Anonymous

The Annals of Otology, Rhinology and Laryngology

Vol. 8

Anonymous

The Annals of Otology, Rhinology and Laryngology
Vol. 8

ISBN/EAN: 9783337375157

Printed in Europe, USA, Canada, Australia, Japan

Cover: Foto ©berggeist007 / pixelio.de

More available books at **www.hansebooks.com**

ANNALS OF
OTOLOGY, RHINOLOGY
AND
LARYNGOLOGY.

FOUNDED BY JAMES PLEASANT PARKER.

H. W. LOEB, M. D., MANAGING EDITOR,
ST. LOUIS.

EDITORIAL STAFF.

T. MELVILLE HARDIE, M. D., CHICAGO.

JAMES T. CAMPBELL, M. D., CHICAGO.

W. SCHEPPEGRELL, M. D., NEW ORLEANS.

G. L. RICHARDS, M. D., FALL RIVER.

GEORGE MORGENTHAU, M. D., CHICAGO.

J. L. GOODALE, M. D., BOSTON.

SEYMOUR OPPENHEIMER, M. D., NEW YORK.

Published Quarterly

By JONES H. PARKER,

St. Louis, Mo.

VOLUME VIII.

1899.

CONTRIBUTORS TO VOLUME VIII.

ADAMS, 281.
Alderton, 10.
Allport, 251.
Andrews, 154.
Ard, 245.
Astier, 41.

BABER, 421.
Bach, 265.
Ball, 160.
Barkan, 170.
Barr, 418.
Bartlett, 300.
Baumgarten, 352.
Baurowicz, 267.
Beck, 1.
Behrens, 95.
Black, 263.
Blake, 280.
Bobone, 378.
Boenninghaus, 249.
Bosher, 265.
Bosworth, 189-194.
Bracken, 283.
Braislin, 156.
Br.eger, 417.
Brindel, 65.
Brose, 247.
Browne, 342.
Buard, 67.
Buck, 97.
Bull, 257.
Bulson, 41.
Burger, 275.
Burnett, 157-244-252-264.
Butler, 283.

CAEN, 58.
Casselberry, 177, 261.
Cheatham, 42.
Cherwin, 42.
Christy, 348.
Clark, 269.
Clement, 66.
Cline, 77, 79, 80.
Cobb, 43, 187.
Collinet, 57.
Coolidge, 182, 346.
Cordes, 351.
Cornick, 99.
Coville, 42.
Cowgill, 268.
Cox, 76, 80.
Cozzolino, 196.
Cryler, 341.
Cube, 351.
Culbertson, 26.
Cuzner, 66.

DAVIDSON, 245.
Day, 84.
DeBlois, 180, 187.
Delle, 58.
Dench, 93, 243, 420.
Doepfner, 281.
Donnellan, 258.
Dufour, 247.
Dunn, 43, 44.

EATON, 176.
Eeman, 416.
Erwin, 9.
Eulenstein, 151, 246.
Evans, 160.
Ewing, 155.

FARACI, 419.
Farlow, 178, 190.
Fischer, 275.
Freudenthal, 168, 293.
Friedenberg, 280.
Fritts, 161.

GALLAGHER, 273.
Gardner, 280.
Getchell, 247.
Gibson, 261.
Gleason, 155, 246.
Glover, 63.
Goodale, 159.
Gottheil, 266, 267.
Gouguenheim, 58, 66.
Gradenigo, 45, 367, 370, 415.
Gradle, 71.
Gradwohl, 190.
Green, 154, 244.
Gruening, 152.
Guye, 413.

HAIGHT, 380.
Haike, 340.
Halsted, 87.
Hardie, 155.
Harris, 341.
Hartman, 265.
Hartmann, 249, 377, 378, 427.
Hartwig, 242.
Hastings, 46.
Havis, 341.
Hengst, 87.
Herman, 280, 284.
Hinkel, 246.
Holinger, 378.
Holmes, 420.
Hopkins, 96, 190.
Hovell, 420.
Hubbard, 180, 187.
Huber, 255.
Hunt, 275.

INGALS, 164, 188, 192.
Ingraham, 67.
Irwin, 64.

JACKSON, 154.
Jacques, 61.
Jakins, 46.
Jansen, 414.
Jenkins, 59.
Jenny, 282.
Jervey, 283.
Joel, 158, 257.
Johnson, 92.
Jones, R., 162.
Jones, C. H., 166.
Jones, W. S., 282.
Jordan, 166.

KATZENSTEIN, 132.
Kayser, 368.
Keen, 238, 282.
Kicer, 59.
Knapp, 153, 382, 408.
Knight, 274.
Koenig, 277.
Koplinski, 168.
Koller, 156.
Kompe, 253.
Korner, 248.
Kummel, 416.
Kuttner, 132.
Kyle, 191, 192, 258.

LAFETRA, 64,
LaForce, 155,
Launois, 46,
Latham, 60,
Lautenbach, 246,
Lederman, 92,
Leduc, 67,
Leland, 185, 186,
Lenhardt, 60,
Lennox-Browne, 344,
Lermoyez, 61,
Lester, 91,
Levy, 299, 349,
Liaras, 46
Libbey, 336,
Loeb, C., 132,
Loeb, H. W., 295,
Logan, 78
Lombard, 42,
Luc, 393,
Lucae, 410
Lydston, 283,

McBRIDE, 364, 414,
McCassey, 71,
McCaw, 28
McConachie, 154, 359,
McConnell, 354,
McDaniel, 71,
Mackenzie, 190,
McKernon, 94, 288,
McLaughlin, 281,
Macewen, 385,
Magenau, 270,
Makner, 354,
Makuen, 179, 183, 185, 190, 391, 274, 287, 354,
Manu, 61,
Marshall, 163,
Martin, 170,
Mayer, 264,
Meltzer, 334,
Meyles, 61,
Milbury, 61, 324
Millener, 157,
Milligan, 371, 420,
Mitchell, 331,
Moline, 62,
Mongour, 67,
Montgomery, 174,
Moore, 273,
Morf, 279,
Moure, 46, 68, 413,
Mullen, 277,
Murray, 263,
Myles, 79, 80

NEWCOMB, 22, 180, 182, 187, 190,
Newman, 157,
Noack, 68,
Northrup, 278,
Noyes, 413,

OHLS, 257
Olliphant, 282,
Onodi, 62,
Oppenheimer, 47, 273, 312,
Orr, 69,
Ostmann, 254,

PACKARD, 244,
Payne, 233, 256,
Peltesohn, 246,
Pfingst, 263,
Pilgrim, 247,
Phillips, 73, 290,
Piaget, 62,
Politzer, 382,
Poole, 353,
Pooley, 93, 95,
Porter, 69, 278,
Potter, 347,
Potts, 242,
Price-Brown, 78, 86, 119,
Pritchard, 356,

QUINLAN, 77, 272,

RANDALL, 242,
Reik, 283,
Reuling, 96,
Rice, 255,
Richards, 69, 90, 91, 108, 158, 274, 288,
Richardson, 286, 287,
Rixa, 62,
Roaldes, 180, 183, 184
Roberts, 256,
Robinson, 258,
Rodman, 71,
Roe, 180, 183, 193,
Ropke, 251,
Rooh, 92,
Rose, 185,
Roy, 354,
Rudloff, 421,
Rupp, 165
Russell, 277,
Rutherford, 47,

SAMPSON, 275,
Sawyer, 165,
Schadle, 161, 344,
Scheppegrell, 238, 348,
Schmiegelow, 364,
Schwartz, 168,
Sendziak, 346,
Shurley, 185, 189,
Simons, 175, 176,
Slack, 274,
Smith, 193, 244,
Solly, 99,
Somers, 55, 161, 257,
Stapler, 245,
Stein, 261,
Stevens, 55,
Stokus, 167,
Stoner, 72,
Stout, 257, 342,
Straight, 81,
Strong, 345,
Stubbert, 99,
Stucky, 125, 230, 278, 288,
Sumner, 342,
Sutherland, 295,
Swain, 183, 188,
Sweeny, 265,

TAULBEE, 169,
Teichman, 151, 248,
Thiesen, 343,
Thompson, 63, 320,
Thorner, 81, 82,
Todd, 335,
Torplitz, 157,
Toussot, 63,
Trask, 170,

UCHERMANN, 422, 428,

VANSANT, 160, 244,
Videl, 163,

WAGNER, 170,
Wallace, 64,
Ward, 185,
Warren, 259,
Wassmund, 340,
Weaver, 70,
Webber, 162,
Wells, 278,
West, 281,
Westbrook, 72,
Whalen, 272,
White, 94,
Whiting, 55, 336,
Williams, 168,
Wishart, 344,
Woolen, 181, 183, 191, 192, 193,
Wright, 258,

ZEIM, 10,

INDEX TO VOLUME VIII.

ORIGINAL COMMUNICATIONS.

ABSCESS, brain, 43, 44, 92, 242, 251.
Abscess, cerebellar, 242.
Abscess, epidural, 43, 44, 92, 233, 288.
Abscess, neck, 157.
Abscess, peritonsillar, 108, 346.
Abscess, peritonsillar associated with diphtheria, 187.
Abscess, peritonsillar complicated by thrombus and septic phlebitis, 185.
Abscess, post-oesophageal, 44.
Abscess, subpetrous, 44.
Accessory cavities of the nose, empyema of, 59, 161, 257, 342.
Accessory cavities of the nose, latent empyema of, 59.
Acoumetry, 364, 370.
Acetanilid in otitis media, suppurative, 335.
Acoustic phenomena of fluid media, 365.
Adenectomy, 344, 421.
Adenectomy, death after, 125.
Adenectomy, hemorrhage after, 62, 125, 170.
Adenectomy, indications for, 160, 264.
Adeno-carcinoma of the nose, 82, 180.
Adenoids, see also tonsil, pharyngeal and deaf-mutism, 247.
Adenoids and ear diseases, 156, 248, 380.
Adenoids, influence of sea climate and surf bathing on, 248.
Adenoids, involution on the reviera, 378.
Adenoids, polyuria and incontinence of urine as symptoms in, 235.
Alveolar hemorrhage, treatment of, 205.
American voice, is it due to catarrhal and other conditions of the upper respiratory tract, 178.
Anesthetic beta-eucain, 353.
Anesthetic mixtures, 72.
Anesthetic, Schleich's, 71, 72.
Angina pseudomembranous due to pneumococcus, 163.
Annulus tympanicus, fracture of, 251.
Antinosin, 157, 282.
Antitoxin, 165, 168, 277, 278, 280, 281, 283, 284.
Antrum of Highmore, see maxillary sinus.
Aseptic methods in surgery of the nose and throat, 96.
Asthma and disease of ethmoid region, 187.
Asthma caused by disease of maxillary sinus, 286.
Asthma, nasal, 256.
Aphonia, hysterical, 275.
Artificial respiration and state medicine, 71.
Atmospheric changes, effect on hearing in chronic catarrhal otitis media, 312.
Atresia of external auditory meatus, 377.
Auditory concussion causing delusional insanity, 26.
Aural polypi and granulation not unfavorable in aural suppuration, 246.
Aural speculum, 244.
Auricle, laceration of, 252.
Auricle, ossification and Roentgen rays, 340.
Auricle, syphilitic perichondritis of, 244.

BACILLUS, Loeffler's in exudations of ulcerations, 63.
Bacillus, Loeffler's, persistence in the mouths of convalescents, 277.
Bacteria, nasal, 258.
Bezold's mastoiditis, see mastoiditis Bezold's.
Black tongue, 266.
Book notices, a treatise on diseases of the ear, 97.
Branchial cyst, 72.
Brain, abscess of, 43, 44, 92, 242, 251.
Bronchus, foreign body in, 182.

CALDWELL-Luc operation, 184.
Cancer of the larynx, 68, 69, 348, 350.
Cautery, abuse of, 258.
Cell, ethmoidal, see ethmoidal cell.
Cellulitis, acute diffuse of submaxillary region, 162.
Choana of tonsil, 265.
Chancre, occlusion of, 158, 161.
Chromo-rhinorhea, 62.
Chorea, larynge 1, 230.
Cleaning the ear in suppurative otitis media, 154.
Cocain, 282.
Cochlea, function of, ——
Cold, taking, 193.
Colorado climate, influence on nasal mucous membrane, 262.
Couzh, reflex, 99.
Croup, pseud membranous, 165, 347.
Cupric electrolysis in ozena, 264.
Cyst, branchial, 72.
Cyst, laryngeal, 164.

DEAFMUTISM, 245.
Deaf'mutism and adenoids, 247.
Deafness, catarrhal, pathology of, 246.
Deafness, electricity in, 157.
Deafness, treatment of, 244.
Deflections of the nasal septum, 287.
Dental septicemia of antrum, 60.
Diabetic ulceration of throat, 293.
Diphtheria, 165, 166.
Diphtheria and membranous croup, 347.
Diphtheria associated with peritonsillar abscess, 187.
Diph heria bacillus, see bacillus, Loeffler's.
Diphtheria, complicated by measles, 64.
Diphtheria, control of, 64.
Diphtheria, death rate, 166.
Diphtheria, diagnosis, 64.
Diphtheria, intubation in, 273.
Diphtheria, quarantine against, 283.
Diphtheria, treatment, 64, 167, 168.

Diplacusis, 151.
Disfigurements, nose and mouth, surgical treatment of, 256.
Diverticula in maxillary sinus, 193.
Dry air in suppurative otitis media, 154.
Dysphoria, 348.

EARACHE in children, 155.
Earache and otitis Media in lobar pneumonia of children, 335.
Ear, cleansing ear in chronic suppurative otitis media, 145
Ear, concussion of, causing delusional insanity, 26.
Ear diseases and adenoids, 156, 248.
Ear diseases, influence of sea climate and surf bathing on, 248.
Ear-drum, affections of, 246.
Ear-drum, tympanic membranes.
Ear external, affections of, 246.
Ear, germs in, inflammation of, 42.
Ear, grippe, 155.
Ear, lesion bilateral, following traumatism, 28.
Ear, middle, see middle ear.
Ear, middle, inflammation of, see otitis media.
Ear, rheuma te diseases of, 422.
Ear, suppuration of, statistics of dangerous complications, 248.
Ear, suppuration of with polypi and granulations, 246.
Ear syringe, infection through, 335.
Ear, traumatism of, 26, 96, 232.
Ear, treatment in whooping-cough, 55.
Ear, tumor of, 46.
Ear, uncomplicated suppuration of, treatment, 30.
Electricity in deafness, 154.
Electricity in stri tures of Eustachian tube, 157.
Electrolysis in ozena, 58, 264.
Empyema of accessory nasal cavities, 58, 161, 257, 342.
Empyema of frontal sinus, 171, 172.
Empyema of the frontal sinus and intracranial infection, 261.
Empyema of maxillary sinus, 43, 184, 257, 279.
Entotic sound perceptions, 55
Epiglottis, fibro-lypoma of, 191.
Epistaxis as an early sign of softening of the brain, 253.
Epistaxis, control of, 263.
Epi-taxis, traumatic, 286.
Epithelioma of larynx, 66.
Epithelioma of maxillary sinus, 73.
Esophagus, foreign body in, 163.
Esophagus, tuberculosis of, 300.
Ethmoid cell, lar e, 62.
Ethmoid cell, sarcoma of, 264
Ethmoid region, disease of and asthma, 187.
Ethmoiditis, suppurating, 324.
Eucain as an anesthetic, 355.
Eustachian tube, electricity in strictures, 157.
Exanthemata, mastoid complications of, 243.
Exophthalmic goitre, surgical treatment, 168, 281.
External auditory canal, affections of, 246.
External auditory canal, section of, 252.
External auditory meatus, atresia of, 377.
External auditory meatus, movable spongy, osteoma of, 151.

FACE, plastic operation upon, 1.
Face, unilateral hypertrophy of, 174.
Facial nerve, 108, 285.
Facial paralysis of otitic origin, surgical treatment, 46.
Falsetto voice in the male, 274.

Fibrolipoma of epiglottis, 191.
Fibrolipoma of tonsil, 187.
Foreign body in bronchus, 182.
Foreign body in nose, 61.
Foreign body in esophagus, 162.
Foreign body in larynx, 269, 272, 245.
Foreign body in pharynx, 162.
Foreign body in respiratory tract, 169.
Formaldehyd in laryngeal tuberculosis, ——
Foreign body in the middle ear, 340.
Frasnotomy speculum, 265.
Frontal sinus, confined suppuration and rupture, 192.
Frontal sinus, disease of, 263.
Frontal sinus, empyema of, 171, 172, 261.
Frontal sinus, infection from maxillary sinus, 60.
Frontal sinus, sarcoma of, 264.

GOITRE, exophthalmic, see exophthalmic goitre.
Granulations not unfavorable in aural suppuration, 246.
Grippe ear, 155.
Gummata of the larynx, 352.
Gumma of tongue, 267.

HAY-FEVER, prevention of, 62.
Headaches, diagnostic characteristics, 71, 283.
Headache, due to centipede in nose, 61.
Headache, nasal, see nasal headache.
Headache, treatment of, 244.
Hearing in chronic catarrhal otitis media, effect of atmospheric changes in, 312.
Hearing, measure of quantitative, 364.
Hearing power, notation, results of, 367.
Hearing, word-spoken and, 42.
Heart dilatation complicating obstruc ion of upper air passages, 320.
Hemorrhage after adenoid operation, 62, 125, 170.
Hemorrhage, traumatic, from the nose and pharynx, 286.
Hemostatic for mucous surface operations, 283.
Hiccough arrested by pressing tongue, 168.
Hypertrophy, unilateral of face, 17.
Hysterical aphonia, 275.
Hysterical larynx, 190.

INCONTINENCE of urine, symptom of adenoids, 255.
Insanity, delusional, resulting from auditory concussion, 26.
Instruments, sterilization with formaldehyde, 283.
Intra-nasal op rations, rubber splints in, 119, 288.
Intubation, 273, 275
Intubation instrument, new, 268, 270.
Iodin treatment of tuberculosis, 67.

LARYNGECTOMY, technic of, 228.
Laryngitis, catarrhal, 274.
Larynx, cancer of, 68, 69, 348, 350.
Larynx, chorea of, 230.
Larynx, cyst of, 164.
Larynx, epithelioma of, 66.
Larynx, foreign bodies in, 269, 272, 274.
Larynx, growths of, 276.
Larynx, gummata of, 352.
Larynx, hysterical, 190.
Larynx, innervation of, 132.
Larynx, instrument for applying nitrate of silver, 352.
Larynx in syringomyelia, 267.
Larynx, myxoma of, 66.
Larynx, paralysis of, 132.
Larynx, polypus of, 65.
Larynx, relation of female generative organs to disease of, 273.

Larynx, stenosis of, 275.
Larynx, traumatism of, 95.
Larynx, tuberculosis of, 67, 268, 273, 349.
Larynx, tuberculosis of, lateral correspondence of, 270.
Larynx, ventricles of, see ventricles.
Lateral sinus, purulent thrombosis of, 43.
Leukoplakia, 163.
Lupus of septum and nasal fossae, 63.
Luschka's tonsil, see tonsil pharyngeal.

MALLEUS, fracture of, 251.
Mastoid cases, fifty-one, 244.
Mastoid complications of exanthema, 243.
Mastoid operation, 152.
Mastoid operation, statistics of dangerous complications, 248.
Mastoid, percussion of, 249.
Mastoid, tuberculosis of, 371.
Mastoiditis, 41, 93, 157, 285.
Mastoiditis, acute suppurative, 233.
Mastoiditis, Bezold's, 152, 153.
Mastoiditis, diagnosis, 245.
Mastoiditis, purulent, 41, 288.
Mastoiditis, treatment, 245.
Maxillary sinus, dental septicaemia of, 60.
Maxillary sinus, diverticula in, 193.
Maxillary sinus, empyema of, 43, 184, 257, 279, 280.
Maxillary sinus, epithelioma of, 73.
Maxillary sinus, infection of frontal sinus from, 69.
Maxillary pneumosinus, 58.
Measure of quantitative hearing, 364.
Meatus auditorius externus, see external auditory meatus.
Middle ear catarrh, see otitis media.
Middle ear, foreign body in, 349.
Middle ear, inflammation, see otitis media.
Middle ear, sarcoma of, 247.
Middle ear, tuberculosis of, 47, 371.
Middle turbinate, see turbinate middle.
Muco-cutaneous lesions, 333.
Mucous surface operations, hemostatic for, 333.
Mouth-breathing, bridle for, 60.
Mouth-disfigurements, surgical treatment of, 256.
Myringitis due to dental pulpitis, 46.
Myxoma, see polypus.

NARES, posterior nares.
Nasal bacteria, 258.
Nasal asthma, 256.
Nasal catarrh in children, 255.
Nasal catarrh, surgical treatment
Nasal chamber, anatomic variations, 341.
Nasal fossae, lupus of, 63.
Nasal headache, 63.
Nasal headache due to centipede in nose, 61.
Nasal mucous membrane, influence of Colorado climate on, 263.
Nasal obstruction, due to anterior arch of vertebrae, 22.
Nasal pressure causing eruptions of face, 263.
Nasal reflex, laryngeal chorea from, 230.
Nasal septum, chancre of, 61.
Nasal septum, deformities of, 262.
Nasal septum, lupus of, 63.
Nasal septum operations, splint for, 333.
Nasal septum, perichondritis of, 161.
Nasal septum, treatment of deflections of, 287.
Naso-pharyngeal disease, anatomic point in, 342.
Naso-pharynx, sarcoma of, 87.

Naso-pharynx, syphilis from arm-to-arm vaccination, 258.
Noise in causation of disease, 46.
Nose, adeno-carcinoma of, 82, 180.
Nose, blue secretion from, 62.
Nose, centipede in, 61.
Nose, deformity of, 159.
Nose, disease of as related to vertigo, 261.
Nose, diseases and trachoma, 10.
Nose, disfigurements, surgical treatment of, 256.
Nose, fossae of, see nasal fossae.
Nose, malignant disease of, 258.
Nose, obstruction of, see nasal obstruction.
Nose, rebuilding of, 95.
Nose, restoration of, 1.
Nose, Roman, correction of, 159.
Nose, sarcoma of, 160, 341.
Nose, septum of, see nasal septum.
Nose, synechia of, see synechia nasal.
Nose, toilet of, 158.
Nose, treatment of, 61.
Nosophen, 157, 283.

OBSTRUCTION of upper air passages complicated by heart dilatation, 320.
Occlusion, posterior nares, 158, 161.
Odor, as a symptom of disease, 71.
Odors and urticaria, 257.
Operations, intranasal, rubber splints in, 119.
Operation, mastoid, see mastoid operation.
Osteoma, moveable spongy, of external auditory meatus, 151.
Otitis media, acute, 82.
Otitis media, chronic catarrhal, treatment by oto-massage, 247.
Otitis media, chronic suppurative, in tuberculous patient, 43.
Otitis media chronic suppurative, opening mastoid in, 382, 385, 391, 408.
Otitis media, chronic suppurative, operative treatment, 93.
Otitis media, chronic suppurative, radical operation in, 246, 382, 410.
Otitis media and earache in lobar pneumonia of children, 335.
Otitis media, chronic catarrhal effect of atmospheric changes on hearing in, 312.
Otitis media, chronic hypertrophic, 155.
Otitis media in infants causing intestinal disturbances, 249.
Otitis media, non-exudative necessity for early recognition and treatment, 247.
Otitis media suppurative, treatment with acetanilid, 336.
Otitis media, suppurative, antinosin in, 157.
Otitis media, suppurative, cleaning ear in, 154.
Otitis media, suppurative, complications, 44, 154.
Otitis media, suppurative, diagnosis, 157.
Otitis media, suppurative, dry air in, 154.
Otitis media, suppurative, nosophen in, 157.
Otitis media, suppurative, treatment, 157.
Otitis media with triple personality, 45.
Otitis, ophthalmoscope in diagnosis of endo-cranial complications of, 45.
Otologic experiences during naval battle of Santiago, 175.
Otology, progress in, 155.
Oto-massage in chronic catarrhal otitis, 247.
Otorrhea with endo-mastoiditis and epidural abscess, 283.

Otorrhea with fracture of the base of the skull, 41.
Otoscopy, magnifier in, 249.
Ozena, bacteriology and histology of, 199.
Ozena, electrolysis in, 58.
Ozena, treatment, 264.
Ozonized air in diseases of the lungs and air passages, 279.

PARALYSIS, laryngeal, see lary x, paralysis of.
Paralysis, posticus 132, 275.
Paramonochlorophenol, 274.
Perichondritis of auricle, syphilitic, 244.
Perichondritis of nasal septum, 161.
Peritonsillar abscess, 186, 346.
Peritonsillar abscess associated with diptheria, 187.
Peritonsillar abscess complicated by thrombus and septic phlebitis, 185.
Perforation of tympanic membrane, 242.
Pharyngomycosis, 289.
Pharynx, foreign bodies in, 162.
Pharynx, influence of turbinate hypertrphy on, 257.
Pharynx, traumatic hemorrhage of, 286.
Pharynx, tuberculosis, 343.
Phenolnatrosulphoricinum, 352.
Phlebitis, of aural origin, 156.
Phlebitis, septic, complicating peritonsillar abscess, 185.
Phonation, 273.
Phthisis, etiology, 280.
Plasmine solution as a cleansing agent, 330.
Pneumosinus, maxillary, see maxillary pneumosinus.
Polypi, aural, not unfavorable in aural suppuration, 246.
Polypus of larynx, see larynx, polypus of.
Polyuria, symptom of adenoids, 255.
Posterior nares, occlusion of, 158, 161.
Posterior turbinate hypertrophy, removal of, 254.
Presidential address, 99, 177, 356.
Program of sixth otological congress, 195.
Pseudomembranous angina due to pneumocccus, 163.
Pseudomembranous croup, 165.
Psedomembranous rhinitis, non-diphtheritic, 255.

REFLEX cough, 90.
Reflex, nasal, causing laryngeal chorea, 230.
Removal of adenoids, see adenectomy.
Removal of tonsil, see tonsillectomy.
Respiratory tract, foreign bodies in, 169.
Rheumatic diseases of the ear, 422.
Rhinitis, acute, 259.
Rhinitis, hypertrophic in females at puberty, 258.
Rhinitis, hypertrophic, histopathology of, 239.
Rhinitis, pseudomembranous, non-diphtheritic, 255.
Rhino-pharynx, see naso-pharynx.
Roentgen rays and ossification of the auricle, 346.
Roentgen rays determining relation of trachea and bronchi to thoracic walls, 280.
Rubber splints in intranasal operations, 119, 268.

SARCOMA of frontal and ethmoidal sinuses, 264.
Sarcoma of middle ear, 247.
Sarcoma of middle turbinate, 58.
Sarcoma of naso-pharynx, 87.

Sarcoma of nose, 160, 341.
Sarcoma of thyroid gland, 279.
Sarcoma of the tonsil, 92.
Schleich's anesthetic, 71, 72, 283.
Septum, nasal, see nasal septum.
Septum operations, splint for, 334.
Siegle's aural speculum, 244.
Sinus, frontal, see frontal sinus.
Sinus, maxillary, see maxillary sinus.
Sinus, thrombosis, 43, 55, 92, 166, 250, 69, 288, 336.
Snare for removing posterior ends of turbina es, 61.
Sound perception, entotic, 55.
Speculum, frasmotomy, 265.
Speculum, Siegle's and aural, 244.
Speech, defective and derangement of the cerebral functions, 287.
Speech and hearing, 42.
Speech, ventricular band, 275.
Splint for septal operations, 334.
Splints, rubber, in intra-nasal operations, 119.
Stammering, 183, 355.
Staphylorraphy, 91.
Submaxillary region, acute diffuse cellulitis of, 162.
Suprarenal gland, 70, 160, 277.
Surgery of ear, nose and throat, aseptic measures in, 96.
Synechia, nasal, 57.
Syringomyel a, laryngeal disease in, 267.

THIOL in nose and throat practice, 278.
Thorner, Dr. Max, 291.
Throat, cough, 265.
Throat, diabetic ulceration of, 293.
Throat disease, spread by milk, 65.
Throat, malignant disease of, 258.
Throat, treatment of, 61.
Thrombosis, purulent of lateral sinus, 43.
Thrombosis, sinus, see sinus thrombosis.
Thrombus, complicating peritonsillar abscess, 185.
Thyroid gland, sarcoma of, 279.
Thyroid tumors, accessory, 344.
Tinnitus aurium, treatment of, 244.
Tone sensation and the cochlea, 47.
Tongue, black, 266.
Tongue, gumma, 266.
Tonsil, acute suppuration of, 186.
Tonsil, chancre of, 265.
Tonsil in systemic infection, 264.
Tonsillectomy, 265, 344.
Tonsillectomy, death after, 125.
Tonsillectomy, recurrence after, 190.
Tonsillitis, acute, pathologic histology, 163.
Tonsil, fibrolipoma of, 187.
Tonsil, naso-pharyngeal, chronic inflammation of, 76.
Tonsil, naso-pharyngeal.
Tonsil, sarcoma of, 92.
Tonsil, submerged, 266.
Tonsil, surgical diseases of faucial, 345.
Trachea and bronchi, relation to thoracic walls determined by Roentgen rays, 280.
Tracheocele, 281.
Trachea, surgery of, 282.
Tracheo-thyrotomy in cancer of the larynx, 68.
Tracheotomy for foreign body, 272.
Tuberculosis, air in, 70.
Tuberculosis, hygiene and dietetics, 66.
Tuberculosis, iodin treatment, 67.
Tuberculosis of esophagus, 300.
Tuberculosis of larynx, see larynx, tuberculosis of.
Tuberculosis of middle ear and mastoid, 371.

Tuberculosis of pharynx, 343.
Tuberculous patients, where to send them, 354.
Tuberculosis, sero-diagnosis of, 67.
Tuberculosis, sero-therapy in, 69, 168.
Tuberculosis, summer care of, 278.
Tuberculosis, treatment, 69.
Tuning forks, plea for accurate definition, 154.
Turbinate hypertrophy, influence on pharynx, 257.
Turbinate, middle, sarcoma of, 58.
Turbinates, posterior ends, new snare for removal, 61.
Turbinates, posterior, hypertrophy of, removal, 254.
Tympanic cavity, acute inflammation of, 244.
Tympanic cavity, hydrochloric acid application, 245.
Tympanic perforations, 242.
Tympanic perforations, closing, 246.
Tympanic ring, see annulus tympanicus.
Tympanum, catheter inflation, 245.

ULCERATION, diabetic of throat, 203.
Urticana and odors, 257.

VALSALVA'S Method, reversed, 331.
Ventricles, eversion of, 68, 164.
Ventricular band speech, 275.
Vertigo as related to nasal disease, 261.
Vocal bands, position after severing recurrent, 275.
Voice, American, see American voice.
Voice, falsetto in the male, 274.

WHOOPING-COUGH, formalin in, 282.

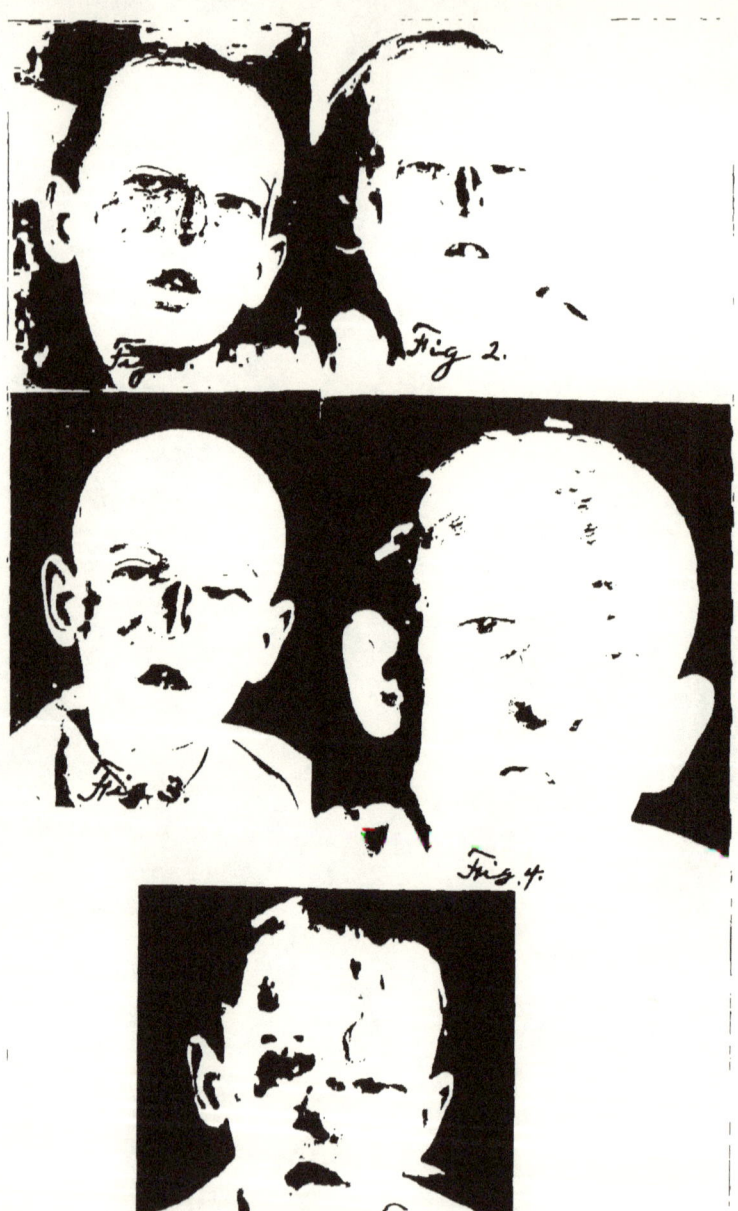

ANNALS
OF
OTOLOGY, RHINOLOGY
AND
LARYNGOLOGY.

A REMARKABLE CASE OF PLASTIC OPERATION UPON THE FACE; FORMATION OF EYE-LIDS AND TOTAL RESTORATION OF THE NOSE.

Dr. Carl Beck,

CHICAGO.

PROFESSOR OF SURGERY POST GRADUATE MEDICAL SCHOOL, AND PROFESSOR OF SURGICAL PATHOLOGY, UNIVERSITY OF ILLINOIS MEDICAL DEPARTMENT.

In studying the modern literature of rhinoplasty, we find that surgeons strive to individualize, and do not attempt to follow a certain typical method. Asepsis in operative work has done away with rapid and brilliant operating, and in its place has come the art of adapting surgical measures closely to the exigencies of the particular case. To express this thought, more clearly, I will refer to the Tagliacozzian method of rhinoplasty. This eminent surgeon laid down strict rules as to the exact number of cuts to be made in a specified time. These cuts, he declared, must be made with certain particular instruments, and the patient must have a particular posture during the operation. If a follower of his method deviated slightly from these prescribed rules, success was not anticipated, and all failures were supposed to be due to irregularities in methods. Modern investigation has proved, that accidental wound diseases are responsible for failures, and since asepsis has come to reign, successes are obtained even by the surgeon, who does not know the method, but has common sense to carry out his plans.

The resection of a joint in exact accordance with a rule laid down by a master would be a slavish and inappropriate method, since the surgeon must resect only what is pathological in his particular case, and what is necessary in order to achieve functional results. In modern surgery, therefore, the tendency is not to lay down certain rules for particular kinds of pathological conditions, as in former times, but to formulate more or less general ideas and indicate general lines of treatment according to which the surgeon must carry out his operation.

Billroth expressed this very aptly in his introductory remarks to the students at the opening of his clinics. "Study closely," he said, "the history and description of cases, more than the disease; experiment, and try to gain surgical tact, and you will become better surgeons than if you carried out the procedure according to the iron rules of masters." These ideas are particularly applicable to plastic surgery. Not one case of defect of the nose is identical with another, and in no case is the method of restoration exactly like that in any other case.

However, we must have, as before stated, some general ideas of restoration, and must make use of the large experience of other surgeons, otherwise, the chances for successful operation will be very slight.

In the following lines I shall endeavor to point out some general rules in describing an interesting case of rhinoplasty. Literally, rhinoplasty means the formation of an entire new nose, but the term is now also used to designate restoration of parts of the nose.

Total loss of nose results from either injuries or disease. At the present time with our conservative methods of treatment, such loss as a result of injury is very rarely found, although the injuries reported during the last war include some cases of considerable destruction resulting from gun shot wounds.

The nose is destroyed as a consequence of disease (mostly syphilis) by progressive ulceration, and while it very rarely happens at the present time in civilized countries, that a case is left untreated until the entire external nose is destroyed, it is a common occurrence in less civilized countries, and in places remote from medical centres.

Total destruction of the nose means destruction of the bony and cartilaginous skeleton, as well as the integument. An interesting case of this kind has come under my observation through the courtesy of Dr. B. Baldwin, to whom I am also indebted for the history, as follows:

The girl is 14 years old, has a healthy mother and two healthy sisters. Her father, a soldier during the civil war, is dead. He was afflicted with syphilis. When the child was 9 years old she showed evidence of syphilis. Not being treated at the beginning, the ravages of the disease caused such destruction of her face that she was kept hidden from the sight of strangers for a long time. This accounts for the progress the disease had made before she ever came under treatment. She was finally brought to this city and given to Dr. Baldwin, Professor of Skin Diseases in the Post Graduate School, for treatment. Her condition as shown by illustration 1. was as follows: In addition to the localization of syphilitic conditions throughout her body, such as gummata of the tibia and other bones—her face showed total destruction of the nose, including the bony septum, leaving only part of the vomer and a little tip of the cartilaginous portion of the septum, fortunately, as will be seen later on, in the plastic. Both lower eye-lids were destroyed, the right one being preserved to about one-tenth of an inch only, and drawn upward in one surface of conjunctiva. She had a large ulcer on the cheek, the left lower eye-lid being entirely destroyed, leaving the cheek, which was also destroyed by the ulcer, in one plane with the conjunctiva. The upper eye-lid on the left side was also partly gone, and the palate destroyed in its central portion. At this time the ulcerations were all in a florid condition, unhealthy, and showing the characteristic tissue of syphilitic granulations.

This pitiable creature cried day and night with pain in the ears and head. Persistent treatment by Dr. Baldwin brought her to a condition in which all the florid symptoms disappeared, and after four months she offered some chance for a plastic operation. Dr. Baldwin tried the plastic himself using the method of forming a nose by the implantation of the finger. He used the middle finger of the left hand, splitting the epidermis of the last phalanx, and separating the skin on the tip of the finger to both sides. Then he took off the nail, and implanted the finger into the freshened border at the root of the nose between the eye-brows, thinking that later on he would use the two phalanges of the finger, and so get a nose. This method is sometimes used with success, as has been proven by a number of experiments, and it gives a fairly good cosmetic result. This experiment failed, and it is not strange that it was a failure in this particular instance, because the use of the Italian method in children, especially syphilitic children, is an extremely difficult task, which rarely gives good results.

In the middle of April she was given to the writer, with instructions to do what he could with her, and he commenced a course of treatment which aimed to remove all traces of syphilis by specific and internal treatment. Unfortunately, from time to time the dreadful disease recurred and interfered with many of the good results of the plastic operations. Nevertheless, now after six months, we have succeeded in restoring the nose and eye-lids and the girl again looks human.

The following steps were taken in the operations:

1st. *Removal of the diseased right conjunctiva, and transplantation of a graft of the mucous membrane of the mouth into the wound.* This first step was taken with the idea of ascertaining if the disease was a mucous infection of syphilis and tuberculosis on the conjunctiva. On this place there appeared some white nodular infiltrations which bore close resemblance to tuberculosis as seen in progressive lupus, and the writer thought it might be such a condition. He had the specimen examined by Prof. Klebs, who found no trace of tuberculosis, so that Dr. Baldwin's decision that the disease was absolutely pure syphilis, was found to be correct. This experiment succeeded only partially, as only a portion of the graft healed on.

2d. *Left lower eye-lid restored.* A tongue shaped flap was dissected from the left temple and turned in the form of a right angle to its pedicle, so that the tongue of the flap formed the lower lid externally, the conjunctiva being dissected and turned back upward into its former normal position, and the border of the dissected conjunctiva and the flap were sutured. This operation succeeded perfectly. I regretted not having exercised the scar of the cheek, as this would have greatly facilitated the further steps of the nose formation. But inasmuch as this was the first step of the operation, and up to this time I had no idea how the girl would react to plastic operations, I thought it best not to remove too much of the tissue, and that I might at some future time replace the scar of the cheek by some healthy integument from the neck.

3d. *Formation of the right cheek and right lower lid.* This time I excised the whole scar of the cheek, dissected the conjunctiva and formed the same kind of tongue-shaped flap on the right temple as before, turning this flap not only as far back as the internal canthus of the right eye, but even over that clear up to the forehead. This was the best step of the whole operation, assuring the union of the flap with the healthy skin on the forehead, and afterward giving to the eye a normal appearance much better than on the other side.

These two operations are blepharo-plastic operations—that is, the restoration of eye-lids. It was no easy matter to restore the cheek in this case, after the dissection of the scar, for though the flap had a wide pedicle, it was long, at its peripheral extremity. While I made a Thiersch graft on the left temple to cover the defect caused by the plastic, I adapted the borders of the right side directly in both instances, and union and healing resulted. But in such cases I prefer adaptation of the borders, as better cosmetic results are obtained, and less time is required for healing.

4th. *The formation of a nose.* This was done in the following manner: The inner lining was formed by two flaps from the cheek, which though narrow, furnishes enough material to insure epidermization. These two flaps did not meet the centre, but were turned as snugly as possible over gauze plugs. The remnants of the nostrils at the aperture were utilized to unite with the small snout-like projection of the cartilaginous septum. A triangular flap from the forehead was turned down and united as after described. The accompanying illustration indicates the condition at this time.

5th. Three weeks afterward the flap was cut off at its

base, and plenty of tissue left to cover the nose at its root. Unfortunately, a portion of the flap died out at this place, and a small opening remains showing root of nose.

6th. An attempt was made to close this opening by bringing the tissue more closely together, but this attempt failed.

In such a case it is not necessary to give a bony structure to the nose, and this would not heal under any circumstances.

Critique.—This case is remarkable for several reasons:
1st. Extent of destruction.
2d. The number of plastic operations necessary.
3d. The manner in which the healing took place.

As to the first, it seemed questionable whether we had the right to undertake the plastic at the time we did. Many of our colleagues disapproved of the attempt, but was it not our duty to relieve this poor girl, if possible? She was bright, and as the final result demonstrated, such a reconstruction was accomplished that she is at least presentable enough to be looked at without abhorrence. The more important question was whether the operation should be undertaken shortly after the eradication of the florid symptoms, or whether we should wait until it was possible to find out if recurrence would take place. I considered this for a time, but decided that as she was in the city, it was best to complete the work before she left for her home. If she were sent home, even with the defect healed, it was to be expected that the influence of the air upon the dried secretions upon the cicatrized formations of the mucous membrane, which itched considerably, would cause her to scratch these itching parts with her fingers, and the results would soon be annihilated. We observed that even under the watchful care of the nurses the child could hardly be prevented from poking her fingers into the large holes of her face, trying to remove scabs, etc. How much worse would this have been if she had gone back to her home in the country, and how quickly would the ulceration and bleeding have brought back the old conditions. These were my reasons for attempting the plastic so soon. In other cases of this kind, especially in full-grown persons, it may be advisable to wait.

The next question of interest is: Why were the remaining operations performed, and why were they so successful? As I mentioned before, I wished to ascertain at first whether any healing would take place, and therefore tried the experiment on the mucous membrane at the corner of the eye-lid. I wished to ascertain whether in this particular case the general rule would hold good that specific cases yield badly to union and healing. The good result proved that she had considerable re-constructive power. I concluded to undertake the restoration of the eye-lids first, because the secretions from this source would certainly have spoiled any operation on the nose. The conjunctiva was turned outward and secreted constantly. Tears, mixed with pus and mucous, constantly ran down the cheeks and these would have interfered with the result of the rhinoplasty. I, therefore, started on the eyes, and in order to find out how much could be accomplished, limited the first operation to the left eye. It healed, and this success encouraged me to try the same methods on the other side. I was sorry afterward not to have extirpated the scar on the left cheek at the same time, as in this way I might have gained healthy skin adjoining the future nose, and might have prevented a little accident, that is, a sloughing off of part of the forehead flap, certainly due to the presence of scar with poor circulation. Had there been healthy skin this circulation would have been much better. This taught me to avoid the same mistake on the right side, and I took a large enough flap to cover all defects. It is important to remember in all plastic operations, that healthy skin should be united to healthy skin or mucous membrane, and if possible, never unite to scar, otherwise, the results of union will be poor.

A very important detail in plastic surgery of the nose is to assure the shape of the nose by the use of some forehead skeleton. This was absolutely impossible in a case like the present, but nature had assisted slightly by leaving a portion of the septum cartilage projecting, thus assuring at least a slight profile of the nose. It was an object of the plastic to make the best use of this projection, first, for the formation of the nostrils, and second, for the formation of a roof. Had I taken a forehead flap only and turned it down in the old classical way recommended by Diefenbach, the nose flap would have shrunk and crumbled and would gradually have atrophied, and perhaps even would have moved with inspiration and expiration. In order to give a certain solid form to such a flap, a lining must be obtained, but in this case it was difficult to get a lining of epidermis. The forehead was already so tense in consequence of the removal of the two temple flaps, that I could not spare the skin for two flaps, one for the inside of the nose and one for the outside. I therefore, decided to use a portion of the sides, that is, the cheeks as inner lining, stripping it off and turning it inward. Even if it did not reach to the central line of the septum enough epithelium was secured to insure epidermization of the inner surface of the nose, thus preventing contraction.

Another important detail is the formation of the external borders of the nostril. In most cases these can be obtained by the dissection of the triangular flaps on the side of the aperture. In most specific cases we find that the wings have been destroyed together, but that portions of them remain attached to the cheek. These portions may be used to advantage by turning them into their old places and they will certainly be preferable to the artificial wings made of a forehead flap.

The third point of interest to which I wish to refer was the method of healing and the restoration of the physiological functions of the nose and eye-lids. Before the plastic, tears ran down her cheeks constantly, producing an eczematous condition of the cheek, but after the plastic her cheeks were dry. The right eye gained almost its normal appearance, but the left eye caused us considerable trouble. There seemed to be a recurrence of the disease in the upper lid, melting away a portion of the same, and keeping it in a swollen condition. The nose very soon became functionally useful as a respiratory organ, and the voice changed immediately in a peculiar way. While her nose was missing, the girl's voice was almost normal, notwithstanding the fact that her palate was missing to a great extent, but immediately upon the restoration of the nose, her voice gained a nasal twang, which would only disappear by filling out the gap in her palate. This change was so striking that I have thought it well to record the same. I shall not attempt to explain the cause.

Although undergoing so many operations, entailing considerable loss of blood, the girl rallied and was sent home, after being presented before the Medical Society where she had been shown once before by Dr. Baldwin when her disease was at its height. She is by no means a perfectly healed subject, where no other restoration or preparation could be performed, but I hope that after a time when she has gained strength and courage, to make some minor operative improvements, which will restore her face perfectly. It is most probable that after a time there will be an abundance of skin where it is now tense. We notice in plastic operations that we can produce skin by massage and by stretching, thus encouraging the growth of epithelium.

TRACHOMA OF THE CONJUNCTIVA AND ITS RELATION TO DISEASES OF THE NOSE.*

By Dr. Ziem,

Dantzig.

The conception of trachoma is undergoing a change. The traditional belief that it is a purely contagious disease, transmissible only from individual to individual, and then only by common use of towels and the like, and that this is the only means of contagion; or, what is only a recent modification of the same ideas, that trachoma is transmitted from some other already infected mucous membrane of the individual, by means of fingers or dirty clothing—all this is, of course, applicable to certain cases, but not to those others which after all compose the greatest number, Professor Feuer and his school to the contrary notwithstanding. Observation teaches that in many parts of the world, especially in the classical home of trachoma—the bottomlands of Egypt and in Syria and Palestine,—trachoma does not appear at all seasons of the year impartially as it ought according to the contagious conception, but that the yearly appearance of the epidemic—or rather endemic, depends on certain general well-defined climatic or telluric conditions. As such the most important are: (1), excessive heat and sunshine; (2), dust and vapor, both of which act directly on the eye itself; (3), mold, especially in malarial districts, (a) swamp air affecting the whole respiratory tract particularly through the nose, from which, either by continuity of tissue or through vascular connections the conjunctiva itself becomes affected; or (b) by means of some agent entering the digestive tract, in drinking water perhaps, so that it is carried to the eye through the circulation.

(1.) Excessive heat and dazzling sunlight, as mentioned above, apply to our subject only in tropical or sub-

---

*Read by title at the annual meeting of the Western Ophthalmologic and Oto Laryngologic Ass'n. at New Orleans, Feb. 10th and 11th, 1899. Freely translated and somewhat abridged by Albert B. Hale, M. D.

tropical regions, e. g. the delta of the Mississippi, the Nile or the Ganges, in the Soudan, Arabia, Persia, on the plateau of Spain and Portugal, in Sicily, lower Italy and Greece, but not in the infected regions of the Rhine, of East and West Prussia, Poland or Scandinavia. In the tropics, in Bagdad for example, 32° north latitude, the bare sand has been known to have a temperature of 78° Cent. and of 55° Cent. in the shade; in the Sahara "where the earth is a fire, and the wind a flame," eggs can be cooked in the sand; even here, eye inflammation due to such causes, are almost never seen among the natives, although they do occur among foreigners. This was the case in the army of Alexander the Great, among the British troops in India, and this evident immunity of the bronze, copper-colored or black native of the tropics, is easily explained by the rich pigmentation of the skin in the face and eyelids, sclera and iris, as well as by the narrower pupil of the resident of the tropics (Lewkowitsch). All of these offer protection against the sun and admit the lowest degree of absorption, a condition which the natives intensify by painting the neighborhood of the eye with some dark antimonial material. Even in our northern latitudes we often see a catarrh produced undoubtedly by some intense (electric) light, although there is seldom any suppuration connected with it. Heat itself, notwithstanding German's theory, at least for the bare-headed native of the tropics, has but little influence on the production of eye-diseases. John Locke, in 1692, called attention to this where he says in: "Thoughts concerning Education:" The heats are more violent in Malta than in any part of Europe; they exceed those of Rome itself and are perfectly stifling, and so much the more because there are seldom any cooling breezes here. This makes the common people as black as gypsies, but yet the peasants defy the sun, they work on in the hottest part of the day without intermission or shelter from his scorching rays. This has convinced me that nature can bring itself to many things which seem impossible, provided we accustom ourselves from our infancy. The Maltese do so who harden the bodies of their children and reconcile them to the heat by making them go stark naked, without shirts, drawers, or anything on the head, from their cradles till they are ten years old.

I agree with German, that the Mohammedan has a most unsuitable head dress in his turban, although I differ from him in thinking that this is because it protects the head too little; I think rather that it, with the senseless habit of clipping and shaving the head, encourages neglect of simple hygienic rules for preserving the natural temperature, and thereby exposes the wearer repeatedly to catching cold. It would be interesting to see whether another headgear or even bareheadedness from childhood (with natural hair, of course) would not of itself reduce here the amount of eye diseases.

Since heat and excessive sunlight of themselves are not accountable for much trachoma or eye inflammations, let us consider a far more important factor.

(2.) The effect of vapor and dust upon the eye. It is quite usual to find an eye patient complaining that some particles of wood (sawdust), or of hay, meal, powdered marble or glass has fallen into his eye during work, although most careful examination fails to detect anything; now all symptoms cease soon after a thorough irrigation with plain water (or salt solution) and everting the upper lid; while without this, as I have often learned by experience, such an irritation may lead to the most pronounced swelling and catarrh. Blennorrhea itself may be simulated, especially if there exists at the same time any nasal catarrh to hinder the free drainage from the eye. One patient I remember, was treated by a colleague radically, even to mercurial inunctions, for nine months, with an advised operation to end up with; whereas, a few days' irrigations under my own care, the nose being equally treated, cured the eye with a vascular keratitis. Investigation showed that the first attack had been due to dust from the street! Often enough false diagnosis is thus given. I remember another case in 1895, when a laborer had suffered from sand blown into his eye a day or so before I saw him, producing a suppurating keratitis, which, in the opinion of a visiting colleague demanded treatment by the cautery, but which after a thorough irrigation to both conjunctivae and nose, and with a bandage, disappeared with the accompanying pain so completely that on the next day he was able to work. It is easy to imagine that even the application of the cautery would not have stopped the inflammation, for the sand had been left in the conjunctival sack to continue its irritation.

Dust of this nature acts disastrously, when there is already a catarrh of the conjunctiva. Another patient, a glazier, with extreme swelling of the upper fornix, morning adhesion of the lids, etc., I saw in 1897. His disease had been called trachoma—although there were no follicles—in this case at the patients' request, I did not attempt a scarification but contented myself with a simple salt solution to irrigate the nostril and conjunctiva, and was very gratified, when I next saw the patient after a month's intermission, to find that a previously existing growth the size of a raspberry, in the upper fornix, had now shriveled to half its former size, either on account of the former radical irrigation that removed all irritation due to the particles of glass, or by reason of the attack on the nasal mucous membrane, on which the trachomatous poison had been destroyed. This patient was subsequently scarified occasionally, but soon after left Dantzig.

I have had similar experiences with workmen engaged on razing the old fortifications of Dantzig, who were often, during dry weather, exposed to excessive dust. In such cases many men who previously had had no eye trouble, and who had been released from military service as quite free from eye disturbance, now began to complain of noticeable swelling and hypertrophy of the conjunctiva, and were subject—as Jacobsohn also remarks—to a more or less intense suppuration, although without follicular growth, with simultaneous swelling and occlusion of the corresponding nostril. A few irrigations to both eye and nose ameliorated the symptoms noticeably, so that they seldom came back regularly for treatment, but escaped to return home, when their task had been finished.

Again, in 1880–82, in Alexandria, old European settlers have often told me, that eye troubles had noticeably diminished, since parts of the city had been paved so that the excessive dust raised by the high winds had been done away with. In Russia I heard the same story. Lemkomitsch, who practiced 14 years in Johannisburg, (South Africa), refers the numerous cases of trachoma among Boers and Kaffirs, who use no such utensils as towels or handkerchiefs, directly to the intense storms which fill eyes, ears, nose and mouth with solid masses of sand. All efforts to rub away this sand are futile or else only half carried out, and the consequence is a great increase of even severe eye trouble. German, in Syria and Palestine, reports that during the hot dry months of September and October, many cases of eye trouble are traceable directly to irritation from alkali dust or from fine vegetable matter; this being a better explanation than his attempted explanation of infection by direct contact. It is a fact, also, that eye diseases diminish as the heat subsides and the rains begin; and that in parts of Syria, better supplied with springs and surface water, and consequently with better vegetation, trachoma is less and milder, than in Palestine.

Prof. Marshall, Zoologist in Leipzig, tells me that he has been in sand dunes during wind storms, and seen the surface so blown about that the clothing would be quite full of sand, the skin, face and hands feel gritty, that the mouth would be choked, and the eyes half blinded by the clouds. On some window panes, he continues, he has seen the glass so beaten by the sand that its transparency was altogether lost! Repeated effects of such sand storms upon organic matter like the cornea can well be imagined! It would be a profitable though a horrible experiment, to see how animals eyes would be influenced by such disturbances, especially in connection with simultaneous attacks on the nose; and to continue it further, whether worse results would come from purely mechanical irritation of itself—the sand being sterilized by heat or natural means—or from impure material—the sand being mixed with dung from domestic animals; that is to say, whether repeated mechanical irritation alone would suffice to produce a trachoma-like inflammation.

The conclusion from all this is that part, at least, of the eye trouble in trachoma districts can be traced to a moderate and repeated influence of dust or sand storms in deserts in Palestine, Hungary, east and west Prussia, etc., in districts that have been in historical times denuded of wood and therefore deprived of proper rainfall, and where therefore, under conditions that did not obtain in Spain or Gaul at the time of Julius Cæsar, or in Palestine at the time of Josephus, trachoma has found a home. It may also be concluded that trachoma has assumed a milder form in parts of Syria better wooded and watered; or in Palestine, famous for its cedars! Does not the psalmist sing of the Cedars of Lebanon which the Lord has planted, in which the eagle builds his nest, and the heron finds his home, which built the temple of Solomon and supplied the fleets of the Phœnicians? To-day reduced to a grove of only 377 trees! In spite of Europe's increasing poverty of woods, lamented by Addison, Reaumur, Buffon and others, it is only since Alex. Von. Humboldt's time that its relation to moisture and moderate temperature has been acknowledged. Even Columbus, the immortal discoverer of America, whose meteorological knowledge has been often granted, ascribed the heavy and cooling rains of Jamaica to her wooded mountain peaks, and noticed that Madeira, the Azores and Canaries had lost in cooling rains since their woods had been so ruthlessly cut down!

The irritants hitherto discussed, act immediately upon the eye itself, although their effect is undoubtedly supported by their action upon the nasal mucous membrane as well. There remains to be discussed other poisons—vapors, emanations—which affect the nose primarily and are carried from there either directly to the eye, or through vascular connections to the conjunctiva and apparatus.

(3a.) Swamp air has been notoriously irritant during all ages. Even Empedocles (450 B. C.) recognized this fact, and a classical example of its effect upon the eye is found in Hannibal's history, who, when crossing the inundated shores of the Arno, fell sick of a fever and lost an eye therefrom. Whether this was really a case of acute trachoma with rapid destruction of the cornea cannot be proven, of course. There are still plenty of individuals who seem to overlook this factor in the genesis of trachoma, and who doubt that trachoma (especially with follicular growths) can be considered an accompaniment of malaria, that it is really a miasmatic infectious disease; although I have emphasized this relationship in 1893. Even such an authority as Professor Raehlmann, in Dorpat, slights the point by saying in a few words that "there can be no doubt of the climatic influences at work in certain localities." This is not enough; he should have said that this factor can be shown to be present in an extraordinary number in almost all of the cases. For example, Finland is the most trachomàtous country of Europe, and here at times eleven per cent. of the surface of the land is swampy or actually inundated. The conditions are about the same near Triest, Wilhelmshafen, and St. Petersburg; in the reedy country of Poland and Galicia, in Liveland about the Dnieper, the Rhone, Po and Arno, the Rhine, the Volga, Tagus, Mississippi and many other rivers, where the outbreaks of trachoma are often associated with acute attacks of malarial fever. Schmidt-Rimpler's *dictum*, that every cold does not lead to trachoma—in so far as it is a scientific assertion, not an epigram—must be accepted, but the same is true of laryngitis, acute middle ear suppuration, or of sinusitis. Unfortunately Schmidt-Rimpler fails to state how he made out that his cases of trachoma had no nasal symptoms. I myself have shown that in most instances simple inspection—either anteriorly or posteriorly—does not suffice either to exclude sinusitis or even simple suppuration itself, since to demonstrate the latter a thorough irrigation is often necessary. In fact, with this irrigation, even after a rhinoscopic examination, pus may be brought away. If then, Schmidt-Rimpler had not practiced this irrigation, we cannot admit the truth of his assertion. This connection between trachoma and nasal infection I emphasized in 1886, as did Nieden in Bochum, (Westphalia,) Vocher in Orleans, Sattler in Leipzig, (298 times in 443 cases or 65 per cent). Kuhnt in Koenigsberg, reports the same results referring to my own investigations, although he is not explicit enough in showing that these investigations were made by me and not by him, or without mentioning directly that with my pump even a higher percentage could have been found. Again, a pupil of Schmidt-Rimpler's, Hoppe in Elberfeldt, declares that in the richly trachomatous Masuren (East Prussia,) he found ozena surprisingly frequent.

All this goes to prove the intimate connection between trachoma and nasal catarrh, and explains, what I had already confirmed five years ago, that many patients with one-sided trachoma will be found to have a nasal catarrh on the same side, and that moreover, not unusually the localization of trachoma in either the upper or lower fornix will be found associated with suppuration of either the frontal or superior maxillary sinus, so that the inflammation must have been extended by means of the vascular anastomosis of the nasofrontal or infraorbital vessels to either the upper or lower lid.

I must mention, too, that in trachoma it is not unusual to find an elevated temperature of the skin of the nose, indicative of inflammation within; in some instances I have even seen radical operations on the lids performed by skilled surgeons, but they ignored the possibility of a simultaneous nasal inspection and consequently the inflammation returned the same as before. I remember one instance very well; the case was presented in August, 1896, to Professor Silex of Berlin. It was a hotel servant who, for admission to military service had nearly been compelled to submit to operation; a year later, after sleeping in a barn for some time, he showed all over that part of the upper lid exposed to view by simple eversion, a mass of trachoma granulations, while he had in addition a catarrh of the eye, and a most decided suppuration of the nose. After he removed from the barn to better lodgings, and when the eye inflammation had been subdued, I even then firmly maintained the opinion that a second operation on the conjunctiva or a radical excision of the tarsus would have no lasting effect, and that to fit him for military service the only proper treatment for the cure of the trachoma—which was causing the nasal catarrh as well—would be to attack the nose at the same time, and to send the patient thereafter to the drier climate in the woods, or to the purer air of the sea.

The same relationship will help to explain the relapses in nearly 50 per cent. of cases of trachoma; even after radical excisions on fornix or tarsus reported by Greff in Berlin, and Hoppe in Elberfeldt. Unfortunately Greff's investigations took no account of this relationship to the nose; those of Hoppe mentioned only cases where ozena was very marked. Kuhnt, contradicting a statement made to me in 1894, in Koenigsberg, says: that he now has better results in treating trachoma, when he attacked the nose at the same time. It is worthy of remark, and of importance too, in connection with Greff's assertion that trachoma runs in families, that occasionally only one member of a family is attacked, although all of them have exactly the same environment; this particular individual will, on careful examination, be found to have a profuse, stinking nasal catarrh, due probably to some preceding fever of an eruptive nature, while all other members of the family, who escaped the fever, have no eye symptoms beyond a simple follicular catarrh, at most. I have seen plenty of such instances, as did my lamented colleague, Scheller.

In cases of this nature a direct contagion from person to person, an immediate contagion of the eye disease, must certainly be excluded. My observation teaches me that the cause is rather a malarial-like infection, excited by dwelling in low, damp, unhygienic rooms, where cattle and all the necessaries of labor are huddled together; here a nasal catarrh begins, continues and spreads—as it does to other organs—to the eye itself, either along the lachrymal passage (no previous disturbance here being necessary, as I in 1895, have shown contrary to Nieden), or by vascular connections, and finally leads to a keratitis vasculosa, and especially to follicular growth, a condition not always demonstrable when the exciting cause is dust alone. This acts in the same way as does any other infectious disease, typhoid or tuberculosis.

This conception, not yet however established in all details, explains the relative immunity, noticed by Swan Burnett, that the negro enjoys against trachoma. The negro, it seems, sweats easily, and is therefore free from malaria. It also helps to explain the freedom Berlin enjoys from malaria and trachoma, not because of its excellent hygiene, as Hirschberg thinks, but because it is built on a sandy and unmalarial soil.

(3b.) Malaria may also be acquired by drinking water (Roe), and in that way through the system affect the eye. This explains my own experiences in Alexandria, where I suffered from African fever; but such a means of infection is not so clear as those I have previously given.

These researches, if pertinent, must give us a clew to the treatment of trachoma. At the very outset I must oppose the well accepted idea that local applications, or radical operations, are the only means of cure. As a matter of fact, the excision of the fornix or tarsus is no radical operation, or else there would not be relapses so often following it (50 per cent, as Greff says). Greff even reports a case where the conjunctiva was practically excised on its entire palpebral area; Hoppe also reports a similar case where scarcely any conjunctiva was left, and yet there were relapses with granulations and pannus. I maintain that all such methods of treatment attack merely the symptomatic conditions; by excising hypertrophies and tarsus, or by expression of granulations by means of Knapp's roller forceps, one removes only the results of the inflammation which, in the one case might be called dust trachoma, in the other, swamp trachoma, while the true means of treatment, according to my argument above, ought to have been to attack the original cause itself, acting in the first case by the irritation of the poisoning dust on the conjunctiva, nose and perhaps adjacent cavities, in the other case through a general systemic poison in the blood itself.

To squeeze out granulations, from which even Greff expects no lasting good, is, I think, analogous to a treatment of typhoid or tuberculosis in which one would cut or squeeze out only the pathological glands of intestine or other mucous membrane, leaving the poison in the blood to pursue its destructive course; or as if a case of pharyngitis with associated nasal catarrh, were treated time and again with the cautery, leaving the inflamed mucous membrane of the nose to continue in the same unhealthy state. To excise the fornix and tarsus appears to even such an authority as Raehlmann in Dorpat, who has seen unfortunate results from the hands of other ophthalmic surgeons, to be no more than an intense bleeding, somewhat more intense than a mere scarification.

In my own cases I have followed the plan of making my treatment rather dietetic and gentle, even when I resorted to scarification; I have irrigated the conjunctival sack, the nose and nasopharynx; I have applied some soothing poultice in gauze to the eye itself; where the cornea was ulcerated I have used heat, as usual; I have on occasions massaged the conjunctiva with some soft substance like lufa sponge soaked in a simple soda solution, followed by cooling applications; I have attempted by means of Drouot's plaster to provoke a mild counter irritation about the nose and nasopharynx, applying it not on the mastoid but rather on the retrolobular region; I have forbidden alcohol, have insisted on a normal condition and even temperature of the head, prohibited the useless washing of it that so often leads to a fresh cold, and ordered stopped the irrational cutting or shaving of the hair, especially in persons like Egyptians who wear turbans or such headgear; I have insisted on warm foot wear, cold or warm foot baths (John Locke, 1692,) or rubber overshoes; I have ordered that most important of all conditions for eye or nose, fresh air and plenty of it, morning, noon and night; and I have tried by abundance of pure water drinking, to overcome any nervous condition which would be so harmful to all pathological states in eye or nose.

By such simple means, in treating persons living under somewhat decent circumstances—hospital attaches, soldiers, etc., I have been able to accomplish very good results, if not a complete cure, and I have at last made them able to work. The lowest classes, those living the life of the poorest animals in cellars or even out-houses, I have treated as best served for clinical purposes, hoping that a not distant future may allow a treatment of them, especially in times of epidemic, in barracks or tents, where the proper surroundings may be preserved. Even here I hope that mild measures will in many cases overcome the seeming necessity for operation.

It is not enough, however, to deal only with the treatment of individuals after the disease is established; we must attack relapses of course, but we must also make every effort to eradicate the cause itself. We must instruct the masses by school methods in the necessity of clean dry dwellings and workrooms, of cleanliness in every sense of the word. When an epidemic prevails we must investigate and overcome dampness and bad air of any kind. We must prevent overcrowding in living and work rooms. For the common good we must establish proper drainage and water supply for all, at public cost; we must prepare proper suburbs for wage workers near large manufacturing plants and by public intervention overcome improper child labor.

On swampy soil, houses must be built in a proper manner—with good foundations (as in St. Petersburg or the

Netherlands.) Cities must have proper sewerage, the country must be drained, the drinking water must be pure; if there is no natural supply of water, cisterns or artesian wells must be supplied. Even in such a great city as Koenigsberg, in 1894, there was no proper drainage, so that the house vats with fecal matter stank unceasingly when carried away in the morning; it is not unjust therefore, to ascribe to such a condition the prevalence there of all kinds of catarrhs. Swamps and shore areas must be dried, but not turned into a barren wilderness so that the opposite effect of dusty wastes be the result. The suitable meteorological conditions must be maintained. Dense woods must be thinned, the sun must have access to hitherto impenetrable forests, so that the climate may be properly changed, as in Finland, the most heavily wooded territory of Europe (57 per cent. forest) where the climate has been made essentially milder, by careful forestry. All this will prove our strongest method of combatting trachoma at its source. If the Mississippi were thinned of its primeval woods along its course, its floods prevented—if forestry laws, as in Switzerland since 1874, in the United States since 1891, were thoroughly established, if woods were built where needed, much would be accomplished to ameliorate the condition of our race.

Mankind has learned during the march of years and centuries, sometimes alas, not to his advantage, that the face of nature can be changed. He must still learn that it is only by a wise change of nature's face in future, that many of the ills—such as trachoma—from which he now suffers, can be blotted out.

NASAL INSUFFICIENCY DUE TO EXAGGERATED PROMINENCE OF THE ANTERIOR ARCH OF THE CERVICAL VERTEBRÆ.*

By Dr. James E. Newcomb, M. D.,

NEW YORK CITY.

LARYNGOLOGIST TO THE DEMILT DISPENSARY AND TO THE OUT-PATIENT DEPARTMENT ROOSEVELT HOSPITAL; INSTRUCTOR IN LARYNGOLOGY, CORNELL UNIVERSITY MEDICAL COLLEGE.

The condition of which I wish to speak very briefly this evening has a relation to the perennial subject of lymphoid hypertrophy in the pharyngeal vault. After all that has been written on this topic, one may well apologize and especially to the members of this Section for daring to re-open the subject. Since its masterly exposition by Wilhelm Meyer some thirty years ago, very little has been added to alter our general conceptions of this morbid condition. It's outlines were so graphically sketched by a master hand that later writers have done little more than to fill in comparatively unimportant details. We have learned that the possibility of certain complications must be borne in mind and the great number of operations done has in certain ways enlarged our range of thought.

In my own work, I have in two instances met with a condition which I have not seen mentioned in any of the text-books nor until recently in any current publication.

In 1895 in a paper read before the French Society of Laryngology, under the title of "A Critical Examination of Certain Rare Cases", Castex said that "numbers of children pass for cases of adenoid disease who have only a scoliosis of the septum or who have the vault of the nasopharynx shortened in its antero-posterior axis or with a wall projecting forward, the latter caused by an exagger-

---

*Read before N. Y. Academy of Medicine Section in Laryngology at stated meeting Jan. 25, '99.

ated hypertrophy of the atlas or of the odontoid process of the axis. In some instances the patient seems to have adenoid disease when he has not, or more rarely may appear free from it when he really has it."

In passing it may be remarked that he says also that adenoids of the posterior wall of the pharynx and not in the vault are more common than is generally taught and that copious hot irrigations with the Weber siphon will prevent the necessity of surgical interference in very young children. Of the truth of the first of these two statements, I am quite convinced; of that of the second, l am somewhat sceptical.

In a paper read before the same society in 1896 by Escat, under the title of "Congenital Stenosis of the Nasal Fossæ and of the Naso-Pharynx," he observes that this stenosis, found especially in individuals of mental deficiency, included among others, cases of prominence of the posterior wall of the pharynx but in such cases this deformity was but an anatomical detail among others which together constituted the general atresia of the naso-pharynx. Escat calls attention to the fact that both Ziem and Delavan were able to induce an arrest of development in animals of the bones of the face and thorax by causing occlusion of the upper air passages. In an examination of 113 children in one of the waif-colonies in France, Balme found 56 cases of adenoids or of other tonsillar enlargements but among these classic deformities there were other deformities manifestly not due to adenoids, such as multiple deformities of the skull and skeleton which constituted stigmata of degeneration.

Some writers with a vivid imagination have tried to deduce from this the conclusion that adenoids themselves are stigmata of degeneration but for the sake of the rising generation of the city in which we live, it is to be devoutly hoped that this latter conclusion is false.

In May of the present year, Mendel (in a paper still before the same Society) gave a brief study of this condition basing his statements on four cases which had been under his personal observation. He divided them into three categories:

1. Those with an osseous projection situated at the level

of the buccal pharynx with the coexistence of adenoids, one case.

2. Those with an osseous projection in the rhino-pharynx with the coexistence of adenoids, two cases.

3. Those with an osseous projection in the rhino-pharynx without the coexistence of adenoids, one case.

CASE 1 was that of a boy aged seven years with obstruction to nasal breathing. There was an osseous projection in the middle line of the buccal pharynx. It did not extend up into the rhino-pharynx and was regularly rounded. Adenoids were present. An especial symptom in this case was dysphagia. Breathing was better at night than in the day time. This paradox is explained by Mendel as due to the fact that in the day time the tonicity of the soft palate kept the latter in contact with the projection thus shutting off the air current from the rhino-pharynx. At night however, the palate was relaxed and so there was free space.

CASE 2 a girl of 8 years with classical adenoid characteristics. Osseous projection in the naso-pharynx; above it were adenoids.

CASE 3 a girl of 4 years. Rhino-pharyngeal projection causing almost total occlusion of breathing through the nose. The vault of the pharynx was free from adenoids.

None of these children presented any suspicion of hereditary or acquired syphilis or tuberculosis. A rather curious fact is that the father of case 1 presented the same anatomical peculiarity as the child but had no symptoms therefrom.

Two pertinent inquiries made by Mendel: should we operate for adenoids in these cases when they are present and if we do, what is the prognosis as to effectual removal of symptoms of obstruction?

In reply to the first, he advises operation. The osseous projection, he says, constitutes two-thirds, so to speak, and the adenoids one-third of the obstruction. If we cannot remove the latter entirely, let us do what we can.

As to prognosis, it is well to caution parents against expecting too much from the operative procedure. They should be informed as to the double condition present and be made to understand that it is partly removable and partly not. Otherwise if the symptoms continue, the physician may be charged with having done a useless operation and one not necessary or he may be accused of having done it improperly.

Some modification of the customary use of instruments may be necessary in these cases for it may be difficult to introduce the usual curettes and forceps into the vault owing to the osseous projection. Smaller instruments than those usually employed may suffice or the finger may be relied upon. As the patient grows older the projection gradually becomes relatively less in size and finally as in the case of the father above referred to, cease to give any trouble whatever.

I have met with two cases in which this condition was present. In neither was there sufficient bony obstruction to prevent the employment of the customary size of forceps or curette. In both the vegetations present covered the mucosa stretched over the projection in the rhino-pharynx and ran up beyond it into the vault. One case occurred some two years ago at the Demilt Dispensary but did not return after the operation and I have not been able to follow it up. The other case was that of a ten year old son of a colleague. In this case, there was a moderate amount of obstruction but the results of curetting have been all that could be desired.

118 West 69th Street.

DELUSIONAL INSANITY RESULTING FROM AUDITORY CONCUSSION.

By L. R. Culbertson, M. D.,

ZANESVILLE, OHIO.

OCULIST AND AURIST TO CITY HOSPITAL; U. S. PENSION BUREAU; C. & M. V. AND B. Z. & C. R. R., ETC.

Mr. F. M., age 38. His family state that he has been somewhat of an epileptic for many years. Examined Sept. 7, 1897. He said he never had any trouble in ears until some ten years ago when a gun was fired very close to his left ear. That there was a loud ringing in ear immediately after and has continued ever since. That hearing in that ear has been affected ever since, and that at times hears loud reports like that of a gun firing. Hearing R. E. Low, conversation normal. L. E. Low, conversation 4 ft. Fork L. E. heard better by Gruber's test. R. E. Watch 1 ft. L. E. Only on contact. R. E. Drum not opaque or retracted. L. E. Drum semi-opaque and congested about manubrium. Tubes open. Slight pharyngitis.

Examination of eyes: Javal astigmometer—1.5 D astigmatism ax. 90 both eyes. Retinoscopy (atropia) each eye —.75 at 90; + 1.5 at 180. Types R. E.—.75 sp $\bigcirc$ + 1.5 cy 90 = 5/5. L. E.— 1. sp $\bigcirc$ + 1.5 cy 90 = 5/5. Pupillary distances equal. No unbalance of oblique muscles. Other ocular muscles not tested. Glasses were given to be worn constantly as I desired to ascertain if ringing in ears and epilepsy could be due to eye strain, or aggravated by eye strain.

He says that he never had noises in the ears or deafness before the gun was fired near him.

His glasses have not given any relief, and taking this fact with the history we can eliminate eye strain as a cause. Eye strain—crossed astigmatism in particular—will cause ringing in the ears and epilepsy.

The concussion of the left auditory nerve from the report of the gun has caused deafness with partial degeneration of left auditory nerve and consequent hallucinations of hearing.

Epilepsy aggravates, and may be one cause of his delusion of hearing.

I append a letter just received from him in answer to mine, asking him if he wore his glasses, and if the noises were as bad as ever.

"DEAR SIR:—

"Yours at hand. You seem to think I am troubled with ringing and noises in my ears.

"Ever since the shot-gun was fired at me passing my right ear, I have been worried by loud noises not in my ear but external noises (such as slamming doors and unlooked for shooting, etc.). It feels as if someone were striking me in, or some place near my ear. On a farm it is not convenient to wear glasses all the time, but I wear them considerably at night.

"If you could imagine my feelings about the hunting season you would about know my trouble.

"Yours truly, F. M."

"BILATERAL AURAL LESION FOLLOWING TRAUMATISM."

By James F. McCaw, M. D.,

Watertown, N. Y.,

OCULIST, AURIST AND RHINO-LARYNGOLOGIST TO THE JEFFERSON COUNTY ORPHAN ASYLUM, MEMBER JEFFERSON COUNTY MEDICAL SOCIETY, CORRESPONDING MEMBER MEDICAL SOCIETY OF THE COUNTY OF KINGS, MEMBER ALUMNI ASSOCIATION OF THE BROOKLYN METHODIST EPISCOPAL HOSPITAL, ETC.

Mrs. S., aged 27 years, housewife, consulted the writer June 1, 1898, for difficulty of hearing, and from her the following history was elicited:

Family history negative. She had always been perfectly healthy with no trouble from her ears until nine months before, when she fell, striking the back of her head. Immediately she was seized with headache, epistaxis, vertigo, nausea, vomiting and severe tinnitus, but retained consciousness. A physician was consulted who prescribed a sedative, when the symptoms gradually abated, and at the end of about four days had entirely disappeared, except the vertigo and tinnitus, which continued, but with diminished severity. At this time her hearing suddenly failed so that she was almost totally deaf during the evening, with slight improvement after a night's rest, but gradually getting worse again toward evening. This phenomenon was noticed for about one week after which there was no appreciable difference in hearing between morning and evening, the deafness being well marked with constant tinnitus. The dizziness gradually disappeared and was entirely gone at the expiration of ten days to two weeks. For about four weeks following this, there was slight improvement of hearing in the right ear. Six months after the accident only slight improvement having taken place, she consulted a specialist and says he removed something from her throat, but without the slightest benefit to hearing. Upon examination, the canals, tympanic membranes and middle ear revealed nothing abnormal. There was found a high degree of impairment of hearing, spoken words were heard only when very loud and close to the ear, and then uncertainly. Watch tick not heard. Perception through the solid media of the skull was very much diminished, especially so for the high tones, which were not heard by air conduction. These findings were all exaggerated in the left ear. Tuning fork lateralized to the right side. Although the writer saw this case when far advanced (nine months after the commencement) he prescribed pilocarpin, potassium iodide and strychnine, but for the month she was under observation there was no improvement. At this time she passed from under observation and has not been seen since.

While this case is apparently one of double labyrinthine trouble, well borne out by the functional examination, the kind and exact location of the lesion or lesions are still matters of conjecture. These cases are exceedingly rare and have not been sufficiently studied to give us much light on the subject. In reviewing the literature at my command I find only one case reported which bears similarity to mine; that is one reported by D. Kaufman in the *Vienna Medical Journal*, 1897, an abstract of which is found in the May number of the ANNALS OF OTOLOGY, RHINOLOGY AND LARYNGOLOGY, 1898. In this abstract it is said, "Kaufman seeks to account for the labyrinthine affection on both sides by the sudden pressure brought to bear through the trauma and the spreading of this to the perilymphatic tracts and thence by means of small consecutive hemorrhages to the parietes of these tracts. In consequence of the hemorrhages there were disturbances of nutrition and deafness."

It seems to me that this theory best explains the findings in my case although probably the disturbances of nutrition had not progressed as far in this case, or more improvement had taken place before I saw it than in Kaufman's, as slight sound perception for low tones was present on both sides. While the writer favors this theory to explain the findings in this case, nevertheless when we consider the extreme rarity of symmetrical traumatic lesions and the complexity of the acoustic apparatus, we must admit there remains still a doubt concerning the accuracy of our diagnosis.

THE TREATMENT OF UNCOMPLICATED SUPPURATION OF THE MIDDLE EAR.*

H. A. ALDERTON.

BROOKLYN.

The treatment of suppuration of the middle ear unattended by any complication whatsoever, such as, *e. g.*, mastoiditis, extradural abscess, sinus-phlebitis, etc., will receive our attention in this article. It is a subject of sufficent importance if it only teaches us never by any chance to be guilty of advising procedures which will tend to bring about these complications, many of which are difficult to diagnosticate and still more difficult to cure. Also, if by the early and judicious treatment of an acute suppuration the too frequent development of the chronic condition may be guarded against. In the vast majority of cases this growth from the acute to the chronic process may be prevented by such treatment.

What is it well to do, in case of impending acute suppuration of the middle ear? First, the causative element should be unmasked and appropriate measures taken to counteract its influence. Should an acute naso-pharyngitis be discovered, it is well to order warm, mildly antiseptic nasal sprays for cleanliness and a mild, mentholated oil to keep down the swelling so that the nasal cavities can be cleared without further danger of forcing infective material into the ear through the eustachian tube. If a tonsillitis is at the bottom of the trouble, it should be treated by gargles, astringent or stimulating applications, incision and constitutional measures, as may be indicated. Teething and dental caries should early receive attention. Unfortunately in the exanthemata, the middle ear inflammation is one manifestation of the disease process and is not dependent upon transmission of infection from the naso-pharynx; the most we can do then is to limit further

---

*Read before the Brooklyn Medical Society Meeting of March 18, 1898.

infection through the eustachian tube, treat the general disease and give such attention to the ear as the general condition of the patient renders possible and advantageous.

Should the attending physician be so fortunate as to be called in during the inception of the attack, he should, providing the intensity of the inflammation does not contraindicate it, very gently inflate the ear by means of the Politzer bag, after having rendered the naso-pharyngeal tract as nearly aseptic as is possible. Inflation should be practiced with the aid of the otoscopic tube so as to use no more force than is just sufficient to open the eustachian tube. This procedure, in conjunction with the use of naso-pharyngeal antisepsis, will often prove adequate to abort an attack.

If seen early and in a sturdy patient, local blood-letting is often of great value. From three to four leeches may be applied in an adult, two in front of the auricle and one or two behind. In a child, one leech in front of the auricle is sufficient. The skin just in front of the tragus, as close to it as is possible, should be surgically cleansed and the leech applied by means of a leech-glass, one aperture of which is firmly pressed against the skin and through the other aperture the leech is prodded until it takes hold, it being necessary at times to incise the skin previously. So soon as a firm hold is taken, the leech-glass is withdrawn and the leech permitted to hang on until the weight of the abstracted blood causes it to loosen its hold. Behind the auricle, the leech should be applied on a level with the orifice of the canal and close to the posterior auricular furrow. Bleeding is to be encouraged by applying moderately warm moist cotton pledgets. After enough blood has been lost the bites should be antiseptically cleansed and the bleeding stopped by pressure, cold, or if need be, by local hemostatics. Blood-letting should not be tried with feeble or anæmic persons.

This should be immediately followed by hot douchings through the external auditory canal. For this purpose any fountain syringe holding about one quart of liquid will serve. The reservoir should be suspended just a trifle higher than the patient's head, force of current not being desirable as it may mechanically irritate the tympanic structures. To the hard rubber terminal of the syringe is affixed a soft rubber tube, of smaller calibre than the orifice of the canal, projecting half an inch beyond the distal end of the terminal for introduction into the meatus. This soft rubber terminal is to be introduced to its full extent in adults and older children but only to just within the orifice in infants. The liquid used should be as warm as the patient can bear with comfort and may be either sterilized water or a solution of boric or carbolic acid in sterilized water. From a pint to a quart of the liquid should be used at each sitting and the sittings should be at intervals of from one to three hours, according to the severity of the symptoms and the age or intractableness of the patient, it never being good practice to use the douche so frequently in infants or intractable children that they are kept in a constant state of nervousness and the ear in a constant state of engorgement from crying, struggling, etc.; here warm applications by means of instillations and hot, moist or dry cloths or the hot water bag are better.

Following each application of the douche the canal should be thoroughly dried out by small cotton tampons, formed on a cotton carrier and then removed from the carrier and introduced by the fingers or forceps, which are left *in situ* for a few minutes and then withdrawn. These tampons should be about one and a half inches in length and fluffy on the distal end, so as not to irritate the drum membrane. The thorough drying of the canal after douching is of great importance; it prevents maceration and exfoliation of the epidermis with the swelling and infection so often attendant upon this condition of the skin; it is also very necessary, if there is a coexisting otitis externa, as a preliminary to the use of medicaments. The drying out should never be done except with the greatest gentleness and caution, the attendant always being required to demonstrate his ability in the presence of the physician, after instruction.

Finally, the orifice of the canal should be stopped by a pledget of cotton between the douchings, so that by no mischance a draught of cold air may enter the ear.

The writer has never seen any benefit accrue from the popular practice of instilling medicated solutions, *e. g.*, laudanum and sweet oil, etc., into the ear, so long as the drum membrane was unbroken, other than might arise from the temperature of the instilled solutions. Benefit is obtained from the temperature of such instillations but mostly in very mild cases that would have done just as well, if not better, under the use of the hot douche of sterilized water; while in the severe cases the temporary benefit derived is more than counterbalanced by the softening of the epidermis, the irritation through the presence of rancid solution and the consequent infection.

Bodily rest is of the greatest importance and should be strenuously insisted upon, all occupation being absolutely interdicted.

A light diet should be ordered and all spicy foods, alcoholic stimulants and the use of tobacco denied.

The bowels should receive attention; a good purgative controlled by an opiate to secure easy and painless stools often doing heroic service.

The normal quantity of sleep should be obtained by the administration of soporifics, where required, combined with laxatives to relieve the bowels.

Two constitutional remedies exert a beneficial influence on the middle ear inflammation, atropine and pilocarpine. The former in small and repeated doses tends to limit the formation of secretion in the tympanum; the latter, to be given to robust persons only, in doses of $\frac{1}{12}$-$\frac{1}{8}$ gr. once or twice daily either by the mouth or hyperdermatically, relieves vascular tension if carried to the physiologic limit. They should, of course, not be given to the same patient.

Those cases in which there is considerable radiation of pain of a neuralgic character around the ear will often obtain great relief from a liniment containing chloroform, opium, belladonna or aconite well rubbed into the skin in the neighborhood of the auricle.

Should all these procedures fail to cause the subsidence of the otalgia and of the constitutional symptoms within a reasonable time and, especially, should bulging of the drum membrane threaten its integrity or the intra-tympanic pressure threaten the life of the contained structures, then paracentesis should immediately be done. With even less delay if the middle ear inflammation is one of the manifestations of an exanthematous disease, as in this case the vitality of the tympanic structures may be destroyed with appalling suddenness.

Having decided upon the performance of incision of the drum membrane, the first step to be undertaken is the thorough surgical cleansing of the external auditory canal. This may be attained by syringing with a warm aqueous solution of tincture of green soap followed in a few minutes by a syringeful of warm sterilized water and thorough drying with cotton tampons. Alcohol, 95 per cent., is then instilled and the canal again syringed with a solution of carbolic acid, 1-40, in sterilized water, the canal again being thoroughly dried out with sterilized cotton tampons. A sterilized speculum is now introduced and under good illumination an incision is made into the most prominent portion of the bulged membrane and extended downward to the lowest limit of the inflamed membrane. If there is a chance for choice the posterior inferior quadrant of the drum membrane is the site of election.

The middle ear should now be gently inflated to drive out the contained secretion, which must be wiped away thoroughly by aseptic cotton tampons. After the canal is entirely dry, the bleeding having stopped, a wick of antiseptic gauze is carried up to the membrane and permitted to lie loosely in the canal and the concha of the auricle, being covered externally by absorbent cotton and the whole kept in place by a few turns of a gauze bandage around the head.

If now at the first dressing, in 6-12 hours, the pain is much relieved and the tympanum seems to have been well drained, the canal is simply dried out with aseptic cotton tampons, after a gentle inflation, and a fresh dressing applied as above. It will often be the experience to find the canal perfectly dry and the drum membrane entirely healed after the second or third dressing. This form of treatment, first advocated by Haug, of Munich, should be persisted in so long as the discharge is not profuse or until symptoms of irritation or retention appear, the dressings being changed at least twice a day.

Should the discharge be so profuse as to keep the dressings continually soaked or should symptoms of irritation or retention appear then it becomes necessary to change the plan of treatment and resort to syringing.

So long as the only desire was to apply moist warmth to the ear, douchings were all sufficient but they have not enough force to wash out secretions from the depth of the canal and their prolonged use encourages tissue swelling and the formation of granulation and polypoid tissue, the most prolific causes of retained secretion and its sequelæ. On the other hand the repeated short, sharp impact of a column of water in syringing, while undoubtedly more irritating to the sensitive structures during its continuance, yet by breaking up and cleaning out the secretions more thoroughly, leaves these structures better able to cope with the effects of the inflammatory process.

There are two syringes on the market which are fairly efficient for lay use, the one ounce hard rubber syringe and the soft rubber ulcer and aural syringe manufactured by the Davidson Company; the glass syringe usually recommended for ear use is an abomination, being too small, without sufficient force and having a terminal poorly adapted for application to the meatal orifice.

Instructions should be given to the attendant as to the correct way in which to hold the auricle during the syringing. In small children the auricle should be held outward, downward and somewhat backward, in larger children and adults it should be outward, backward and upward, the lumen of the canal thus being most perfectly straightened. The terminal of the syringe is then introduced into the orifice of the canal and the stream directed inward and slightly upward, in adults also somewhat forward.

At least a pint of liquid should be used at each sitting and the syringings repeated every 2 to 4 hours according to the quantity of the discharge.

Any mild antiseptic may be added to the sterilized water but this is not so necessary if the canal is thoroughly dried after each syringing; the only purpose an antiseptic serves being that of preventing infection of the skin surface, as the solution never reaches the middle ear. The middle ear is not contaminated from the canal if the latter is kept clean and free from decomposing secretions.

In treatment by syringing, as well as with gauze, it is of vital importance to provide for thorough drainage from the middle ear cavity by such a free incision in the drum membrane that all parts of the cavity are in communication with the outlet. For this reason, whenever the perforation seems inadequate, the opening should be enlarged by carrying the incision from under the posterior fold around the periphery of the drum membrane to the antero-inferior quadrant, being careful to incise the membrane only and not to disturb the ossicular chain by entering too deeply. This is the more important as cases are on record in which, because of dehiscences of the bony inner tympanic wall, the bulb of the jugular has been penetrated with alarming, if not fatal, hemorrhage. If the tympanic attic is involved, the incision is to be extended through Schrapnell's membrane and even up into the attical space. This latter procedure is especially indicated since Forns has shown that normally there exists, in the majority of cases. no communication between the tympanum and the attic, membranous folds completely dividing these spaces from each other.

Following syringing, the canal should be gently and thoroughly dried and the orifice stopped with absorbent cotton, the saturation of which by secretion at any time previous to the expiration of the regular interval indicates the necessity for another syringing.

At least once a day, after syringing, gentle inflation by means of the Politzer bag should be practiced, extruded secretions being removed in the usual way. This manipulation empties the tympanum of secretion, prevents the formation of adhesions and improves the hearing: the chance of forcing infected material into the mastoid cells not being worthy of consideration when contrasted with the benefit to be derived, and this seems especially problematical since Forns' experiments and since Politzer has shown that the same process exists in the mastoid process. "In all cases of suppurative otitis media in which there did not exist during life any trace of inflammation of the mastoid process, and where an autopsy was made, we constantly found pus in the mastoid cells." (Ann. des mal. de l'or., 1892, No. 5.)

It is the writer's custom never to insufflate powders in cases of acute suppuration. The powders never reach the inflamed surface, the perforation being too small, and they tend to pack into the fossa between the drum membrane and the antero-inferior canal wall; thus interfering with cleansing, forming a nidus for decomposing secretions and, most important, obstructing the perforation and causing retention. If the discharge is so slight as to make the powders of any use then the antiseptic gauze wick carried to the fundus of the canal answers the purpose much more satisfactorily, as it not only absorbs and disinfects the discharge but also keeps it away from the skin lining of the canal, thus providing against the disturbance of the epidermal cells so frequently seen when powders, such as boric acid, are used.

Any other conditions which arise during the course of the treatment must be treated according to ordinary surgical principles; thus granulations obstructing the perforation must be removed, etc.

Under this method of treatment most of the acute cases will naturally proceed to perfect recovery with closure of the perforation.

Should the process become chronic under treatment or should the patient present himself while suffering from a chronic otorrhea, then the treatment must be regulated according to the conditions found at the examination.

First and foremost it is necessary that access to the tympanic cavity be perfectly free. To this end all granulation tissue must be removed and its recurrence prevented. The opening through the drum membrane must be sufficiently large to permit remedies to penetrate easily into the tympanum. It is the writer's custom whenever the perforation is at all small to excise, under cocain anesthesia, a large segment of the drum membrane in the region of the pre-existing perforation. No hesitancy should be felt in undertaking this procedure, the regenerative power of the membrane being equal to bringing about a closure of the perforation in the course of healing in the majority of cases. In case the perforation is or has been made sufficiently large, the treatment may be continued with the knowledge that the remedies will probably reach the diseased tissues, if rightly applied.

Next in importance is cleanliness. This is attained by syringing, at home, with any mild antiseptic solution, preferably boric or carbolic acid in boiled water, as in the acute stage; the syringing being always assisted by thorough drying out of the canal and stoppage of the orifice with cotton. At the office, the writer first syringes with a weak solution of tincture of green soap in water, removes this by a syringeful of boiled water, inflates the ear, drying out any expelled secretion, then exerts gentle suction by means of the Siegel's otoscope to remove any secretion clinging to the crevices in the tympanum, drying out the same with cotton. The ear is then syringed with a mild carbolic solution and again gently dried, following which lukewarm alcohol is instilled and left in place for a few moments. The alcohol is thoroughly removed and the ear is in shape to receive such medication as may be indicated. Of course, if Schrapnell's membrane is perforated the attic should receive attention. For syringing the attic either Hartmann's canula, attached to an Alpha syringe by means of a slender rubber tube, or a canula devised by the writer may be employed; the latter instrument only being applicable when the ossicles have been thrown off by nature or removed surgically.

If the recesses of the tympanum or attic are found to be shut in by bands of fibrous tissue or by granulation tissue, these obstructions must be thoroughly removed by the curette, under cocain anesthesia. In the case of granulation tissue the base from which it grows must be scraped and cauterized with chromic acid.

If the bony wall of the tympanum or the ossicles are found to be carious, the carious tissue must be removed to secure healing and to prevent recurrence. It is advisable in contemplated removal of the ossicles to consider the functional ability of the ear, where the hearing is nearly normal.

Occasionally it is well, in such cases, to hold one's hand providing drainage is good and the patient can be kept under observation at longer or shorter intervals. Nature here sometimes brings about remarkable cures from the standpoint of the conservation of the hearing. But where the hearing is much reduced and always where symptoms of retention are occasionally present, it is invariably the better practice to remove the carious ossicles, after a preliminary period of non-surgical treatment has proven ineffectual, thus securing adequate drainage from the attic and insuring against recurrence of the suppuration.

While many cases are cured by the removal of carious bone or ossicles, there are a few cases in which the seat of the disease is so deeply situated, *e. g.*, in the antrum, that the measures recounted merely ameliorate the condition and fail to cure. In these and also following removal of the ossicles, the writer has derived much benefit from the use of the canula devised by himself for syringing out the mastoid antrum. It consists of a Hartmann's canula whose distal end is prolonged first upward and then backward, so that the aperture points directly toward the aditus ad antrum when the canula has been introduced into the tympanic attic. The vertical distal portion measures about one-third of an inch in height. It serves not only for syringing out the antrum but also for introducing liquid remedies and powders. It is sometimes difficult to empty the antrum of liquid after syringing; the writer has succeeded in overcoming this feature by carrying a narrow strip of gauze or cotton up into the aditus and then by inclining the head forward siphonage completely empties the cavity so that the cotton introduced afterward comes away perfectly dry.

In those very rare cases in which the measures detailed above fail to induce a cure, the writer recommends the use of his antrotome to make a counter opening into the antrum, through which opening syringing, the application of medicaments and, in a limited degree, curettage can be instituted. The writer has modified his technique in the use of this instrument in that he now lifts out the membranous canal before applying the instrument, instead of, as formerly, slitting the posterior membranous canal wall to introduce it into the tympanum. Burnett thinks that this instrument, while very ingenious, might endanger the lateral sinus in certain cases, but as the cases in which the lateral sinus lies so far anteriorly as to be in danger are very rare and as the drill hugs the posterior canal wall extremely close and is only 3 mm. in width, this danger would seem to be very small. If the drill did extremely rarely wound the sinus it would do more than the chisel or spoon does in the radical operation now so popular for the cure of the same condition.

The writer does not feel any great enthusiasm over the so-called "radical operation," especially since reading Stetter's article reporting a number of recurrences or failures to cure in cases operated upon at Halle and elsewhere. Carious or necrotic tissue should be removed wherever situated but with as little disturbance of healthy parts as is consistent with the attainment of the object sought.

In the above description of the treatment of chronic suppuration of the middle ear only the broad general principles have been indicated. Purposely, because every case must be a law unto itself beyond these generalities. The procedures described above, or similar ones, must be undertaken in every case where called for, while the particular plan of treatment pursued outside of these is a matter of individual preference or personal idiosyncrasy on the part of both physician and patient. Some cases will respond nicely to the dry treatment by means of boric acid powder, aristol, iodoform or Haug's naphtholated gauze, etc.; others do better under the use of solutions of nitrate of silver; others, under alcoholic instillations, either pure or combined with boric acid, salicylic acid and glycerine; sulphate of zinc solutions act well in some cases and so on and on.

To describe which particular remedy is indicated in each particular case would take more time than we have at our disposal to-night and warrants study by itself.

ABSTRACTS FROM CURRENT OTOLOGICAL, RHINO-LOGICAL AND LARYNGOLOGICAL LITERATURE.

I.—EAR.

Otorrhoea For Three Years—Fracture of the Base of the Skull—Operation—Recovery.

1. ASTIER. *(Rev. hebdom, de Laryngologie, d'Otologie et de Rhinologie,* No. 45 Nov. 5, 1898.) Read before the French Society of Laryngology May, 1898.

A man of 57 came to the office of the author with paralysis of right half of his face, and of the entire right side. It was stated by the house physician that the patient had fallen down a stairway. He was suffering for three years from otorrhea. After much guessing about the different possibilities the mastoid process was opened and the skull trephined. A fracture was discovered, but no abscess of the brain. The patient recovered completely, and his otorrhea was cured. *Holinger.*

A Case of Mastoiditis.

2. BULSON, A. E. *(Journ. Amer. Med. Assn.,* Jan. 7, 1899.) The interesting features of the case reported are as follows: The development of a well-marked sinus phlebitis, with probable thrombosis, after apparent recovery from abscess of the mastoid antrum which had been thoroughly evacuated; the disappearance of the phlebitis and probable thrombus without the development of other pyemic manifestations; the development of an epidural abscess somewhat remote from the site of the carious process.

The importance of thoroughness in mastoid operations is urged, the operation including not only a free opening of the mastoid antrum and obliteration of its pneumatic structure, but careful search for and removal of small areas of diseased or carious bone. *Scheppegrell.*

Some of the Special Germs in Inflammation of the Middle Ear, with an Interesting Case.

3. CHEATHAM, W. *(Medical Record*, Oct. 1, 1898.) Of 102 cases investigated by Milligan, eight were tuberculous. Zaufel and others state that in acute otitis media due to cold and catarrhal extension from the nasopharynx, the pneumo-bacillus of Friedlaender and the diplococcus of Fraenkel play an important part. The characteristics of the bacillus of Friedlaender are as follows (W. C. Parkes in *British Medical Journal*):

1. Non-motility. 2. Polymorphism. 3. Decoloration when stained by Gram's method. 4. The presence of a well-developed capsule, especially when taken from the heart blood of the inoculated mice after death.

The cultural characters are as follows: 1. Whitish semitranslucent, sticky growth in bouillon. 2. Aerobic and anaerobic growth in gelatin, causing no liqufaction. 3. Whitish, moist, raised growth on slanted gelatin, the growth sliping to the bottom of the tube after four or five days. 4. Abundant gas production in glucose-gelatin stroke cultivation. 5. Slimy and almost transparent growth on agar and blood serum. 6. Formation of acid in dilute lactose bouillon. 7. Coagulation of milk when an acid reaction (in four cases before the ninth day, in one case on the eleventh day.) 8. Brownish abundant growth on potato.

In a case of acute suppuration of both middle ears, an examination showed the presence of a pure culture of the diplococcus of Weichselbaum. *Scheppegrell.*

Some Relations Between Hearing and the Word Spoken.

4. CHERVIN. *(Rev. Hebdom. de Laryngol., d'Otol., et de Rhinol.*, No. 42, Oct. 15.) The author speaks about an anomaly which he discovered not very exceptionally amongst children; especially girls, and a few adults. They cannot distinguish between some consonants, for example t and k, s and ch or our j. The author thinks this is a hereditary imperfection, which must be overcome by education. In a case he cites the mother of the child made the same mistake—no wonder the child did.

Holinger.

A Study of Thirty Cases of Antral Empyema.

5. COBB, F. C. *(Boston Med. and Surg. Journal*, Dec. 1, 1898.) In the first case reported, a rubber had been introduced into the antrum through a dental cavity and set up the inflammatory process. In the second case, it proved to be due to a foreign body which rested between the alveolar process and the antrum and was found to be a twelve-year molar.

Seven cases were due to acute catarrhal conditions and were characterized by severe rhinitis, followed by a discharge of muco-pus from one side only. Of the total number of cases cited, four were undoubtedly syphilitic and yielded to antisyphilitic treatment. In one case only was there evidence of malignancy. Ethmoidal disease was associated with antral suppuration in seven cases.

Scheppegrell.

Chronic Suppurative Otitis Media in a Tubercular Patient—Brain Abscess—Trephining Through the Mastoid.

6. COVILLE ET LOMAARD. *(Annales des Malad. de l'Oreille, du Lar., du Nez et du Phar.*, No. 11, Nov. 1898.) Man of 28 was suffering from an old otitis media, and tuberculosis. A mastoid operation became necessary. Three days after the first operation a cerebellar abscess was evacuated the presence of which was not suspected before the operation. The patient made a comparatively good recovery but died twenty days after with symptoms of weakness and of consumption. *Holinger.*

Purulent Thrombosis of the Lateral Sinus. Epidural Abscess Extensive Subperiosteal Abscess With Oedema of the Scalp, Face and Neck—Operation—Recovery.

7. DUNN, Richmond. *(Archives of Otology*, Vol. XXVII, No. 6.) A young man, aged 17, with a history of scarlet fever and repeated attacks of right middle ear inflammation. The whole scalp was very sensitive to touch and general edema extended from nape of neck to eyebrows. This edema was greatest in amount, over the occipital region and on the right side. There were irregular chills, high fevers, profuse sweats followed by temperature sinking to normal. Occasionally the right ear discharged fetid pus.

The neck was swollen in front and particularly on the right side as far down as the clavicle.

The right mastoid antrum was opened and fetid pus found, curetting backward an epidural abscess was found. The lumen of the sinus was filled with clot which was cleared out toward the bulb and then posteriorly till a free flow of blood was obtained. Subsequently a large collection of pus formed beneath the scalp in the occipital region. Incisions were made and the parts drained.

Paralysis of the right external rectus was present for a time, then disappeared. *Campbell.*

Purulent Mastoiditis Complicated by Epidural, Subpetrous and Post-Oesophageal Abscesses; Death Presumably From Internal Hemorrhage.

8. DUNN, RICHMOND. (*Archives of Otology*, Vol. XXVII, No. 6.) A man, aged 38, for five weeks had suffered from severe pain in the left ear. At no time had there been any discharge. On account of swelling of adjacent tissues the m. t. could not be clearly seen. There was swelling over the mastoid and partial left facial paralysis. The mastoid was opened and practically the whole process was infiltrated with pus and granulations.

For 10 days the patient did well but complained of insomnia and some stiffness in the left side of the neck. A swelling now appeared about the upper end of the sternomastoid. Thinking that an abscess had formed an incision was made but no pus found. The skin was now reflected from the post-mastoid region and the sinus exposed. An epidural abscess containing about one drachm of greenish pus was opened. This abscess lay between the sinus and the inner table of the skull following the lateral along its course about half of an inch and the sigmoid for about the same distance.

Patient has continued to complain for a number of days of great pain when attempting to swallow. He refers the pain to the back of the larynx. Examination of pharnyx and larynx was impossible on account of patient's inability to open his mouth. Examination of external ear canal shows its lumen very small and the upper wall extremely sensitive to the slightest pressure. Patient stated that he had drawn from his nose into his throat a lot of badly-tasting matter and this suggested the possibility of pus in the middle ear escaping by way of the eustachian tube.

Under chloroform the operation wound was again examined; while probing in the upper posterior angle of the auditory canal an abscess was opened from which considerable pus escaped. While under the anæsthetic patient coughed up a considerable amount of pus. The pain in swallowing persists and he experiences great pain when the neck is moved. The swelling under the sterno-mastoid has increased. In making pressure over the swelling in the neck a great quantity of pus was seen to flow from the mastoid antrum. This abscess being opened, was found to be situated below the inferior face of the petrous portion, around the styloid process and foramen. Two weeks later the abscess in the neck burst into the œsophagus and much pus was expectorated. The discharge diminished greatly but the patient did not regain his strength, before death he passed by the bowel much clotted blood probably as a result of rupture of a vein into the post-œsophageal abscess. No autopsy was made. The mistake was made in not recognizing at first a case of Bezold's mastoiditis.

Campbell.

Otitis Media With Triple Personality.

9. ERWIN, A. J. *(Journ. Amer. Med. Assn.,* Jan. 7, 1899.) The otitis media presented no unusual characteristics. The three personalities which were entirely separate and distinct, each with its own facial expression, language and style, are of much interest and are important from a psychologic standpoint. *Scheppegrell.*

The Value of Ophthalmoscopic Examination for Diagnosis of Endo-Cranial Complications of Otitis.

10. GRADENIGO. *(Annales des Malad, de l'orielle du Larynx, du Nez et du Phar.,* No. 12, Dec., 1898.) In half of the cases we find lesions of the papilla. Only exceptionally we find papillitis in mastoiditis without endocranial complication. In extradural abscesses around the sinus the papillitis may be the only symptom. It is advisable to examine each patient for this symptom. The papillitis does not give any indication of the location of the abscess. The recovery of the papilla after operation may

be regarded as indication of complete result of the operation. *Holinger.*

Noise as a Factor in the Causation of Disease.

11. HASTINGS, ROBT. W. (*Journ. Amer. Med. Assn.*, Dec. 24, 1898.) The author refers to the effect of noise on the ear itself, its effect on existing diseases, its power to originate disease, and its effect on the health of the community. No excitement is more frequent, none more persistent than noise in neurasthenia. In illness it not only prevents sleep, which is an important factor in the therapeusis of disease, but also produces irritation upon the nerve centers. It is a factor in the causation of certain ear diseases; in the course of many acute and systemic diseases, and in many nervous diseases, chiefly neurasthenia.
Scheppegrell.

A Case of Inflammation of the Membrana Tympani Apparently Due to Dental Pulpitis.

12. JAKINS, PERCY. (*Journ. of Laryngology, Rhinology and Otology*, Nov. 1898.) The patient, a girl of the age of 10, was suffering from acute myringitis, which the writer felt was due to a disease of the two upper central incisors. Both nerve canals were drilled and a quantity of decomposing matter was evacuated and the inflammatory condition of the membrane speedily disappeared. *Loeb.*

Tumor of the Entrance of the External Ear.

13. LANNOIS. (*Rev. Hebdom. de Laryngol, d'Otol. et de Rhinol.*, No. 42, Oct. 15. Read before the French Laryngol. Soc. May, 1898.) A woman of 75 was suffering from this tumor for 12 years. The tumor had grown rapidly during the last 6 months before death. It proved to be an adeno-epithelioma, of very mild malignity. The case is published as unique. *Holinger.*

Surgical Treatment of Some Facial Paralysis of Otitic Origin.

14. MOURE AND LIARAS. (*Rev. Hebdom. de Laryngol. d'Otol. et de Rhinol.*, No. 51, Dec. 17, 1898.) The authors sum up their careful work in the following conclusions: Facial paralysis in old otorrhoea may be due to compression by granulations or sequestra. The compression is

difficult to locate. The fact that there is electric reaction of degeneration does not prove that we should not operate, the nerve may even then recover. It is sufficient to remove all diseased tissue along the fallopian canal, especially above the oval window, and in the attic. *Holinger.*

Tuberculosis of the Middle Ear.

15. OPPENHEIMER, S. (*Medical Review of Reviews*, Nov., 1898.) The tubercle bacilli may be conveyed to the ear through the eustachian tube by the blood vessels and lymphatic system, and from the external auditory canal through a perforation of the membrana tympani. In practically all cases of tubercular infection of the tympanic cavities, contagion is conveyed to the part by way of the eustachian tube.

The morbid changes after infection may be either acute or chronic, the latter being more frequent. The symptoms vary with the progress of the disease and the amount of destruction present. The first indication is frequently the sudden appearance of pus in the external auditory canal, without any evidence of the disease being present previously. Examination shows the tubercle bacilli in the pus.

Enlargement of the periauricular glands, due to the tubercular infection, is sometimes the first symptom present. The nodules in the middle ear are about the size of a mustard seed, scattered irregularly over the entire mucous membrane. When touched with a probe they are found to be hard and resistant. The tubercular process may remain latent in the aural cavity for a considerable time without apparently causing serious trouble, but extension of the disease usually occurs sooner or later and pressages a fatal termination. An early diagnosis is important and offers a chance for the removal of the diseased area in a limited number of cases. General remedies are important.

Scheppegrell.

Tone Sensation with Reference to the Function of the Cochlea.

16. RUTHERFORD, WM. (*Lancet*, Aug. 13, 1898.) Our auditory sense excels all others in the power of distinguishing different qualities of sensation. It can distinguish tones varying in pitch from 16 to 20 vibrations to 40,000 or

even 60,000 per second. A good violinist can detect about 50 differences of pitch within the range of a semitone, implying that he can feel a difference due to half a vibration. But in some highly trained ears the sensitiveness is even greater, so that difference of pitch due to a quarter or even one-sixth of a vibration can be felt throughout the ordinary scale (Luft). How are we enabled to appreciate differences of pitch? As we all know, Helmholtz supposed that the same nerve cell cannot give us two different sensations even of the pitch of tone and therefore a separate nerve-fibre and cell must be impressed to enable us to hear tones differing by, it may be, only the fraction of a vibration. This was his extension of doctrine of specified activities to explain the different qualities of tone sensation. To render his theory feasible he supposed that the cochlea must contain an immense number of minute resonators capable of severally responding by sympathetic vibration to all the tones we can hear. Each cochlear resonator was supposed to stimulate its own special nerve-fibre and this in turn its own cell in the brain. When he advanced that theory in 1862, the structure of Corti's organ was so imperfectly understood that Corti's pillars were supposed to be the nerve terminals. They increase very slightly in length from base to apex of cochlea (from 0.06 mm. at the base to 0.1 mm. at the apex), yet Helmholtz deemed that small difference sufficient for his theory. But when it was ascertained that no pillars of Corti occur in the bird's cochlea, and that in the mammal they are mere supporting structures apparently unconnected with nerves, the theory of their function as resonators had to be abandoned.

But the resonance theory was not allowed to drop. Hensen came to its support with a new suggestion. He showed that the fibres of the basillar membrane increase slightly in length from base to apex of cochlea and that they may play the part of resonators for tones of widely different pitch, although the shortest fibres are only about $\frac{1}{5}$ mm. and the longest about $\frac{1}{3}$ in length. But the fibres are so heavily damped by layers of cells cemented to them on both surfaces of the membrane and along their whole length that they are ill-adapted for resonance. Hensen thought that the basillar fibres run parallel with each other and that they lie in a single layer like piano strings;

but in the bird's cochlea there are two layers of fibres and irregular fibres running between them. It is impossible that systems of fibres so arranged can severally act as resonators. A nightingale would scarcely be able to learn its songs if it had to rely on the individual vibration of basillar fibres to enable it to appreciate tones of different pitch. But another argument seems cogent against Hensen's theory. If Corti's membrane had consisted of a single layer of tense parallel fibres, each cemented to the hairs of a series of Corti's cells and varying much in length throughout the cochlea, Helmholtz and Hensen would not have been so unreasonable in ascribing stimulation of Corti's cells to a segmented vibration of the superjacent fibres. But the sound waves first encounter the comparatively lax membrane of Corti and after affecting the hair cells reach the heavily damped subjacent basillar membrane, which is surely in a most unfavorable position for sound analysis.

In Ziehen's "Physiological Psychology" Corti's membrane is substituted for the basillar membrane in alluding to Hensen's theory. No one has ever ascribed a power of sympathetic response to the fibrils of Corti's membrane. It is not a tense membrane; its fibrils are in laminæ and in the bird it is semi-gelatinous. Nor has anyone ascribed to the short rod-like hairs of Corti's cells the power of selective and sympathetic vibration. They are, according to the writer, the mere transmitters of all vibrations, whether simple or complex, to Corti's cells. Therefore, the minute structure of the cochlea places insuperable difficulties in the way of Helmholtz and Hensen's theory and very likely if Helmholtz had known the structure of the cochlea in mammals, but more especially in birds, as described by Retzius, the resonance theory would never have been proposed.

But there is a physical defect in the theory of Helmholtz which he entirely overlooked; for example, if we sound three tuning forks, giving the notes of a common chord—say C, E, G—each fork produces a simple tone due to a series of simple vibrations. The three sets of vibrations are compounded and enter the cochlea as a compound vibration. But although we have the compound sensation of a harmony we can discriminate its three constituent tones. Helmholtz inferred that the cochlea must analyze the compound vibration and that three sets of terminals and at least three different sensory cells in the brain must be affected. Harmony, however, depends on the relative frequencies of vibrations. It is strictly determined by the mathematical ratios of numbers. In the common chord of C, E, G, the numbers are 264, 330, 396 vibrations per second. Helmholtz never supposed it possible for the cochlear resonators to send these or any such numbers of impulses along their respective nerve fibres. He abandoned the mathematical cause of harmony at the cochlea and imagined that it could be accounted for by the specific activities of cells in the auditory centre. When he published his book on "Tone Sensations" in 1862 he could not have been impressed by a passage in Fechner's "Psycho-physics," published two years previously (1860), containing these significant words: "The first, the fundamental hypothesis is that the activities in our nervous system on which the sensations of light and sound functionally depend are—no less than the light and sound themselves—to be regarded as dependent on vibratory movements."

Helmholtz, in his book on "Sensations of Tone," gives no evidence that he ever sought to determine how many impulses per second can be transmitted along a nerve; on the contrary, he shows quite clearly that he dismissed the subject without experiment. This statement may be verified by examining page 253 of the third edition of his work translated by Ellis and published in 1875. Yet eight years before that edition appeared, his own assistant, Bernstein, had made his researches on the nerve impulse by means of his differential rheotome and had ascertained that the electrical change which accompanies the transmission of a nerve impulse lasts only 1/1400th of a second (0.0007 second) even when the nerve is powerfully stimulated. Therefore, even when tested in this manner, 1400 perfectly discrete impulses can be transmitted per second in a frog's sciatic nerve. But Bernstein's method of experiment, though highly ingenious, is far too crude to inform us of the limit to the frequency of the impulses that a nerve is capable of transmitting. The question is, if a nerve can transmit, say, 1400 perfectly discrete impulses per second, is there any reason why it should be supposed incapable of transmitting 14,000 or even 40,000 per second to produce the highest audible tones? It may be imagined that if Bernstein had made his experiments before Helmholtz published the first edition of his book he could not have abandoned the mathematical cause of harmony at the cochlea, but must have believed that the vibrations produced by the several tones of a chord are continued into the brain. The writer believes he would have adopted this view of the matter even though he adhered to his theory of cochlear analysis.

Eighteen years ago (1880), while thinking over the great difficulties that beset the resonance theory, it occurred to the writer that the telephone might by analogy help observers to understand better the action of the cochlea. The telephone transforms sound vibrations, however complex, into electrical currents of corresponding frequency, amplitude, and wave forms and these in turn are re-transformed into sound vibrations similar to those received. The theory of hearing which he was led to propose is that there may be no analysis of sound in the cochlea; that all the hair cells may be affected by every sound, simple or complex, and through them the sound waves translated into nerve vibrations of corresponding frequency, amplitude and wave form; that in the sensorium the nerve vibrations give rise to sensations varying in quality with that of the incoming impulses. This theory has been termed the telephone theory of hearing, and no doubt it is a convenient term to distinguish it from the resonance theory though it should be understood that the proposer never regarded the hair cells as transforming the mechanical vibrations of sound into an entirely different mode of motion analogous to electricity. The essential point in the theory is simply this, that there is no analysis of sound waves in the cochlea. Theoretically a single hair cell, nerve fibre, and sensory cell should suffice to give us all the different sound sensations; but probably the great number of auditory terminals and sensory cells renders one far better able to feel slight differences in the quality of sounds. This theory of tone sensation has the advantage of carrying the physical cause of harmony onto the auditory centre, and the validity of the argument advanced in 1886 was afterwards proved by Hermann by experiments on the production of differential tones. When two discordant tones are simultaneously produced a beat is heard, and if the vibrational differences between the two primary tones be sufficiently great the successive beats give rise to the sensation of a third tone whose pitch is the vibrational difference between the two primary tones; for example, if the primary tones have respectively 880 and 1056 simple vibrations—namely, notes A^1 and C^2—the pitch of the beat tone is 176 simple vibrations—namely, the note F.

Hermann produced the two primary tones from two tuning-forks, with a third fork at rest, but capable of consonating to the beat-note if it had been produced objectively, but the third fork remained unaffected, although the observer heard the beat tone loudly. He therefore concluded that if the beat-tone failed to excite a reasonator outside the ear it could not affect any supposed reasonator in the cochlea; consequently it must be a purely subjective phenomenon arising from the conflict of vibrations in an auditory centre. That experiment, which has been fully confirmed, proves that the auditory nerve transmits to the sensorium vibrations of the same frequency as the sound waves and that they produce in the auditory centre harmony or discord, according to their relative numbers. Hermann's conclusion from these experiments is "that there is no alternative but to drop the Helmholtz hypothesis of resonators in the ear, although so elegant."

Although the telephone theory of the action of the cochlea enables us to understand the production of sensations of harmony and discord it still leaves much that is in the highest degree obscure. The discrimination of pitch is a great difficulty. We all know that it varies greatly in different persons. Those who have little or no ear for music have been found unable to observe any difference of pitch between C and E of the scale, although there is an interval of 66 vibrations between them. On the other hand, a musician with an extremely good ear may by practice become so sensitive to pitch that he can detect a difference due to one-fifth of a vibration even when the vibrational frequency is 1000 per second. That fact is all the more remarkable, because when the vibrational frequency rises above 500 per second the individual vibrations cannot be detected, although a change in the sensation can readily be felt. The accurate and refined appreciation of pitch seems therefore to imply an exceptionally perfect molecular state in the cells of the auditory centre; but how it is that the consciousness is so sensitive to changes in vibrational frequency when the individual vibrations are no longer perceptible must remain a mystery.

The writer considers that he was justified in arguing 12 years ago that our sensations of harmony and discord cannot be reasonably explained unless we believe that the auditory nerve transmits to the brain impulses of the same frequency as the sound vibrations. The proof of that has since been furnished by the experiments of Hermann on the differential tones already referred to. But it by no means follows that the theory of cochlear analysis is thereby disproved; on the contrary, it merely compels the upholders of the Helmholtz theory to admit that if certain cochlear terminals are affected by vibrations of a certain frequency they must transmit to the brain impulses of the same frequency as the sound vibrations. Although the validity of the argument on this point be granted it still remains an open question whether certain auditory terminals are impressed by tones of a certain pitch or whether they are all impressed by audible tones of every pitch. If the observations made by Gruber, but especially those by Stepanow, are reliable and free from fallacy we should be obliged to conclude that the loss of the apical half or so of the cochlea does not involve deafness to tones of low pitch. The results of their observations are therefore hostile to the resonance theory and favorable to the telephone theory. On the other hand, the two cases observed by Bezold and that by Moos and Steinbrugge show that deafness to tones of a high pitch may be associated with nerve atrophy in the basal turn of the cochlea. It is very difficult to look upon that association as a mere coincidence and apparently more reasonable to regard the basal portion of the cochlea as necessary for the appreciation of tones of high pitch. But if a corresponding relation cannot be shown between the apical portion of the cochlea and tones of low pitch the upholders of the resonance theory must feel that they are supported by only one of the two legs that are necessary in this case.

But if it appears impossible to harmonize the resonance theory of the cochlea with anatomical structure and with certain pathological observations how does it stand with the telephone theory? If it be true, as Gruber and Stepanow would have us believe, that the basal turn of the cochlea can enable us to hear all tones, notwithstanding destruction of the apical half, it seems reasonable to expect that the apical portion should still enable us to hear all tones, notwithstanding nerve atrophy in the basil portion. Bezold's cases, however, appear to show that the basal portion is necessary for the appreciation of high tones. We are therefore landed in the dilemma that pathological observations on the apical portion of the human cochlea are opposed to the resonance theory and favorable to the telephone theory, while those of the basal portion favor the resonance and oppose the telephone theory. In Bezold's second case only about two octaves of audible tones remained in the middle register; the loss of high tones was ascribed to nerve atrophy, but no cause could be assigned for the loss of low notes. On the telephone theory one is compelled to ask why there should have been deafness to many low and so many high tones, although nerve fibres enough remained to enable the patient to hear two octaves of the middle tones. If evidence of high tension in the labyrinth had been recorded in that case an explanation could have been found on the lines of Bernett's experiments, but in absence of that evidence the writer is bound to say that he does not at present see how that case is to be explained on the telephone theory.

Whatever be the true theory it is important to note that tone defects are far less common at the lower than at the upper end of the scale and that in the middle of it they are exceedingly rare. Politzer states that in cases of aparrent tone defects in the middle register the use of resonators generally shows that there is only diminished sensitiveness instead of total deafness to the particular note. At first thought one might be inclined to regard gaps in the scale as evidence in favor of the theory that each auditory fibre is concerned in the sensation of a particular tone, but in the well-known case reported by Magnus, where the tones F, F sharp, G, G sharp, A sharp and B were inaudible, ankylosis of the stapes and calcification of the orbicular ligament were found after death and assigned as the cause. *Loeb.*

Entotic Sound Perception.

17. SOMERS, L. S. *(Medicine*, Dec., 1898.) Entotic sound perception which is due to causes in the ear itself refers to sounds, such as the pulse-beats in the surrounding arteries and the rushing sound of the blood, the latter being especially strong when there is increased resonance of the ear, as in closure of the meatus or tympanum, or when fluid accumulates in the tympanic cavity, during increased cardiac action, or in hyperesthesia of the auditory nerve. Of the various causes, anemia and hyperemia are first enumerated. Local treatment is important, but constitutional measures are frequently equally so. Each case should be treated according to its indications.

Scheppegrell.

Treatment of the Ears in Whooping Cough.

18. STEPHENS, G. A. *(Jour. Amer. Med. Assn.*, Jan. 7, 1899.) The author claims that paroxysms of coughing diminished or ceased after treating the ears, which often in this disease have a serous or purulent secretion, or are merely sensitive to pressure. After washing the ear morning and night with warm boricated water, he paints the external passage with the following:

   Cocain hydrochlorate - - - 1.25 gm.,
   Glycerin - - - - - - 16 gm.,
   Van Swieten's solution - - - 20 drops,
   Aq. dest. - - - - - - 15 gm.

Scheppegrell.

A Contribution to the Clinical Stages and to the Technique of the Operation for Sinus Thrombosis.

19. WHITING, New York. *(Archives of Otology*, Vol. XXVII, No. 6.) The author deals with,

1. The Clinical Stages of Sinus Thrombosis.
2. The Technique of the operation for the relief of the same.

The course of sigmoid sinus thrombosis may be conveniently designated for the purposes of clinical classification as comprising three stages, characterized by local and systemic manifestations; the anatomical appearances of the sinus wall, the pathological changes in the clot and the signs of circulatory obstruction may be denominated as local factors; while rapid and excessive fluctuations of temperature frequently repeated rigors, peripheral or centrel metastases, etc., embrace the essential systemic symptoms.

The clinical stages are designated:

FIRST STAGE: The presence of a thrombus, parietal or complete (chiefly composed of fibrin, red blood cells, exfoliated endothelium, leucocytes and homogeneous protoplasmic cells) not having undergone disintegration and accompanied by slight or moderate pyrexia, rigors being usually insignificant or absent.

SECOND STAGE: The presence of a thrombus, parietal or complete, which has undergone disintegration with resulting systemic absorption, characterized by frequent rigors and pronounced septico-pyemic fluctuations of temperature.

THIRD STAGE: The presence of a thrombus, parietal or complete, which has undergone disintegration with systemic absorption, accompanied by rigors, rapid and great fluctuations of temperature, and central or peripheral embolic metastases terminating usually in septic pneumonia, enteritis or meningitis.

The diagnosis of sinus thrombosis in the first stage is seldom made preliminary to the operation for mastoiditis and its detection follows, as a rule, the recognition of extension of carious disease, through the inner table along the course of the sigmoid groove.

The only safeguard against encountering the increased gravity of the second stage is to operate immediately upon the recognition of the first stage. The transitional period between these two stages is usually brief, and its completion is commonly heralded by a sharp rigor.

THE TECHNIQUE OF OPERATION UPON THE SIGMOID SINUS.

The usual curvilinear mastoid incision is made extending from one inch below its lip to a point one-half inch above the temporal ridge. A second incision is made beginning at the centre of the first and extending backward two inches or more toward the occipital protuberance.

The mastoid antrum and pneumatic spaces are opened and along its posterior border the bulging convexity of the sigmoid groove is seen. The sigmoid groove may be quickly opened with curette or broad-beaked rongeur but under no circumstances must a chisel be employed.

The most accessible part of the sigmoid groove for opening is the knee and descending portion. The knee lies at about the level of the supra-meatal spine and usually ½ to ⅔ of an inch posterior to it. When the groove has been opened further exposure of the sinus is most readily accomplished with the rongeur.

When evisceration of the thrombus is complete the bleeding is best controlled by pressing a wad of closely folded gauze *upon* the distal extremity of the opening in the vessel. Bleeding from the jugular bulb is controlled by packing gauze *into* it. The removal of the gauze 2 or 3 days later is never followed by hemorrhage of any significance.

The shock attendant on the more serious of these operative procedures is frequently very pronounced. Intravenous injection through the median basilic vein of 16 to 24 ounces of normal salt solution at a temperature of 105° to 108° F. or an injection into the bowel of this salt solution at a temperature 110° to 115° F. should be employed.

At the moment of opening the sinus wall, the foot of the operating table should be elevated; *firstly*, because the blood pressure in the sinuses of the dura mater is measurably increased and the likelihood of admission of air to the open vein reduced to a minimum; *secondly*, in order to maintain the equilibrium of the general intracranial fluids, which might be seriously disturbed by suddenly induced anemia of the brain, consequent upon the very copious bleeding of the sinus operation. *Campbell.*

II.—NOSE AND NASO-PHARYNX.

Remarks on Nasal Synechias.

20. COLLINET. (*Rev. Hobdom. de Laryngol, d'Otol et de Rhinol.* No. 49, Dec. 3, 1898.) The etiology of this affection is more complicated than appears at first sight. Besides the usual point after galvanocautery operations the author mentions congenital synechias, synechias after diphtheria, scarlet fever, and measles, chronic and acute rhinitis, especially in lymphatic individuals, etc. The author describes an instrument to

dissect synechias. It consists of two parallel knives, the distance of which may be regulated with certain screw or wedge devices. Several layers of gauze in the opening are advised for after treatment. They have to be changed every day. *Holinger.*

Endothelial Sarcoma or Angio-sarcoma of the Middle Turbinated Body.

21. DELLE. *(Rev. Hedom. de Laryngol. d'Otol. et de Rhinol.* No. 10, Dec. 10, 1898.) A case of sarcoma of the middle turbinated body is reported. It had progressed through the ethmoidal cells, and the base of the skull, producing meningitis. The patient died without treatment. *Holinger.*

Maxilliary Pneumosinus of Cystic Origin.

22. CAEN, FREMONT. *(Rev. Hebdom. de Laryngol. d'Otol. et de Rhinol.,* No. 45, Nov. 5, 1898.) This term was introduced by Lermoyez. A man of 42 had at different times a most serious toothache connected with much swelling of the cheek, which disappeared after a few days. These attacks recurred. A doctor incised the tumor during an attack and it collapsed letting some air escape. It filled up as soon as the incision closed. The two first molars were incompletely extracted long before, leaving stumps of the roots. These were extracted by the author and the alveoli drained into the sinus. A complete cure followed.
Holinger.

Interstitial Electrolysis with Copper Electrodes in Ozena.

23. GOUGUENHEIN ET LOMBARD. *(Annales des Malad. de l'Oreille Larynx du Nez et du Phar.,* No 11, Nov. 1898.) The conclusions of the authors are quite encouraging. Interstitial electrolysis with copper electrodes has a certain influence on ozena and its manifestations. The bad odor disappears. How it acts, it is difficult to say. The formation of oxychloride of copper and its influence are doubtful. The good results seem to persist for a long time, yet it cannot be said that they were permanent. Accidents are rare with a good technique. The dangers have been exaggerated. *Holinger.*

A Bridle for the Control of Mouth-Breathing.

24. JENKINS, N. B. *(Medical Record, Vol. LIV.)* The device is made of four strips of cotton webbing, one inch wide, and commercially known as "stay binding."

Scheppegrell.

Latent Empyema of the Accessory Cavities of the Nose.

25. GOTTLIEB KICER. *(Laryngoscope,* February, 1899.) The following observations are the result of 200 post mortems:

Empyema of the Maxillary Sinus existed in 39 cases. In nine of these the affection was bilateral. In 16 (6 bilateral) there was pus. In 13 (1 bilateral) a muco-purulent secretion was found. In 10 (2 bilateral) a sero-purulent secretion was present. In 42 post mortems one case of non-purulent secretion was found.

Empyema of the Sphenoidal Sinus existed in 29 post mortems. Seventeen of these were bilateral. In 14 (1 bilateral) pus was found; in 9 (3 bilateral) muco-purulent secretion and in 7 (2 bilateral) sero-purulent secretion was present. In 40 post mortems there was one case of non-purulent secretion found.

Empyema of the Ethmoid Cells existed in 7 post mortems. One of these was bilateral. In 4 there was pus, in 2 muco-purulent secretion and in 2 sero-purulent secretion. In 33 post mortems one case of non-purulent secretion was found.

In the 200 post mortems 105 contained the products of empyema in one or more of the accessory cavities. Thus in the maxillary sinus in 75 cases (37 per cent) in the sphenoidal sinus in 66 cases (30 per cent), in the ethmoid cells in 39 cases (19 per cent.), in the frontal sinus in 29 cases (15 per cent).

Of the 200 post mortems 59 were found with empyema in one or more of the accessory cavities, viz.: In the maxillary sinus in 19 per cent, in the sphenoid sinus in 15 per cent, in the frontal sinus in 7 per cent, and in the ethmoid cells in 3.5 per cent.

Concomitant Empyema was found in 21 post mortems. In 7 of the 19 cases of frontal sinus empyema, pus was also found in the maxillary sinus of the corresponding side. In two of these the frontal sinus accumulations were bilateral.

The greenish-yellow pus referred to by Killian as characteristic of empyema of the frontal sinus was seen in only three of the 13 cases; it was also found twice in the maxillary sinus and in the sphenoid sinus, and once in the ethmoid cells. This is very natural, as the coloring of the pus is due to the chromogenic bacteria.

In 11 of the post mortem examinations all of the accessory cavities contained a non-purulent secretion, and if the post mortem diagnosis were considered, the above facts might have been determined as the cause of death. A post mortem rhinoscopic inspection suggested empyema in about one-half of the subjects. The case-records indicated that not one of these cases had been diagnosticated as such. *Loeb.*

Dental Septicemia of the Antrum.

26. LATHAM, V. A. *(Jour. Amer. Med. Assn.,* Jan. 21, 1898.) A patient of eight years developed symptoms which finally suggested involvement of the antrum. One of the symptoms which led to this diagnosis was "slight crepitation"(?), which caused the author to believe that the antrum was involved. The temporary second molar was extracted and no necrosis was found, but evidences of roughened bone felt. The danger of injuring the cusps of the developing teeth is referred to. The cavity was drained of all pus and washed with a hot saline solution. The case made a good recovery.

The second case of empyema in a patient of 23 years.

Scheppegrell.

Infection of the Frontal Sinus from the Maxillary Sinus.

27. LENHARDT, (Havre.) *(Rev. hebdom. de Laryngologie, d'Otologie et de Rhinologie,* No. 41, Oct. 8, 1898.) Read before Soc. Franc. de Lar., May, 1898. As a rule we find the suppuration of the frontal sinus with inflammation descending from the upper to the lower sinus. This avenue may be passed in the opposite direction and the author reports two cases which prove that the inflammation of the frontal sinus may rise from an inflammation

of the maxillary sinus. The proof is based on the principle *post hoc, ergo propter hoc*. *Holinger.*

A Case of Syphilitic Chancre of the Nasal Septum Resembling Pseudo-sarcoma.

28. LARMOYEZ. *(Annales des Malad. de l'Oreille du Larynx, du Nez et du Phar.*, No 12, Dec., 1898.) A very interesting case indeed. The case only became clear when a secondary roseola appeared the day before an extensive operation on the nose was intended. *Holinger.*

Automatic Curvature of a Snare with Flexible Loop used to Remove the Posterior Ends of the Turbinated Bodies.

29. MANU. *(Annales des Malad. de l'Oreille du Larynx, du Nez et du Phar.*, No, 11, Nov. 11, 1898.) The author found difficulties in removing the posterior ends of the turbinated bodies. He invented a new instrument. It is a snare with each end of the steel wire fixed for itself in a screw apparatus, which allows of twisting the wire lengthwise. By more or less twisting the one wire or the other the loop may be bent in a direction wanted.

Holinger.

Chronic Headache Caused by a Centipede in the Nose.

30. MEYJES, W. P. *(Journal of Laryngology, Rhinology and Otology*, Nov. 1898.) The headache which was confined to the right eye was combined with a slight mucopurulent secretion from the corresponding side of the nose. To lessen the hyperemia and swelling boric acid was administered, and after a heavy fit of sneezing a small insect which turned out to be a chilopod (centipede) was expelled alive. *Loeb.*

The Relation of the General Practitioner to the Rhinologist, and the Treatment of some Affections of the Nose and Nasopharynx.

31. MILBURY, F. S. *(Brooklyn Med. Journal*, Nov. 1898.) By working harmoniously, the general practitioner and the specialist may be of incalculable benefit to each other in the practice of medicine. Many affections, which on superficial examination appear to have no connection

with nasal diseases, are speedily cured by the correction of some pathologic condition of the nose. *Scheppegrell.*

A Case of Blue Secretions of the Nose (Chromo-Rhinorrhee.)

32. MOLINE. (*Hebdom Rev. de Laryngologie d'Otologie et de Rhinologie*, No. 44, Oct. 29, 1898.) A lady of 25, was treated for acute otitis media. During the treatment she became seriously ill with indefinte symptoms mostly resembling a severe attack of grippe. Three months later she complained that at times when she was blowing her nose blue masses were mixed with the usual grayish secretions of the nose. The author at different times saw these blue lumps of mucous and had an examination made by Dr. Engelhardt with a completely negative result. The author excludes hysteria for different reasons. He comes to the conclusion that in one of the accessory cavities some unknown process must have its seat. *Holinger.*

A Case of a Rare Anomaly.

33. ONODI. (*Hebdom, Rev. de Laryngol., d'Otol. et de Rhinol.*, No. 52, Dec. 24, 1898.) The anomaly which is nicely demonstrated in a drawing consists in a very large ethimoidal cell. The measurements are: height 2 cm., width 3 cm. and length 4 cm. It looks like a large frontal sinus betwfen the bony sheets of the roof of the orbit.

Holinger.

Grave Hemorrhage After Adenoid Operation in a Hemophile.

34. PLAGET, GRENOBLE. (*Hebdom, Rev. de Laryngol., d'Otol. et de Rhinol.* No. 45. Nov. 5, 1898.) Serious hemorrhages have been reported from different sources. This one teaches a new lesson because it deals with a condition which may be diagnosed beforehand. The hemorrhage which brought the patient in great danger was stopped at last with tamponade of the nose and naso-pharynx. *Holinger.*

The Prevention of Hay Fever.

35. RIXA, A. (*Journal Amer. Med. Assn.* Jan. 21, 1899.) The author advocates irrigation and sterilization of the nasal chambers in order to prevent hay fever by

rendering the parts aseptic. Two weeks before the onset of the disease, the nasal cavities and post-nasal spaces are irrigated by a harmless antiseptic solution by means of a douche and an atomizer. Hydrozone of a strength of eight to 25 per cent. is preferable. In obstinate cases, boracic acid, menthol and glyco-thymoline in a four per cent. solution of eucain is prescribed.

[Eucain, cocain and other local anesthetics are dangerous agents to place in the hands of the patient, and should never be prescribed. The abstracter has repeatedly called attention to this, and recently in an article entitled "The Abuse and Dangers of Cocain," published in the *Medical News*, Oct. 1, 1898.] *Scheppegrell.*

Headaches from Nasal Causes, with Some Illustrative.

36. THOMPSON, J. A. (*Journal American Med. Assn.*, Jan. 14, 1899.) A review of the clinical history and a description of the symptoms of headaches due to pathologic conditions in the nasal chambers. In some obscure cases, only the result of the treatment will confirm the diagnosis. The correction of the abnormal condition of the nasal chambers is indicated in the treatment of these cases.

Scheppegrell.

Clinical Study of Primary Lupus of the Septum and the Nasal Fossæ.

37. TOUSSOT. (*Hebdom, Rev. de Laryngol., d'Otol. et de Rhinol.*, No. 45., Nov. 26, 1898.) The historical introduction to this article is very short. Had he known the publications of Siebenmann, this article might have been more valuable. The author publishes eleven observations. The diagnosis of this affection is not very difficult. For differential diagnosis syphilitic ulcerations must be borne in mind. The treatment is curetting of the parts and cauterization with lactic acid or galvano-cautery.

Holinger.

III.—MOUTH; PHARYNX; DIPHTHERIA.

Remarks Concerning the Presence of the Short Bacillus of Loeffler in the Exudations of Ulcerations.

38. GLOVER, JULES. (*Rev. Hebdom, de Laryngol., d' Otol. et de Rhinol.*, No. 53, Dec. 24, 1898.) After adenoid operation

the patients often suffer from acute tonsilitis, which sometimes becomes quite grave. The short bacillus of Loeffler has often been found in the false membranes, which however could not be considered diphtheritic in the present sense of the word. *Holinger.*

Control of Diphtheria.

39. W. K. JACQUES. (*Jour. Amer. Med. Assn.*, Nov. 26, 1898.) The fear of cholera and epidemics of yellow fever has caused state officials to spend thousands of dollars in questionable methods for the protection of the people. Diphtheria, however, which is absolutely preventive has received little or no attention from these officials. The best authorities on the subject believe that if children were given the highest scientific protection there need not be a single death from Klebbs-Loeffler diphtheria. We have the knowledge and the remedy; all we lack are the individuals in official life to apply them for the protection of the public. *Scheppegrell.*

Diphtheria Complicated by Measles.

40. LA FETRA, L. E. (*Medical Record*, Oct. 8, 1898.) A girl of six years developed nasal diphtheria extending to the pharynx, which was successfully treated by the injection of 1500 units of antitoxin and local antiseptic irrigation. *Scheppegrell.*

Diagnosis and Treatment of Diphtheria.

41. IRVIN, JAS. S. (*Va. Medical Journal*, Sept. 23, 1898.) A careful review of the diagnosis and treatment, including a detailed review of the local and constitutional symptoms, differential diagnosis and bacteriology of the disease. The laboratory technique for developing and examining the bacillus is referred to and the importance of prophylaxis emphasized. *Scheppegrell.*

Diphtheria and Its Modern Treatment.

42. WALLACE, C. H. (*Medical Fortnightly*, Nov. 1, 1898.) Antitoxin should be used not only in cases in which a diagnosis of diphtheria has been made, but also in suspicious cases. *Scheppegrell.*

Spread of Throat Illness by Milk.

43. X. (*Public Health*, Oct. 1898.) During 1897 there was at Surbiton a well-marked outbreak of follicular sore throat and as a result of personal inquiry it was very soon found that milk supply was the apparent factor in distribution. A visit to the premises was made with no results, and a close inspection of the cow-sheds and of the employes was also made. A sewer was being laid in the road outside the farm, and for a limited period there was a smell to be noticed; but it was unlikely that this could have affected the milk (since it was not kept stored on the premises near the road) as to render it in any way dangerous to health. A man whose business it was to milk the cows, was examined and found to be out of health, and suffering from well-marked tonsilitis and gastrodynia, and further with suppurating whitlows on the hands. He immediately discontinued any employment with the cows and no further case occurred. There were considerably over 30 cases of illness, some very severe and probably many more.

Scheppegrell.

IV.—LARYNX.

Histological Examinations of Polypus of the Larynx.

44. BRINDEL. (*Rev. Hebdom. de Laryngol., d' Otol. et de Rhinol.*, Nos. 46 and 47. Nov. 12 and 19, 1898.) The author gives a review of the literature on this subject. His own statistics contain 33 cases.

They are:
Myxoma, 11, 33 per cent.
Papilloma. 9, 28 per cent.
Angio-myxoma, 5, 15 per cent (2 cysts).
Angio-fibro myxoma, 2 (1 cystic).
Fibro-myxoma, 2.
Fibroma. 2.
Adenoma, 1.
Chondro-fibroma, 1.
All were pseudo cysts.

In 29 cases the position of the polypus was given; 14 times the polypus was situated on the left vocal cord, 12 times on the right and twice at the anterior commissure. In one patient a big polypus was located in Morgagni's pouch. the anterior third of the vocal cords was usually the place of origin. Out of 29 patients, 21 were men, 8 women, 7 were between 20 and 30, 10 between 30 and 40, 10 between 40 and 50, 2 between 50 and 52. Not one case was found below 20 years. The occupation of the patients had no influence. The polypi were with one exception all removed through the mouth. The microscope alone gave proof of the benign or malignant nature of the polypi. *Holinger.*

Myxoma of the Larynx.

45. CLEMENT, PARIS. (*Rev. Hebdom. de Laryngol, d'Otol. et de Rhinol.*, No. 42, Oct. 15, 1898.) Benign tumors of the larynx are met with very exceptionally. Usually they are papillomata. The author reports this case because it is a myxoma. It occurred in a tenor singer of 42. The following conclusions are taken therefrom:

1. Myxoma of the larynx was for a long time little known although the examination is easy.
2. It does not necessarily interfere with respiration.
3. The voice is changed in timbre and extent for singing.
4. It does not necessarily need to be operated upon. A good operation causes no interfernece with the function of the organ. *Holinger.*

Tuberculosis; Its Hygiene and Dietetics.

46. CUZNER, A. T. (*Journ. Amer. Med. Assn.*, Dec. 17, 1898.) The author believes that tuberculosis is a disease of malnutrition of the tissue elements of the body; that the bacillus while it has considerable work to do and has well-marked influence over the course of the disease is not the primary factor or cause. He believes if taken in the early stages, before the system has become exhausted by its inroads, it can and has been cured in numerous cases. Climate and hygiene are valuable, but the most important part in the treatment of tuberculous patients is the diet.

Scheppegrell.

A Case of Epithelioma of the Larynx.

47. GOUGUENHEIM AND LOMBARD. (*Annales des Malad. de l'Oreille du Larynx., du Nez et du Phar.*, No. 12, Dec.,

1898.) The tumor was very moveable, and pedunculated. It was removed with the hot snare. The patient made a good recovery. *Holinger.*

A Practical Mode of Administering Iodine Hypodermatically in the Treatment of Pulmonary Tuberculosis.

48. INGRAHAM, C. W. *(Medical Record,* Oct. 1, 1898.) The formula recommended is as follows:

℞ Iodine - - - - 1 2 gr.
Bromine - - - - 1 14 gr.
Phosphorous - - - 1 100 gr.
Thymol - - - - 2 3 gr.
Menthol - - - - 2 3 gr.
Sterilized oil - - - - ℥i.

The commencing dose is 15 minims, while the maximum daily dose is one fluid drachm. The injections are never made oftener than once daily. *Scheppegrell.*

Treatment of Tubercular Laryngitis.

49. LEDUC. *(Annales des Malad, de l' Oreille du Larynx. du Nez et du Phar.,* No. 11, Nov. 1898.) The author demonstrated at the congress in Moscow a simple method of inhaling powders into the larynx and trachea of patients with tubercular laryngitis.

He uses:
Diiodoform 8.0.
Hydrochlorate of cocain 0.08.

2 to 6 aspirations a day. Sometimes he adds to that morphine 0.04. The laryngeal symptoms disappeared very readily. *Holinger.*

Acute Oedema of the Larynx.

50. MAGUAN. *(Rev. Hepdom. de Laryngol., d'O'tol. et de Rhinol.,* No. 52, Dec. 24, 1898.) A case of acute edema of the larynx is reported in a man of 59. After three days the symptoms disappeared after sprays and an external application of counter irritants. *Holinger.*

Sero-Diagnosis of Tuberculosis.

51. MONGOUR AND BUARD. *(Echo Med.,* Dec., 1898.) Sero-reactions were obtained in four cases of tubercular

pleurisy and nine of tuberculosis. In 21 other patients they also discovered latent phthisis by this means.

Scheppegrell.

Tracheo-Thyreotomy in Cancer of the Larynx.

52. MOURE. (*Rev. Hebdom. de Laryngol. d'Otol et de Rhinol.*, No. 43, Oct. 22, 1898.) A rather gloomy view is taken about operative interference in cancer of the larynx. The greatest number of patients die shortly after laryngectomy, and those who live are in such a condition that they would better die. The author would replace this operation by tracheo-thyreotomy. He cites two observations of epithelioma of the vocal cords with tracheo-thyreotomy, recovery during five years. But against those there are two observations with recurrences, the fourth after a second operation cured since seven months. The method of operating is then described. 1. Tracheotomy from the lower margin of the cricoid cartillage downwards. A Trendelenburg canua is introduced. Next is splitting of the larynx from the lower margin of the hyoid to the upper of the cricoid. All within the same incision of the skin, which reaches from above the hyoid to the sterum. This gives a very good access to the larynx where any tumor is easily destroyed with the thermocautery. The wound is then closed by sutures. *Holinger.*

The Eversion of the Ventricles.

53. NOACK, ALEX. (*Rev. Hebdom. de Laryngol., d'Otol. et de Rhinol.*, No. 44, Oct. 29, 1898.) The author sums up the contents of his paper in these four points:

1. The spontaneous eversion of the mucous membrane of Morgagni's ventricles, which was admitted by certain authors, is not possible on account of the fixation of this membrane to the underlying structures.

2. The acute inflammation of the mucous membrane of the lower parts of the ventricular bands may look like an eversion.

3. The so-called eversion is due to the presence of a tumor which originates in or on the mucous membrane. The inflammation connected with this tumor must come into consideration.

4. To verify the diagnosis the histologic examination of the tumor must be made, as Dr. Moure shows.

Holinger.

A Case of Laryngeal Cancer.

54. ORR, C. J. *(Medical Review*, Dec. 3, 1898.) A man of 53 years, otherwise robust, complained of hoarseness and some slight dyspnœa. The dyspnœa became so aggravated that a tracheotomy was performed. About 20 days later the patient died from asphyxia, due to obstruction of the trachæ by a clot of bloody mucus. The unusual symptom in this case was the very small degree of pain present.

Scheppegrell.

Treatment of Tuberculosis.

55. PORTER, W. *(Medical Review*, Dec. 3, 1898.) A valuable contribution to the therapeusis and therapeutic indication of tuberculosis. *Scheppegrell.*

Consumption: Cases that Should Go to Colorado and Cases that Should Stay at Home.

56. RICHARDS, G. L. *(Trans. Fall River Med. Society*, 1898.) Suitable cases with means to live in comparative comfort are greatly benefited at Colorado, Arizona and New Mexico. Others had better remain at home and there are many places near by, such as Sharon and Princeton where patients may derive much benefit.

Scheppegrell.

Some Statistics Upon Sero-Therapy in Tuberculosis.

57. STUBBERT, J. E. *(Trans. Amer. Climatologic Assn.*, 1898.) A report of 82 cases treated by sero-therapy. Of 36 incipient cases, the bacilli disappeared in 30 per cent. and decreased in 41 per cent., and there was general improvement in over 94 per cent. Of 42 moderately advanced cases, the bacilli decreased in 30 per cent. and there was improvement in 71 per cent. In the far advanced cases, there was general improvement in 25 per cent. In cases of laryngeal tuberculosis, the ulcerations healed in 50 per cent.

The author believes that sero-therapy is more satisfactory than treatment by any one drug, but admits that we have not yet found a specific for tuberculosis.

[Compare with Dr. Weaver's article on "Specific Action of Air on Consumption," an abstract of which is published in this issue.—*Scheppegrell.*]

The Specific Action of Air in Consumption.

58. WEAVER, W. H. *(Journal Amer. Med Assn.,* Jan. 14, 1899.) The method described is as follows: The patient is directed to take as full an inspiration as possible and then close the glottis and bear down as if straining so as to increase the tension of air in the upper chest. In performing the act in this manner, the greater effort is expended in the muscles of the abdomen, and consequently the air is forced upward.

In order to increase the interest in the treatment, the nebulizer should be given the patient, and the importance of getting the antiseptic nebulizer well down into the air cells by the method described above impressed upon him. Tuberculosis of the lungs in the early stage is the most curable of any of the chronic diseases, and in the last stage the most hopeless.

A table of 56 cases shows the following results: 19 first-stage cases cured; of 28 second-stage cases, 16 were cured, 7 improved and 3 died; of 9 third-stage cases, no cures, six improved temporarily. Leaving out the hopeless third-stage cases, 80 per cent. were cured.

Scheppegrell.

V.—MISCELLANEOUS.

Notes on the Therapeutic Uses of the Suprarenal Gland.

59. BATES, W. H. *(Medical Record,* Oct. 8, 1898.) In a case of exophthalmic goitre a decided improvement followed its application. One tablet representing eight grains of the gland was taken three times daily.

Scheppegrell.

The Diagnostic Characteristics of Headaches According to Their Origin.

60. GRADLE, H. *(Journ. of American Med. Assn.,* Nov. 18, 1898.) Among nasal diseases, the forms more likely to lead to persistent and unusually severe headaches are the more intense suppurative inflammations of the sinuses, in which cases nasal symptoms can always be elicited. Nocturnal aggravations are not uncommon under these circumstances. Moderate but very persistent headache is occasionally complained of in nasal stenosis of any kind, but only by distinctly neurotic subjects.

It is less generally known that inflammatory conditions of the pharyngeal tonsil are sometimes the source of continuous headache, usually not very severe in both children and adults. *Scheppegrell.*

Odor as a Symptom of Disease.

61. MCCASSEY, J. H. *(Medical Age,* Dec. 27, 1898.) Most diseases have characteristic odors and the sense of smell should be exercised in making differential diagnoses.
Scheppegrell.

Artificial Respiration in Relation to State Medicine.

62. MCDANIEL, E. D. *(Journ. Amer. Med. Assn.,* Dec. 3, 1898.) A plea for the scientific application of artificial respiration. *Scheppegrell.*

Schleich's General Anaesthesia Not a Success

63. RODMAN, H. *(Medical Record,* Oct. 1, 1898.) The general consensus of patients is that the Schleich mixtures are not unpleasant to inhale, but in this respect it has no advantage over chloroform. In the administration of these mixtures the reflexes are lost early, especially the conjunctival reflex, thereby depriving the anæsthetist of one of his most important safe-guards. Patients gag and vomit to almost the same extent as with ether or chloroform, which is also contrary to what has been claimed for these mixtures.

In view of these unfavorable results, the Schleich mixtures have been practically abandoned at the Mount Sinai Hospital, and one of its strongest advocates on their introduction into this country is thoroughly disappointed with their results. *Scheppegrell.*

Bronchial Cyst, With Report of Case Treated by the Injection Method.

64. STONER, H. H. (*Medicine*, Nov. 1898.) After cleansing with a two-per-pent. carbolic solution, an ounce of Lugol's solution was injected into the cavity of the cyst and manipulated in such a manner as to bring the solution in intimate contact with every portion. Two weeks later the tumor was reduced to a mere nodule, and fourteen months later had shown no tendency to refill. *Scheppegrell.*

Anæsthetic Mixtures in General Anæsthesia, with Special Reference to the Sehleich Mixtures.

65. WESTBROOK, R. W. (*Brooklyn Med. Journal*, Nov. 1898.) A conservative article on an important subject. The author believes that the Schleich mixture should be given an extended trial. The condition to be desired is a "safe anæsthetic in the hands of a safe man," and such a man will never fail to realize that anæsthesia always means a step towards death. *Scheppegrell.*

REPORT OF THE FOURTH ANNUAL MEETING OF THE AMERICAN LARYNGOLOGICAL, RHINOLOGICAL AND OTOLOGICAL SOCIETY.

HELD AT PITTSBURG, PA., May, 1898.

BY ROBERT C. MYLES, M. D., NEW YORK CITY,

SECRETARY OF THE SOCIETY.

(Continued from Vol. vii. No. 4.)

Primary Epithelioma of the Antrum of Highmore, with Histology of a Case, and two Camera Lucida Drawings. By Wendell C. Phillips, M. D. (Abstract.)

Tumors of the superior maxillary bone have attracted the observation and taxed the skill of surgeons for many years. In the earlier years, all tumors in this region were supposed to originate in the antrum of Highmore— even those that we now know to be primarily located elsewhere. Primary sarcoma of the antrum is not so rare, and it would seem that the earlier observers did not carefully differentiate between sarcoma and epithelioma. A careful research of literature has brought to light a few authentic cases of primary epithelioma of the antrum. Short abstracts of these have been made.

Morel states that epitheliomata may originate under the periosteum, or in the spongy portion of the maxillary bone, or they may originate in the antrum—starting in the epithelial layer covering the mucosa, or in that lining the glands. He cites no cases. Reclus reports two cases operated upon by M. Verneuil, English reports one of epithelial carcinoma in the antrum. Verneuil and De Gaetano each report one case, as well as Reinhard.

The case reported by Dr. Phillips, is as follows:

J. G., German, 58 years of age, came under treatment

[NOTE.—Through an oversight in the proof reading, the fact that the report was prepared by Dr. Myles, and furnished to the *Annals* by him was not recorded in our last number. For this omission apology is hereby made.]

March 7th, 1897; is of heavy build and ruddy complexion, weighing 215 pounds. Has always drank beer and light wines, and used tobacco. Six years ago remembers having had pain in the region of the right antrum; several teeth in the right upper jaw, which were in a state of decay, three years previously had been extracted. The pain had continued, and one and one-half years later, an opening had been made into the antrum through the alveolar process, through which opening there had never been much discharge of pus or blood. This opening had never closed. Four months ago, he noticed a growth around the opening which had rapidly increased in size. There was found to be a large cauliflower-like excrescence projecting from the alveolar opening. It was about two inches long from before backward, and three-quarters of an inch broad. It appeared to be a large mass of granulation tissue. Careful examination revealed a pedicle, which extended into the antrum, and which bled when touched with a probe. There was a sensation of fulness, with some pressure in the region of the antrum, but no external swelling or bulging, and no severe pain. Transillumination revealed a dark area over the entire region of the antrum. There was no glandular enlargement. The nasal cavity upon that side was quite normal—no polypi, no excessive secretion. The eye did not protrude, and was normal in every way. Believing the growth to be made up of polypoid or granulation tissue, he was informed that an operation would be necessary for its removal. The operation was performed at the Post-Graduate Hospital, March 15th, 1897, under ether. The large protruding mass was removed by a cold wire snare, after which a probe was passed into the antrum, which was found to be completely filled with the same kind of tissue. The opening was enlarged by means of curettes and gouges, until large enough to admit the finger, and the entire mass was removed. Hemorrhage was excessive.

Special pains were taken to curette every portion of the antrum; and the large opening made it quite possible for this to be accomplished. There was no indication of extension of the disease into adjacent tissues or sinuses. The cavity was thoroughly cleansed with bichloride solution, and carefully packed with iodoform gauze. The patient made an uneventful recovery, and after about six weeks the packing was discontinued, and the wound allowed to close up. From this time, to March 14th, one year after the operation, he has been examined once a month. There are now no visible signs of recurrence, no pain or tenderness, nor glandular enlargement, and no loss of appetite or flesh. His weight is now 218 pounds. The opening into the antrum is still entirely closed; the eyesight is good, and there is no fetid or purulent secretion.

Microscopical examination of the growth was made by Dr. Wright, and it was found to be an epithelioma. That the growth was primarily from the antrum, there could be no doubt. Its gross appearance was certainly unlike epithelioma which, together with the fact that primary epithelioma of the antrum is almost unknown, had led to the diagnosis of a benign growth. Dr. T. M. Prudden also examined the slides, and entirely coincides with the views of Dr. Wright.

Two camera lucida drawings were exhibited, the first showing the region where the epithelial joins the edematous portion of the growth; the second, being a high-power drawing, showing the infiltration at one point of the epithelial cells into the edematous tissue. That there has been no recurrence, is probably due to the apparent incipiency of the growth, enabling its thorough removal. The polypoid degeneration of the mucous lining of the antrum, had no doubt, existed for a long time. Had there been extension into the adjacent sinuses, especially the ethmoidal and sphenoidal regions, or had the bony walls of the antrum become infiltrated or destroyed, or had there been extensive glandular enlargement, with a cachectic diathesis, the results would, no doubt, have been very different. By thorough removal of the entire mass, however, it is to be hoped that all traces of the epithelioma have been obliterated. And the absence of recurrence after one year and two months have elapsed, would seem to bear out this conclusion. The large opening, allowing such thorough curettment, is also believed to have contributed to the successful termination.

Dr. Wendell C. Phillips, New York:—Mr. President and fellows: I have only a word to add, and that is to emphasize one point, that the growth, which I removed seemed to be made up of epithelial cells ingrafted upon the surface of a mucus polypus in the antrum of Highmore. The literature on this subject is very meagre. It is impossible to gain very much information from books or published cases. It is a matter of speculation as to how and where these growths originate. If this was a true primary epithelium, ingrafted upon the surface of a mucous polypi, it goes far to prove that benign growths may and often do become malignant.

Chronic Inflammation of the Pharyngeal Tonsil, with Little Hypertrophy, by Chas. N. Cox, Brooklyn, N. Y. (Abstract.)

Universally conceded that lymphoid hypertrophy at vault of pharynx of sufficient degree to interfere with respiration is productive of harm.

In the typical case, where obstruction is a marked feature, no question as to the advisability of complete removal of growth.

Purpose of this paper is to call attention to those cases in which there is chronic inflammation of pharyngeal tonsil with little or no hypertrophy. The condition is analogous to that often found in the faucial tonsils.

It is one of the most frequent causes of nasopharyngeal catarrh in children. Patients subject to frequent attacks of "cold in the head," which is, in many instances, not a coryza, but an acute inflammation of the nasopharyngeal tonsil. Increased liability of such subjects to infection. That condition known as a "cold" is more often the result of entrance into the system of pathogenic organisms carried into the air-tract by dust, than that of exposure to cold or damp air.

Frequency of so-called "bilious attacks" in children suffering from this disease, even where there is little hypertrophy. Believed by the author to be a septic process due to infection. Disease attended with more or less general debility and lack of tone, most cases anemic. Eustachian tube and middle ear frequently involved. Sometimes a very small collection of hypertrophied lymphoid tissue will impair the resonating power of nasopharynx to such a degree as to prevent the production of the finest quality of tone. Removal of this little hypertrophy is sometimes followed by most gratifying results in patients who are professional singers. Many cases get well with simple local and general treatment.

Author is not an advocate of indiscriminate and reckless scraping out of every naso-pharynx that presents itself, but maintains that in certain cases of chronic inflammation of the pharyngeal tonsil, even where there is little hypertrophy, removal of that little is the quickest and most efficacious method of treatment, and the one attended with the most lasting results.

Discussion—Dr. Lewis C. Cline, Indianapolis, Ind.:—I wish to congratulate Dr. Cox on his very interesting paper. I would like to ask him to explain a little further as to his method of operating. I think in all cases we should satisfy ourselves that we are going to do our patient good by operating before we decide to do so. The time for indiscriminate operating in every case is past. We all know some who operate on every patient whether or not there is anything to operate upon. Of course, if we have something which is obstructing the nasal passage, we should generally remove it. I frequently remove adenoid growths by degrees, not using anesthesia but a little cocaine, and then use the forceps and remove a small portion of the growth. I think I get as good results as those who use anesthesia and remove the entire growth at once; although, of course, I know there are cases where it is necessary to use anesthesia and remove the growth at one sitting; but the gradual method in private practice will produce just as satisfactory results. In curetting we should be very careful about scraping in too deep.

Dr. F. J. Quinlan, New York City:—I thoroughly disagree with the last speaker in making sectional operations. I think that if we have a growth which interferes with respiration, the voice or with the hearing, it should be removed at one operation in its entirety. I do not approve of the galvano-cautery in these cases. I was not here when Dr. Cox read his paper, so I do not know what stand he took on this question; but it seems to me that the use of the galvano-cautery is a dangerous measure, further, that it seems to me as non-surgical. I do not think there is anything more surgical than the knife, even if you do get a little bleeding occasionally. The worst case of hemorrhage I ever saw resulted from the use of the galvanic snare. It kept me busy all night and the following day. One other objection to sectional operations done without ether in children is that they terrify the child causing it often to interfere with the movements of the operator.

Dr. Price Brown, Toronto:—I agree with the last speaker. I think if possible the whole operation should be done at one sitting. I think this is particularly so in the case of children. In these cases, I often do the digital operation using the finger only. It can be done very well in this manner. You can get a much better idea of the condition existing within the naso-pharynx from the touch of the finger than when the impression has to pass through the curette. In older persons I have used the forceps, but have never found it a satisfactory instrument. I very much prefer the different modifications of Gottstein's curette. Still, even with this instrument, you can rarely complete an operation at a single sitting—a second being required several days later to render the removal thorough. I might mention one case I had in connection with this discussion. I removed almost the whole mass at one sitting. The patient was a prominent singer, and would not permit me to remove any more at that time. The result of the operation was some improvement, but she did not have a perfect voice. Finally three years later she came to me again and I removed the rest. Her voice soon recovered, and it is now practically normal.

Dr. Jas. E. Logan, Kansas City:—It has not been my privilege to see many cases of chronic hypertrophy in children. I thought the paper referred more especially to those cases found in adult life.

So far as the operation is concerned, I do not approve of the galvano-cautery. The curette is much better, and better still is the ring knife of Schulz.

I think chronic hypertrophy in the adult is due largely, if not altogether to diseased pharyngeal tonsil in the child.

Dr. Lewis,C. Cline, Indianapolis, Ind.:—Of course I use anesthesia in some of my cases for the removal of adenoids. I use it whenever it is necessary. In the majority of cases I do not use it. I do not see where the objection lies to taking off a part of the growth at one operation, and the balance at subsequent operations. I believe with a little bit of training and the right kind of forceps that better results can be produced in this manner without danger of complications. It is my experience that parents submit more readily to the gradual operation. I have had very little trouble with children. Cocain, in my experience, is all that is necessary. I have been criticised three or four times for doing the gradual operation, but I believe the doctors who criticise me most have never tried my method. I must insist that in many cases removal by the gradual method is preferable.

Dr. Robert C. Myles:—I heartily endorse this prolonged discussion of a subject of so much importance. There have been two very radical views expressed this afternoon, and we have listened to some diversified ideas as to the proper method of procedure. I think that both methods should be employed. I have had cases in which, for many reasons, Dr. Cline's method was pursued, and in others in which a more radical procedure was followed. If one will take the specimen shown by many operators as having been removed from the rhino-pharynx, and place it under a microscope, very frequently it will be found to consist of muscular tissue, fibrous tissue and sometimes even periosteum. The tissue is not adenoid tissue. As a matter of fact, every case should be studied in itself. Certain cases will require one operation, and some another. What is required is a scientific comprehension of the necessities of each case. There is no doubt in my mind that for the more extensive operations, the use of the anesthetic is advisable. One important factor in these operations, is the use of a proper sized curette, and a sharp one. Passing it well up until it strikes the septum, and then carrying it backward while the attendant holds the head of the patient in a straight position, one can often remove a very large amount of growth with one sweep of the curette. So far as ether is concerned in the operation upon children, it has been my experience that the child dreads the ether more than the operation. Chloroform is preferable, but dangerous.

Dr. Cline:—Do you use chloroform?

Dr. Myles:—There is a strong objection to the use of chloroform in New York City and I use it only on rare occasions.

Dr. Cline:—Do you operate without anything?

Dr. Myles:—Usually with crystals of cocain or a strong solution of ether gas or chloroform.

Dr. Lewis C. Cline:—I do not wish to take up much more time in the discussion of this question, but this discussion recalls one in which I took part. At the last meeting of the Mississippi Valley Medical Association, I read a paper upon this subject and upon its conclusion several gentlemen proceeded to criticise the gradual method. After a heated discussion the participants were about equally divided in opinions. After the discussion, one of the gentlemen said to me that if the operation was on his own child he would not permit the radical operation, but in the hospital work, where you wanted to get rid of your cases rapidly, that he would always remove the growth at one sitting. I think either in hospital practice or private practice we can get results by my method if we have the right kind of an instrument and go at it right. As Dr. Myles says many children will submit to an operation, either with curette or forceps, rather than to submit to anesthesia. That has been my experience.

Dr. Chas. N. Cox, Brooklyn, N. Y.:—Mr. President, I have been very much interested in the discussion although it got pretty far away from the subject of my paper, and for the benefit of some gentlemen who did not seem to understand just what my subject was, I will say that I took the liberty of changing the title to "Chronic Inflammation with Little Hypertrophy." I am not prepared to dispute the statement of Dr. Logan that all of these cases must be the result of original diseases in childhood. As to my methods of operating, I think perhaps the personal equation enters very largely into all operations. Some surgeons may be so skilful as to remove these slight hypertrophies with the forceps without anesthesia without causing a great deal of pain to the patient. I have sometimes in adults, without anesthesia, removed the growth during several sittings. In children I never do. I always operate with anesthesia, usually chloroform, patient in horizontal position, using a Gottstein curette, or some modification thereof. In addition to that, I always pass my finger well in to see that I have completed the operation and not left any fragments or small portions.

Dr. Howard Straight, Cleveland, Ohio: It is easy enough to decide as to the necessity for removal in the marked cases, and the authorities all agree that adenoids should be removed when they obstruct; but I have been treating many cases in which the hypertrophy was slight for a number of years by radical operation, and in the majority of cases I have secured good results. I do not want to go on record as favoring the indiscriminate scraping of the nasopharynx of every patient who comes under my charge. I know, of course, there are cases in which no operation is necessary; but I believe they are not nearly so common as many suppose. Curetting the cavity of the naso-pharynx, like curetting the cavity of the uterus is sometimes difficult to determine whether it is necessary to curette, and I have often been compelled to make a post-nasal digital examination. If I find a much thickened membrane I curette lightly. I never found very much in the literature of the subject to bolster me up in my plan except a slight reference in McBride, but the paper of Dr. Cox altogether justifies me in what I have been doing for a number of years.

Dr. Max Thorner, Cincinnati:—I wish to say that I entirely agree with Dr. Cox and Dr. Straight. It seems to me that this question ought to be settled once for all. It is not many years ago that some of our text-books and monographs on the subject said that if there was a small amount of adenoid tissue, which does not cause obstruction, you need not remove it; and many operators were of the opinion that it is not absolutely necessary to remove at an operation, every trace of adenoid tissue. The point which I wish to emphasize is this: It is not the amount of adenoid tissue which should be the guide as to whether or not an operation is necessary, but it is the disturbance which that adenoid growth may cause. This is particularly the case in people who must make professional use of their voices, as speakers and singers, and often in children where the reflex disturbance may be great, though the amount of lymphoid growth may be entirely out of proportion to the trouble caused by it.

Acute Suppuration of the Middle Ear. Jas. E. Logan, M. D., Kansas City, Mo.

On motion, discussion on this paper was postponed until Thursday morning.

End of afternoon session.

Adeno-carcinoma of the Nose, by Max Thorner, M. D., Cincinnati, Ohio. Abstract.

The author refers to the confusion regarding the nomenclature of nasal tumors. In his paper he wishes to speak about such neoplasms only as are histologically to be classed as adenomata and adeno-carcinomata. The case under consideration is one in which a typical adenoma of the nose developed into an adenocarcinoma.

The case reported is that of a farmer aged 47, white, who was referred to the author by Dr. V. T. Churchman, of Charleston, W. Va., Sept. 16th, 1898.

About one year ago he noticed some obstruction in the nasal cavity which gradually increased until breathing through that side was absolutely impossible. Four months after he noticed the trouble, the doctor removed a large growth from the nose with a snare, after which breathing was again free for about one month. Then the same trouble reappeared.

Another large portion of the growth was again removed. He was free for about two weeks, when breathing was impeded again, and two weeks later the left nasal cavity was entirely closed. Operations were repeated at intervals of about one month with moderate hemorrhage, and for most of the time were not very painful. His only complaint is obstruction to breathing. Has not lost weight. Appetite good.

The following is the condition upon entrance into the hospital: Man of medium size, fairly well nourished, nothing abnormal to be seen about his face. Hearing in left ear diminished; the left side of nose entirely obstructed by a growth which extends from the vestibule backward, and fills completely the space between the choana and the eustachian tube; color greyish-red, surface uneven and resembling somewhat a mass of cauliflower, is soft and bleeds upon touch; origin can not be ascertained, but it seems to come from the middle meatus, which is completely obliterated. Septum free from growth; no glands enlarged. A large portion was removed with a snare. Microscopic examination proved it to be typical adenoma. Two and a half weeks later, nose was again obstructed, and much was again removed. Microscopic examination confirmed the first diagnosis. On October 22d under chloroform anesthesia an enormous amount of masses were removed from the nose and post-nasal space with snare and curette. Pieces from the size of a filbert to that of a small walnut were removed, all being very friable. Microscopic examination of these portions was made by the pathologist of the hospital, Dr. Freiberg. He reported as follows: "The surface of the growth is not papillary but smooth. Lying in a well developed stroma of young connective tissue, abounding in easily stained nuclei, is seen an enormous aggregation of tubuli of various conformation. Some of these are fairly straight with lumina of small calibre, others convoluted in their course and others still short with large dilated lumina, reminding one of cystic formation. Here and there is to be seen a typical collection of epithelial cells without evident lumen. The tubuli are lined with a tall cylindrical epithelium whose nucleus is large and very easily stained. I have been unable to detect anything like cilia of these epithelia. Taken altogether the picture reminds one forcibly of the malignant adenoma of the uterus—I should call it malignant adenoma. On October 29th, one week after the operation, the growth returned, and a more radical operation by temporary resection of the upper jaw was suggested to the patient. The patient refused operation and left the hospital.

A few months after the patient returned home he began to decline. The growth had to be removed every few weeks. On April 25th Dr. C. wrote that he had operated upon the patient eight or ten times since his return, and that the operations had grown to be very painful. General health very bad, sallow complexion. The septum and right side of the nose involved; eyelids edematous. On April 25th some more masses were removed and sent to the Johns Hopkins Pathological laboratory. It was reported to be a typical case of adenoma changing into an epithelioma. The patient was not seen for one week, and then his nose was double its size and purple; eyes swollen, protruding and bloodshot, was only able to swallow soft and liquid food. Patient died on June 12th. The following is the history of the family physician who attended patient during his last few weeks: The growth broke through the wall of the nose at its bridge, from where hemorrhage took place; the left eye was finally destroyed, and the growth in left orbit measured two and a half inches at the time of death, bleeding at all times. No hearing for ten days preceding death, mind entirely destroyed the last five or six days.

This is a case of malignant disease of the nose, the duration of which is about two years. The question arises whether this was an adenoma that underwent carcinomatous changes, or whether it was not a case of benign tumor in addition to which there developed later a carcinoma. Adenoma of the nose is looked upon by many as a benign tumor; however, all authors agree upon the possibility and even on the probability of an adenoma becoming malignant. Pathologists and clinicians mention the manifest malignant tendencies of adenoma of the mucous membranes, and speak of a form of adenoma of the uterus as adenoma malignum. In a still further advanced stage, when the epithelial elements assume the shape of a dense cell conglomeration, we are in the habit, according to Ziegler, to call such a growth an adeno-carcinoma. And with such an occurrence, no doubt, we had to deal in this case.

Dr. E. W. Day, Pittsburg, Pa.:—I would like to report a case of naso-pharyngeal sarcoma, in which the results were not good. The patient died from the effects of the operation. The patient was a boy, H. L., aged 14 years. When he first came to my office there was marked protrusion of the right eyeball and cheek. A mass was protruding from the right nostril. The history given by the patient was as follows: Two years previous there had been some trouble in the right antrum which had been treated by a dentist, with all appearance of cure. The following year a growth was observed to block the right nostril, and an attempt at removal was made by a physician which resulted in a very copious hemorrhage that was with great difficulty controlled. A small section of the growth protruding through the nostril was removed for microscopic examination and a diagnosis of round-cell sarcoma was made. As death was inevitable in any case, the father gave permission for an operation, and the tumor was removed by excision of the superior maxillary. The growth had two attachments, one broad and firm attached to the basilar process of the body of the sphenoid, the other thin and ribbon-like coming from the antrum. When the growth was torn loose from these attachments a very copious hemorrhage of venous blood resulted. The hemorrhage was controlled very quickly by packing the cavity with pledgets of gauze: but in this short space of time an immense amount of blood had been lost, and the shock to the patient was very great. Free stimulation and intravenous injections of a large quantity of the normal salt solution was used without effect, and the patient survived but a few hours after the operation. The growth was quite large, some eight inches in circumference, and heart shaped with the apex at the nose. In the right nasal fossa, the turbinates disappeared from pressure, and the antrum was almost obliterated from pressage on the inner wall. In this growth, extending through the center, was a very large venous cavity which would admit the small finger, which led direct to its attachment at the sphenoid and had almost direct connection with the cavernous sinus at the base of the brain. The excessive hemorrhage had come from this large cavity.

Dr. Robert C. Myles:—These cases do not often occur, but when presented afford a series of interesting consequences. A few years ago a lawyer was sent to me by a general surgeon for diagnosis of antrum disease. He had been operated upon by a general surgeon and by a dentist. A clinical diagnosis of a malignant growth was made. I removed a section and Dr. Prudden pronounced the specimen a sarcoma. In the operation by Dr. Wyeth a portion of the superior maxillary and pterygoid bones were removed, and an artificial jaw was inserted. With the exception of a rather disagreeable taste in his mouth, the patient is doing very well. Last year I operated on a woman 50 years of age for a growth in the faucial tonsilar region, which three pathologists stated was sarcoma, the fourth was not certain. It extended into the muscular tissue, a very unfavorable symptom. The tumor was removed by deep dissections, exposing the constrictor muscles; this was over a year ago, and the woman at the present time is entirely well. I think if the case had progressed a short time longer, perhaps for a month or more, it would have been beyond recovery. Some pathologists claim that these cases of recovery could not have been sarcoma.

Dr. Price Brown, Toronto:—I had a case of this kind to report at Baltimore three years ago, and I speak of it again to-day because the man is still alive without any return of the disease. This was the case of a young man, 19 years of age, who had had what was supposed to be sarcoma and had been operated upon several times with excessive bleeding, but there was a rapid recurrence after each operation. Pathological examinations were made and the disease was pronounced round cell sarcoma. He was advised to go to Baltimore, which he did, and the final operation was performed in one of the hospitals there. He recovered, and came back to Toronto. About a year afterward he came to me, and, upon examination, I found a large, solid, hard, growth in the right nasal passage. It was about half way back and sessile, and seemed to be growing from the septum vault and the external side. He was bleeding considerably at the time, and I saw an operation must be done soon, and that it could not be operated upon with a snare. I used the cautery. The first time it caused considerable bleeding and I plugged the nostril. At the second operation, two days later, the hemorrhage was immense. I had to get professional assistance to help stop it. I then used electrolysis, passing one needle through in front and the other behind the soft palate, both of them penetrating the tumor. The seances were continued as long as the patient could bear the pain, and were repeated several times at intervals of one to two days. This had the effect of removing or lessening the amount of the mass, but it became less red in appearance. It practically limited the blood supply. I then again used the galvano-cautery as before. There were a number of operations, perhaps twelve or fifteen sittings in all. Finally, the whole of the growth was removed. All the operations were done under the influence of cocain. Twice over I had microscopic examinations made of sections, These pronounced it to be fibroma.

Dr. Thomas H. Halsted, Syracuse, N. Y.—About three years ago, a man fifty-five years of age, consulted me because of an exophthalmos of the right side. The eye was pushed upward and outward. There was some pain over the antrum and some swelling. On examining the nose, clear yellow pus was seen coming from under the middle turbinated body, and quite a large mucous polypus, originating apparently from the region of the hiatus semilunaris. Trans-illumination showed the right antrum and pupil perfectly dark, the left translucent. He gave a history of an ulcerated tooth with antral suppuration following, dating back eight years. I diagnosed empyema of the antrum, and opened through the alveolus, finding comparatively little pus, but the antrum filled with a red, friable mass, which under the microscope was shown to be carcinomatous. The microscopical examination of the polypus showed it to be a simple mucous polypus, non-malignant. The point I desire to make is this. Had this polypus been removed say before the exophthalmos appeared and before the malignant disease of the antrum was discovered, the case might have been looked upon as a benign growth undergoing, after operation, malignant change.

At the last meeting of the Society, I reported a case of naso-pharyngeal sarcoma in a child fourteen months of age.

Sarcoma of the Naso-Pharynx, with Report of a Case. By D. D. Hengst, M. D., Pittsburg, Pa. (Abstract.)

Sarcoma is met with more frequently in the naso-pharynx than carcinoma. Up to 1889, Bosworth has collected nineteen cases of sarcoma, which appeared in the naso-pharynx. The disease appears earlier than carcinoma. Ten of the above cases occurred between the ages of one and thirty years, and of the ten, five were between ten and twenty years. The male gender was more frequently attacked in the proportion of one to three.

The character of the secretion is somewhat diagnostic. It is of a sero-mucous character and quite ichorous and offensive. Epistaxis is often present. The general health is naturally much impaired, though no special cachexia is seen.

Difficult deglutition and dyspnœa, with impaired healing all depends on the size of the growth. The origin of the tumor is usually from the basilar process of the occipital bone, beginning in the deeper layer of the mucous membrane lining the pharynx, and growing in the form of lobulated rounded growth. Infiltration extends in all directions. The color of the growth depends upon its vascularity and composition. The only positive way we can reach a diagnosis is by microscopical examination of a section.

Carcinoma is usually more firm to the touch. Enlargement of the cervical glands generally occurs early in pharyngeal malignant disease, and is observed in about one-third of the cases.

Prognosis—The earlier in life the more rapid the course of a sarcoma. A small rounded celled tumor runs a much more rapid course, than a spindle celled variety. A mixed growth grows less rapidly. Generally the prognosis is exceedingly unfavorable.

The author prefers the cold snare in removing the tumor, and slits the palate if necessary, to get at the growth. In using the cautery snare there is danger of exciting inflammatory reaction of a severe character in the surrounding healthy structures.

The external operation should only be performed when all other means are out of the question, and it gives promise of prolonging the life of the patient, if the patient's condition is fair at the time of operation.

The case reported by the author was a young man 14 years old. Difficulty in breathing, dysphagia, headache, dullness of hearing, epistaxis with an icherous irritating discharge from the nostrils were prominent symptoms. The enlargement of the cervical glands could be detected.

There was bulging forward of the right side of soft-palate, presenting almost the appearance of an acute phlegmonous tonsillitis. A semi-solid mass could be felt almost completely filling the post-nasal space, and was attached to the walls of the pharynx and soft palate.

It was decided that an external operation was the only method advisable. Dr. Buchanan performed the operation by making an incision transversely from the zygomatic process of the malar bone, skirting the lower border of the orbit to its inner and lower angle. This incision was continued downward between the nose and cheek and along the upper lip to the median line, dividing the lip. After the soft parts were separated a wire saw was drawn beneath both bones, by means of a blunt hook. Another saw was passed under the nasal process of the superior maxilla.

The soft palate and uvula were split in the median line, and a drill hole was made well back in the middle line of the hard palate, through which another wire saw was drawn into the nostril. The bones were sawn through in a few seconds, and the superior maxilla lifted from its bed. The growth was cut away with scissors after extensive dissection, though it was impossible to remove it entirely. After the parts were brought together few signs of the operation remained.

Shortly after the operation, however, the naso-pharnyx again filled with the mass, and the patient succumbed, twenty-five days after the surgical treatment. The microscope showed the growth to be a small round-celled sarcoma.

AMERICAN LARYNGOLOGICAL, RHINOLOGICAL AND OTOLOGICAL SOCIETY— EASTERN SECTION.

WASHINGTON, JANUARY 28TH, 1899.

ABSTRACT REPORT.

The meeting was called to order by the President, Dr. Charles W. Richardson, who delivered a brief address of welcome.

The following papers were then read and discussed:

Reflex Cough.—Dr. Geo. L. Richards, Fall River, Mass.: —The author defined reflex coughs as those in which no demonstrable lesion is found in the air tract, from nose to terminal air vessicles. The term includes all those coughs, extra respiratory in origin, which are due directly or indirectly to irritation of the fibers of the pneumogastric nerve, whether peripheral or central. The cough of pressure origin, as from aneurism and large thyroid and new growths outside of the larynx and trachea, and included in the term "reflex." A cough of nasal origin is in a sense not reflex since the nose is a part of the respiratory tract, yet as it is not supplied by the pneumogastric but communicates with it through the vaso-dilator nerves from the superior cervical ganglion of the sympathetic, it seemed to him proper to call nasal coughs reflex within the meaning of the above definition.

He believed that reflex cough is produced by an irritation passing along sensory fibers to the point of explosion in regions supplied by some other nerve, and where there is no direct communication travels by way of sympathetic or some other ganglia and in this way also accounted for the distant nature of some of the reflexes producing cough.

He stated that reflex and nervous cough is characterized by some or all of the following: Sudden appearance, rythmical in character, free intervals when no signs of cough, expectoration absent or slight in amount, no fever or marked constitutional disturbance, may come at regular intervals, may continue for years or stop at any time, or eventuate in other symptoms, stops when person's attention is fully occupied, most marked when under observation, usually absent at night (always if purely nervous), absence of physical signs in the respiratory tract and its tone is various—sometimes hacking, bellowing, shrill, croupy or metalic. He then considered in detail various reflex coughs having their origin centrally, or in the ear, nose, or more distant organs of the body.

Further Experiences in Operative Procedure in Staphylorraphy.—Dr. Jno. C. Lester, Brooklyn:—The author referred to his former paper on this subject, read at the Pittsburg meeting, May 11th, 1898, in which he endeavored to show that it is entirely possible to operate in extensive defects of the soft palate after the age of puberty, not only with the hope of complete closure of the cleft, but also with restoration of function. He stated that recent experiences had confirmed him in the opinions then stated, that so much could be accomplished in selected cases, and, reported a case recently operated upon with success. The patient was a female aged 29 years with an extensive fissure in the soft palate which extended to and included a small portion of the hard palate and, the neighboring parts of the pharynx and nose were considerably diseased.,

The essential features of the operation were: The introduction of sutures as the first step in the procedure; the employment of scissors curved on the flat for denuding the edges; the use of large sized silver wire (No. 24), for sutures, and the employment of a Miles' tonsil punch for denuding the angle and that portion of the cleft included in the hard palate; and, the use of the writer's serrated tonsil curette for further preparation of the tissues.

The advantages claimed for this modification were the possibility of controling hemorrhage by the introduction of the sutures as the first step in the operation; the case and acuracey with which silver wire can be introduced and the length of time such sutures can be allowed to remain in the tissues.

Discussion.—Dr. Richards asked at what age Dr. Lester would prefer to operate and whether in dealing with a child of five or six he would operate or postpone procedure for several years.

Dr. Lester replied that he would prefer to operate after puberty and that if he had a hard subject to deal with he should do a tracheotomy first.

Sarcoma of the Tonsil.—Dr. Root, New York:—Dr. Root reported a case of sarcoma of the tonsil occurring in a young man only 23 years old. Family history was negative. Patient's illness began one year ago with an ordinary cold. Deglutition was painful and left tonsil became enlarged and developed into an abscess which was several times lanced, but it remained permanently enlarged and there was some swelling and tenderness externally. Examination showed a large mass involving the whole of the left tonsil, presenting a hard and somewhat irregular surface, slightly injected but not ulcerated. Externally the chain of lymphatics were to some extent involved. Microscopical examination of small portion of the growth enabled the author to diagnose small spindle called sarcoma. Two rather extensive operations were performed during which the tonsil, the enlarged glands and a portion of the inferior maxilla were removed. The patient is convalescent.

Discussion.—Dr. Johnson advocated the operation described and suggested the use of the toxines for those patients who would not submit to operation and referred to a case reported several years ago which he cured by this means.

An Unusual Case of Sinus Thrombosis and Epidural Abccess Complicated by Malaria—Operation—Recovery.—Dr. M. D. Lederman, New York.

The patient was seen on April 2nd with a severe ear ache of the right side. Two weeks before he had had a similar attack which was followed by a thin watery discharge from the external auditory canal. A few days before entering the hospital the second attack came on and his attending physician had incised the drum because of its bulging and of a painful tender mastoid. The patient's condition became worse, however, and he was admitted to the hospital where he was treated with warm douches and the Leiter's coil. On the day of admission temperature was 104 and between that time and April 18th it varied from normal to 105.2. Phenacetin was used in the first few days and when on April 9th the plasmodium of malaria was found in the blood, injections of quinine were used. On the 17th ophthalmoscopic examination showed optic neuritis on the right side and when on the 18th a mastoid operation was performed the antrum and mastoid cells contained pus and granulation tissue and the lateral sinus was thrombosed for at least an inch and a half and pus flowed freely when it was opened. The patint made a good recovery.

Two Operations for Mastoiditis with Unusual Features.— Dr. Thomas R. Pooley, New York.

The first case was one of acute mastoiditis which was operated upon and the operation was followed by persistent vomiting lasting for 48 hours and becoming at one time quite serious.

The second case was a mastoid operation performed in an acute case and followed by inflammation of the frontal sinus with escape of pus through the nose.

Discussion on the Operative Treatment of Chronic Purulent Otitis.—Dr. S. Macuen Smith, Philadelphia.

Dr. Smith stated that in acute middle ear disease and especially that following or occurring during epidemic influenza he thought it best to open the drum head early and that a free incision should be done instead of a simple paracentesis. When mastoid symptoms are evident, he operates early and thinks that trouble is apt to occur as a result of that conservatism which allows complications to arise before operating. He thought that no harm could be done by at least performing the Wilde's incision and that that could be done as a preliminary step to further operation if necessary. He remarked upon the rapidity with which granulation tissue presents itself in even acute cases.

Dr. Edward B. Dench, New York.—Dr. Dench began with the discussion of the two papers just read and took exception to Dr. Lederman's administration of an antipyretic in a case of suspected mastoid disease. He thought that in such a case it was essential to study the course of the temperature and that we should not administer anything which might mask it. He thought that the case should not have been left until optic nuritis developed, but should have been operated upon much earlier, for he believed that no damage results from a perfect aseptic operation.

Referring to the use of the ophthalmoscope to assist in a diagnosis in cases where cranial complications are suspected, he stated that his own experience had been that in very many cases it was of no use as there had been severe complications without evidence of optic neuritis or cloudiness of the disc.

He objected to the use of the Wilde's incision without further operation upon the bone.

Dr. Dench then described the development of the Schwartze-Stacke operation and its value in the treatment of chronic as well as acute middle ear suppurations. He stated that he had always been an advocate of the operation for the removal of the ossicles in chronic otitis media suppurative in preference to the radical Schwartze-Stacke operation, but a recent review of his records had shown him that by the simple method he had had but 58 per cent. of cures, whereas by the radical method the percentage had reached 64.

Dr. J. F. McKernon, New York.—Dr. McKernon emphasized what Dr. Dench had said regarding the use of antipyretics and also concerning the Wilde's incision. In reference to the acute cases of otitis media he advocated an early and very free incision of the drum head.

In the treatment of chronic purulent otitis where there has been no previous mastoid involvement and where we can examine the middle ear he believed in the removal of the ossicles through the canal.

Dr. J. A. White, Richmond: Dr. White said in regard to optic neuritis as being a certain means of diagnosis of cerebral complications, or of a thrombus complicating chronic otorrhea, he doubted if it was at all reliable. He stated that he had seen these troubles without neuritis, and had seen a neuritis that led him to perform a mastoid operation without finding any complications.

He referred to the difficulty frequently encountered of getting patients to submit to an operation for chronic otorrhea. He thought it made little difference whether you performed an ossiculectomy or a radical mastoid operation, so that the result secured was perfect drainage, but he believed that this was more perfectly attained by the radical operation.

Dr. Pooley:—Dr. Pooley differed with Dr. White as to the value of ophthalmoscopic examinations in these cases, and stated that it was almost a postulate that where you have optic neuritis, and can exclude certain intraocular diseases, that you must have a cerebral disease to produce that form of inflammation of the optic nerve generally known as choked disc. He believed that whenever you find a purulent inflammation of the middle ear associated with optic neuritis, there is pretty sure to exist an inflammation of the brain.

Dr. Lester:—How do you distinguish between optic neuritis, due to a constitutional disease, and one due to otitis?

Dr. Pooley:—Optic neuritis, when due to a lesion in the brain, is of that character known as choked disc; that is, there is intense swelling of the optic nerve head, and if that be associated with disease of the middle ear we naturally associate the two as cause and effect. Of course, if we accept this as proving that there is cerebral disease, we have not absolutely determined its nature or location. It may be due to sinus thrombosis, disease of the meninges, or to any other of the conditions that arise from purulent inflammation of the middle ear. It may also occur with cerebral tumor, but since that so rarely occurs with the other symptoms that lead one to suspect brain trouble from otitis it need scarcely be considered.

The Rebuilding of a Nose Without the Insertion of an Artificial Bridge.—Dr. T. P. Berens.

The author reported a case of long-standing deformity resulting from a severe injury to the nose in early childhood. The nasal bones were pushed apart, turned outward and flattened so that the bridge of the nose was very broad and flat. The nasal speculum revealed the columnæ well over to the right side, largely occluding the right nares, while the body of the septum was deflected to the left, and its upper half was adherent to the walls of the vestibule.

Under anesthesia the soft parts were divided from the septum and the cartilaginous septum as well as the "bony nasal spine" of the superior maxilla were broken so that the nose was made quite pliable and was easily molded. A perforated cork splint was inserted in each nostril and a plaster cast of a normal nose, obtained from a friend, was bandaged firmly in place as an external splint. This splint was removed after three days and a Fox eyeglass clip substituted.

Self Inflicted Wounds in Both Eyes, Both Ears, Tongue and Larynx, Leading to Thrombosis of the Sinus Transversus.—Dr. Geo. Reuling, Baltimore:—The patient, a man 41 years of age, suffering from mental depression and religious hallucinations had sharpened the end of a medium sized painter's brush handle and repeatedly stabbed himself. Both eyes had been perforated in the ciliary region resulting in prolapse of the corpus ciliary and iris in one eye and detachment of the retina in the other; both membranes tympani had been perforated and particles of tympanical tissue were hanging from the fissures; the tongue and left tonsil were also marked by superficial abrasions. The patient died on the seventh day after the injury, and post mortem showed an infectious thrombus of the transverse sinus, which had probably resulted from infection produced by the wounds in the middle ear and labyrinth.

The Necessity for Antiseptic and Aseptic Methods in the Surgery of the Ear, Nose and Throat.—Dr. Woolsey Hopkins, New York:—Dr. Hopkins read a very excellent paper in which he urged upon all specialists the necessity for adopting more rigidly the methods in daily use by the general surgeons. He referred to the neglect of simple antiseptic methods in many special clinics and hospitals both here and abroad and referred to the case with which modern methods of cleanliness can be adopted both in office and hospital work.

BOOK NOTICES.

A TREATISE ON DISEASES OF THE EAR.

BUCK, ALBERT H. Third revised edition. New York, William Wood & Co., 1898.

This, the third revised edition, marks the transition of the work from the domain of the manual into that of the treatise. The printer and the binder have combined to make a very presentable book. The arrangement of the subject matter in the chapters, while still open to objection as in the earlier editions, is so in a much less degree.

In chapter I, on General Diagnosis, the writer condemns the speech test as being very unsatisfactory. By attention to the sound values of spoken or whispered words and by the use of residual air (that left in the lungs after a forced inspiration followed by an ordinary expiration) the speech test can be and is made by many aurists the most reliable and accurate test for variations in hearing that we have at our command and is far and away more to be depended upon than the average patient's impression.

The reviewer cannot agree with the writer in that "It is only in exceptional cases that we derive valuable information from the tuning-fork test." The writer himself seems a little uncertain as to his convictions on this subject, for on page 1 he states, "Although in the vast majority of instances he will derive no material aid from the procedure, he should, nevertheless, apply the vibrating tuning-fork to the patient's forehead or vortex," and on page 9, "In an enormous majority of cases of impared hearing that come under our observation, these (affections of the sound-conducting apparatus) are precisely the regions which are mainly involved, and we, therefore, expect, with considerable confidence, to hear the patient say—when the tuning fork is applied to the vertex—that the sound of this fork is heard more loudly in the defective ear." The writer fails to mention certain well-known tests, systematically employed by many aurists, e. g., Schwabach's.

The forehead mirror depicted on page 14 does not compare in freedom of action with Hartmann's. The writer gives the impression that he only exceptionally looks through the central aperture of the mirror; this seems to be the only correct method of use, to the reviewer.

Most aurists would object to the recommendation, "that in the majority of cases the bowl may be used both as a reservoir from which to fill the syringe and as a receptacle for catching the water that runs out from the ear," page 29.

On page 33 the writer states that there are generally good grounds (in the exanthemata) for believing that the primary source of the disease (of the middle ear) must be placed in the pharynx. Rudolph and Bezold, as the result of post-mortem examinations on eighteen temporals of patients dead of measles, deny this and present evidence to show that the middle ear is affected primarily, in the same manner as the pharynx and the skin.

Chapter III constitutes a very valuable epitome of the more common symptoms of ear diseases, and, if the reviewer is not in error, is an addition to previous editions, as apparently also is chapter X.

The writer will hardly bring many aurists over to his preference for the removal of impacted cerumen by instruments rather than by syringing, except in certain cases, page 110.

Dr. Robert Lewis, Jr., has furnished, in very little space, a very readable, instructive and conservative account of naso-pharyngeal conditions, chapter XIII.

The reviewer has not had so fortunate an experience as the writer, who says that "the incus usually comes away as soon as its chief support—the malleus—has been removed" (p. 360), in doing an ossiculectomy for the relief of chronic suppuration of the middle ear. In fact, the removal of the incus has usually been the most difficult step of the operation.

We read, p. 369, "when the foot-plate (of the stapes) goes, the hearing also vanishes. This, at least, must be the rule." It will seem to many of those who have artificially removed the stapedial foot-plate that some other element, than the bare removal of the foot-plate, must be taken into consideration to account for the deafness in any particular case.

The writer well says, p. 427, "Acute suppurative mastoid inflammation, if left to itself or if treated in the impotent fashion which prevailed even as recently as thirty years ago, is a disease remarkably full of disagreeable and dangerous possibilities."

On page 440 occurs the statement that "In the light of this experience, therefore, I have been forced to draw the conclusion that there is only one thoroughly safe course of acute suppurative inflammation of the mastoid cells, viz.: *to expose to view a small area not only of the sigmoid sinus, but also of that part of the dura mater which lies in the vicinity of the posterior end of the antrum.*"

The chapter on Periphlebitis and Infective Thrombosis of the Sigmoid Sinus, new, is especially timely and one of the best in the book. A mistake has been made on page 463 in attributing to Whiting the discovery of the significance of puffiness of the eyelids as a sign of interference of the flow of blood through the cavernous sinus: S. Phillips and J. W. Stirling wrote of this long before.

Chapter XXIII, on the diseases of the auditory nerve, seems to the reviewer the weakest in the book and the only one that seems to have been inadequately handled from the standpoint of the results of modern research.

Taken all in all, Dr. Buck has given us a work that is pretty truly representative of American Otology as it is practiced to-day by those aurists, conversant with the publications of domestic and foreign arthorities in this field of surgery, ready to accept that which is true and to reject that which is false. He is to be congratulated on having produced a book worthy of his reputation as a teacher and practicer of aural surgery. HENRY A. ALDETON.

ANNALS
OF
OTOLOGY, RHINOLOGY
AND
LARYNGOLOGY.

VOL. VIII. MAY, 1899. No. 2.

PRESIDENTIAL ADDRESS BEFORE THE AMERICAN LARYNGOLOGICAL, RHINOLOGICAL AND OTOLOGICAL SOCIETY, 1899.

BY S. E. SOLLY, M. D.,

COLORADO SPRINGS, COLO..

I have the pleasure of welcoming you to the fifth annual meeting of our society and to congratulate you upon its growth and prosperity. The trite saying that it fills a long felt want is certainly applicable in speaking of the utility of our institution. That you have done me the honor to elect me your president, I regard as evidence of the recognition of that class of practitioners who, while treating a considerable number of nose, throat and ear cases, do not entirely confine their practice to that specialty. The fact that there are so many hundreds of men who are doing good work in these branches all over the country who are not pure specialists is an indication of the recognition of the value in the general practice of medicine, of the knowledge and skill in treating these diseases. So many cases of nasal and laryngeal disease are overlooked and neglected on account of the ignorance of this department of medicine among general practitioners; and also too often among medical consultants. In the larger cities, of course, it is natural and best that certain men should devote themselves exclusively to the treatment of the diseases of the upper air tract; but in less populated places it is desirable that the general practitioner should have a more thorough knowledge and capacity for the treatment of these cases than he usually has, and this Society which allows such general practitioners to stand alongside the specialists is fulfilling a great purpose. Both classes of physicians can learn from each other; and our gatherings serve to keep the specialist in touch with general medicine which is essential to his best success, and on the other hand, enables the general practitioner to acquaint himself, and keep abreast, with the progress of this special art.

The subject to which I wish particularly to call your attention today is the unity of the respiratory tract; that while a nose and throat specialist should know all that can be learned about the upper air tract, he should also be competent to make a diagnosis of pulmonary disease, and be acquainted with the general signs of an early or latent tuberculosis and more frequently than is customary, to extend his investigations below the larynx. On the other hand, the medical consultant who is an expert upon affections of the chest, should either examine, or cause to be examined, the upper tract in many more cases than he usually does.

It is not necessary in addressing you, for me to explain the anatomical and physiological relation of the nose, larynx and lungs, or to insist on their frequent connection in pathological conditions. The facts are admitted, but practical and intelligent action taken upon these facts is much more rare than perhaps many of you suppose. As explaining why I feel justified in addressing you on this subject I must be pardoned for speaking egotistically for a brief space. I have practiced in Colorado for nearly twenty-five years, where I have had under my care and generally prolonged observation a very large number of persons affected in some portion of the respiratory tract, who for the most part were subjects of tuberculosis. During this time I have never ceased to rejoice that previous to coming to Colorado, while practicing in London, I studied and worked under that great laryngologist, Sir Morrell Mackenzie, and that I have since endeavored to keep up with the progress of the specialty. The reason is because I have found so many cases of pulmonary disease in which the condition of the nose and throat was such that it had evidently influenced the causation of the lung affection; or that it was interfering with its progress toward recovery and if left unrelieved was liable to make the cure incomplete, or occasion a relapse. On the other hand, in cases coming to me for treatment of the nose and throat I have not infrequently found that there was also disease in the chest.

For some years in my first examinations of patients coming to me with pulmonary tuberculosis I did not examine the nose and throat, unless there was some special reason for doing so, but for the last few years I have made this examination a routine practice, as I have found that many patients who do not complain of nose and throat symptoms nevertheless have affections of this region which are important to recognize and often to treat. I believe, therefore, that in all cases of chronic pulmonary disease, the upper air tract should always be inspected.

Not infrequently patients in whom I have found abnormalities of the nose, but which had not attracted the attention of the physicians who had sent them to Colorado, have only been consciously inconvenienced by these affections after residence in Colorado. Probably because the dryness, dustiness and the diminished pressure of the air increased the turgescence of the mucous membrane and especially of the turbinate bodies. This by diminishing the lumen of the nasal passages causing more or less stenosis to arise where there was an exostosis or divergence of the septum which at sea level had caused no important obstruction; such obstruction having previously given rise to discomfort only when the mucous membrane was swollen by a passing catarrh. Such cases complain of chronic catarrh in Colorado and are comfortable still when they return to sea level.

As a rule it is best to remove the septal prominences because when those possessing them take cold they are more liable to have secondary affections of the respiratory tract lower down. The same rules, of course, apply more or less to other interferences with perfect nasal breathing, such as chronic hypertrophies of the turbinates, polypi, etc., which have escaped notice until the influence of the air of Colorado has called the physician's attention to them.

It has been my practice for some years to take a careful history of the patients previous to their development of tuberculosis and I find that in the cases in which there is a history of frequent attacks of bronchitis and catarrhal pneumonia there generally exists more or less nasal stenosis. The number of those in whom nasal abnormalities are present would appear to be still more frequent among the cases that are subject to naso-pharyngitis, tonsilitis and laryngitis.

The influence of nasal obstruction upon catarrhal affections of the chest, it is conceded, is largely due to mouth breathing during attacks of nasal catarrh and to exposure to damp, chilly weather. As I have said marked nasal obstruction in these cases being only temporary, has not attracted the attention of their attending physician. Where pulmonary catarrh has resulted, no doubt a soil was created for the growth of the bacillus in the lower respiratory tract, while in those who have had tonsillitis it is likely that the bacilli have found an easier entrance than usual from the tonsils into the lymph vessels.

Again, in cases of unilateral nasal obstruction a catarrh of the nasal pharynx has probably allowed a readier lodgment of the bacilli in the naso-pharynx because of a sluggish air current behind the obstructed side and an entrance into the system through the glandular structure in this region. There are also cases of catarrhal affections of the upper air passages in which there may be no stenosis which nevertheless may have afforded a ready entrance for the tuberculosis. I believe that the treatment of these affections has enabled those in whom the tuberculosis has become arrested to return to their homes and occupations with much less danger of renewed catarrhal attacks and so much lessened liability to recurrence of the tuberculosis.

As an aid to the arrest of pulmonary tuberculosis, treatment of diseased conditions of the nose and throat are often of the greatest service. The cough is frequently and noticeably diminished, particularly, of course, where there has been an elongated uvula, or a hypertrophy of one or more of the tonsils. For these reasons I believe that where a chronic affection is discovered the entire respiratory tract should be investigated and that in about 50 per cent. of the cases more or less treatment of the upper air tract is called for. It is very difficult at present to prove the truth of this belief by statistics. Dr. E. F. Ingals made a statistical inquiry into the relation of nasal disease to pulmonary tuberculosis which he reported before the British Medical Association at their meeting in Montreal two years ago.* Any communication from a physician of such wide experience and eminence in the treatment of the nose and throat is entitled to the greatest respect and consideration. He offers an analysis of 14,953 cases. Of these, 1,272 had phthisis; 4,714 had nasal disease without phthisis; 237 had nasal disease with phthisis; 6,058 had neither phthisis nor nasal disease. In considering the question of nasal disease in the causation of phthisis, he very properly writes that it is necessary first to ascertain what is the proportion of nasal disease to the general population. He estimates that 46 per cent. is the ratio of well-marked nasal disease to the population, and he concludes from his own statistics that the proportion of cases of phthisis among those with nasal disease is less than the proportion of cases of phthisis to the whole population. He, however, excepts atrophic rhinitis, in which disease he finds the proportion is greater. In arriving at his conclusions he says that Delavan, in an examination of some 2,000 skulls, found exostosis or deviation of the septum in 50 per cent. He himself estimates that 25 per cent. more of the population had other nasal affections. He, therefore, concludes that about 75 per cent. of the human race have nasal disease. He further says that it is believed that 12 per cent. die of pulmonary tuberculosis, while the number of those who recover is not known, but in 25 per cent. of the autopsies made upon people dying from other diseases there had been found evidence of old pulmonary tuberculosis. So, it may be inferred that 37 per cent. of the human race suffer from pulmonary tuberculosis as against 75 per cent. who have nasal disease, and it is also probable that many cases of

---

*The Relation of Nasal Disease to Tuberculosis. By E. F. Ingals. *British Medical Journal*, Nov. 13, 1897.

both diseases escape record. Dr. Ingals expresses his astonishment at these results, and admits they are opposed to his clinical conclusions.

For my part, I cannot but believe that there is some fallacy in his statistics which is not at present apparent, and one naturally demurs to his deduction that disease of the upper air passage exerts a deterrent influence upon pulmonary tuberculosis.

Dr. W. Freudenthal thus comments on Dr. Ingals' paper:

"Dr. Ingals says that of these 38 per cent. with tuberculosis comparatively few suffer from nasal disease. Thus, for example, of his 830 cases of pulmonary tuberculosis only 237, or about 28 per cent., showed some nasal trouble. 'Of the 237 cases which make up this 28 per cent. I find that 168 consisted of exostosis and deflection of the septum, which * * * is present in 50 per cent. of all persons of the European race; *therefore, many of these would have had no possible influence* in causing the pulmonary tuberculosis.' I fail to see the logic of Dr. Ingals' conclusions. Because 50 per cent. of all Europeans have deflections of the septum, must we exclude them from our statistics? Are deflections of the nasal septum to be considered normal because so many civilized people have acquired them?

"Deflection of the nasal septum is a pathologic condition which also tends to produce post-nasal catarrh, and I consider it a very important etiologic factor in favor of our theory. But Dr. Ingals goes on to exclude other possibilities by saying: 'Further, my records show that of all the cases of pulmonary tuberculosis, 1,272 in number, only 27 of the patients, or about 2 per cent., complained of having had any previous nasal disease, which is 4 per cent. less than the normal average.' His position must be very weak if he is forced to fall back on such arguments. Were we to be guided by the complaints of the patient we would, for instance, still have to treat many cases of persistent headache as malaria and fill the patient with quinin and similar drugs, as we formerly did. We would

---

*Annals of Otology, Rhinology and Laryngology. p. 173. February, 1898.

never be justified in removing polypi, etc., in cases of asthma because the patient does not complain of his nose."

Dr. Freudenthal goes on to refer to his examination of the nasal passages of 500 patients in the New York Hospital for the Ruptured and Cripples, in which he found numerous cases of serious nasal disease among those who did not complain of their nose or throat. Again, he says of 75 consumptives at the Montefiore Home. 38 complained of their nose or throat, and of the remaining 37 who did not complain no less than 23 showed abnormalities in their nasal passages, and in only 14 were there no marked changes at all.

In the same article he expresses the belief that the majority of cases of permanent tuberculosis receive their initial infection from a diseased mucous membrane in the naso-pharynx. His article is well worth reading, and furnishes sound, though not absolutely conclusive, evidence in support of his theory.

He believes that partial obstruction of one nostril allows germs to settle upon the same side of the naso-pharynx with comparatively little disturbance, by air currents. Some observations of my own lend support to this theory. In a paper read by me before the Colorado State Medical Society, at their meeting in Denver, June, 1894,* I reported the results of an analysis of 200 cases of pulmonary tuberculosis. These cases were not in any way selected, but taken in the order in which they presented themselves, to me for their first examination, running back from a recent date. Of these 200 chest cases, 33 had also nasal disease, 38 also laryngeal, and 23 had both in addition to their lung affection, making 56 with nasal and 61 with laryngeal complications: only clearly marked cases being noted.

The nasal cases were divided: 1st. into those in which the disease was most evident or was known to have commenced, or existed solely or mostly on one side more than the other, and these again into right or left; and, 2d, into those in whom the amount of disease or obstruction appeared about equal, and in whom there was no clear evi-

---

*The classification was made from the side upon which the chief cause of the disease appeared to be, rather than where its worst symptoms might happen to be exhibited later in its course.

dence of its having begun on one side more than the other, these being grouped for convenience under the head of median.

For the purpose of the inquiry as to the relation of the nasal disease to the pulmonary, the cases were again classified under right, left and median, with regard to their pulmonary disease. The side of the chest in which the symptoms first appeared, or in which it was most manifest deciding the heading, while those in which the disease was about equal on both sides, or the side of origin was unknown, or had apparently begun on both sides at once, are classified under the head of median.

Defined in this way, it was found that of the nasal cases in 24 the disease appeared to have originated upon the right side and 14 upon the left. Of the 24 cases of right nasal disease the right lung was primarily or chiefly affected in 17, the left in 1, both equally in 6. Of the 14 cases of left nasal disease, 8 were primarily affected in the left lung, 4 in the right and 2 equally in both. So that in the 38 nasal cases where the disease could be classified right or left, 65.8 per cent. had their lung disease primarily or chiefly upon the same side as the nasal disease. It is interesting to note that the case of right nasal disease in which the lung disease was on the opposite side, was one where stenosis of the right nostril had been caused by fracture of the septum, and of the 4 cases of left nasal disease where the pulmonary disease was on the right side; 1 also was known to have been a traumatic stenosis.

I have recently made a similar analysis of 100 cases of pulmonary tuberculosis, taking them in the order in which they presented themselves for examination. Of these I find 54 had also nasal disease. Of these 54 cases 31 had septal deformities, causing more or less stenosis. The remaining 23 had naso-pharyngeal catarrh without marked stenosis on either side, and in whom there was no septal disease giving rise to obstruction.

Of the 31 cases with nasal disease the stenosis was present and most marked in the left nostril in 21, and in the right nostril in 10. The left lung was chiefly or solely affected in 14; the right lung was chiefly or solely affected in 17. There was laryngeal disease in 29 per cent. of the 31 cases. Of the total 100 cases, 25 had laryngeal complications, of which 21 had tubercular laryngitis and 4 simple chronic laryngitis. Five of the laryngeal cases were without nasal disease. Of the 4 cases of chronic laryngitis, all had nasal complications; and of the 21 cases of tuberculous laryngitis 16 had nasal complications.

Of the 31 cases of septal deformities there was marked stenosis, the deformity obstructing the nostril on the same side of the body as the lung which was the first and usually the most seriously affected, in 20 cases: that is, in 64.5 per cent, very nearly the same proportion as in the previous analysis of 200 cases just referred to, in which it was 65.8 per cent. In considering the cause of this curious symmetry between the nasal and pulmonary affection I wrote as follows:

"The facts would appear to suggest an underlying cause in a common deficiency of resistance to disease or injury on the same side of the body, which was probably caused by a pre-natal or post-natal imperfection of development or growth, either congenital or acquired by circumstances or habit: this developmental deficiency being one of nutrition or innervation, or both."

While this may possibly account in a measure for the nasal obstruction, and also for the progress of the pulmonary disease, yet I am inclined to agree with Dr. Freudenthal's theory, and to think that the bacilli in many cases find the readiest point of entrance in the naso-pharynx behind the obstruction. These facts and the theories deduced from them, appear to me, to enforce the wisdom of treating the respiratory tract as a whole.

THE FACIAL NERVE IN ITS RELATIONS TO THE AURIST.*

By Geo. L. Richards, M. D.,

FALL RIVER, MASS.

OTOLOGIST AND LARYNGOLOGIST TO THE FALL RIVER AND EMERGENCY HOSPITALS. FELLOW OF THE AMERICAN LARYNGOLOGICAL, RHINOLOGICAL AND OTOLOGICAL SOCIETY, ETC.

I have chosen this subject to present to you because I have thought for some time that an insufficient amount of attention was being given to this nerve in aural literature. Facial paralysis is more common as an aural complication than the current literature of the subject would lead us to expect or else my personal experiences have been especially unfortunate.

That the subject is a neglected one from the text book standpoint is easily proven by a reference to the new American Text Book of Diseases of the Eye, Ear, Nose and Throat, where no reference to the subject of facial paralysis can be found in either text or index. When it is recalled that the editor of the aural division of this book is one of our most distinguished American aurists, the contempt in which the subject is held is apparent. With the exception of the text books of Politzer, Bacon, Barr and Bishop those books which treat of the subject, do so in such a manner as to imply that the accident of facial paralysis is not of so much account after all. The index of Dr. Dench's book on the ear contains no reference either to the facial nerve or to facial paralysis but there are several short references to the nerve and its paralysis in the text, giving its anatomy, its possibility of affection in purulent and inflammatory affections and of damage in mastoid operations, ending with the astonishing statement that injury of the facial is not a serious accident as its

---

*Read before the American Laryngological, Rhinological and Otological Society at its annual meeting in Cincinnati, Ohio, June 2d and 3d, 1899.

function is in most cases restored in from three to five weeks under the use of the faradic current.

While I am willing to concede that affections of the facial nerve bear no relation in severity or danger to others that may occur in the aural region or be associated with the facial trouble, yet I regard facial paralysis as a by no means trivial condition; on the contrary it is a most distressing one, producing what Bishop well calls a shocking deformity. The possibility of its occurrence is always to be borne in mind by the aurist, avoided whenever possible and always treated as a serious condition. To the aurist in a large metropolitan district the occasional case of facial paralysis occurring in connection with some ear operation may not matter, but to the dweller in a small city, like myself, the presence of a person or two in the city with face drawn to one side, unable to laugh or properly close their mouth, to whistle or to close their eyes is an advertisement not to be desired. Especially is this the case if, for any reason, the doctor is given the credit of having produced the condition. Then it is freely given out that doctor so and so operated on the ear and ever since the face has been crooked. Perhaps the operation was the result of some urging on the part of the physician and the patient now wishes it had not been done. This condition of facial paralysis is a walking advertisement of the worst kind, visible to all men. Most of our aural surgery is absolutely concealed, but this part of it, or complication of it appears on the face where he who runs may read. It was said in a derisive way by the assistants in the eye department of the Halle Clinic that all of Schwartze's mastoid patients had facial paralysis. This was very far from the truth, though it did occur now and then. Schwartze was always on the lookout for the accident and warned his students to be.

References to anatomy before this society are no doubt entirely superfluous yet I can not refrain from a brief review of the topographic anatomy of the facial nerve before going on to point out conditions in which I regard it as of concern to the aurist. Arising from the medulla it enters the internal auditory meatus with the auditory nerve, lying first to its inner side and then in a groove upon it and being connected with it by one or two filaments. At the bottom of the meatus it enters the aqueductus Fallopii and follows the serpentine course of that canal through the petrous portion of the temporal bone, from its commencement at the internal meatus to its termination at the stylo-mastoid foramen. It is at first directed outward towards the hiatus Fallopii, where it forms a reddish, gangliform swelling, the geniculate ganglion. and is joined by several nerves; then bending suddenly backward, it runs in the internal wall of the tympanum, above the fenestra ovalis, and at the back of that cavity passes vertically backward to the stylo-mastoid foramen. Here it divides into main branches after passing through the substance of the parotid gland and supplies the muscles of expression of the face. Within the aqueduct it gives off the tympanic branch to the stapedius and the chorda tympani. Topographically it is in close relation to the foramen ovale. the distance from the canal at this point being but a few lines. It is also in close proximity to the deeper mastoid cells and close to the inner tympanic wall above the foramen ovale. At this point the bony partition is often very thin and may be wanting. Hence it follows that as a result of chronic processes, especially those involving destruction of tissue, the facial nerve can be injured as a result of pressure or by involvement of its neurilemma in the pathological process. This is the more likely to be the case if the process has lasted some time, though it may occur in connection with acute troubles, otitis medi acatarrhalis or suppurativa. Panzer has shown a preparation from a case of acute tympanitis in a child where the facial canal showed a defect and the perineurium and nerve fibers were affected by inflammation. This shows how facial paralysis may arise in children from acute tympanitis.

Politzer regards transitory facial paralysis more common than has been supposed. It has been observed in simple non-perforative catarrh by Wilde, von Tröltsch, Politzer and others. Weiss has reported a case of perforation of the membrana tympani with a knitting needle which was followed by facial paralysis. Cartaz reports a case of paralysis from pressure in connection with acute otitis, cured after paracentesis and another case occurring in connection with the acute earache of influenza.

Forty-eight hours after this began facial paralysis supervened. Although all the ear symptoms subsided the facial paralysis was very rebellious, requiring a long course of electricity before final recovery.

In most cases of facial paralysis coming on in the course of an ear trouble an old suppuration is probably at fault. Bezold observed this in 1 per cent. of all o. m. p cases. This has usually lasted some time and is accompanied by evidences of caries and probably granulation formation, as in the following: Female child of 6 years old, with running ears and one-sided facial paralysis, which was said to be of recent origin. Both middle ear cavities were filled with granulation tissue. This was removed with sharp curette and the facial paralysis improved, but did not entirely disappear during time child was under observation. I think the paralysis due to pressure on the nerve, the facial canal being laid bare at some point, or else involved in the chronic carious process affecting the rest of the middle ear.

Intra-tympanic operations, ossiculectomy and curettage have been growing popular of late years as a cure for old foul suppuratives. In operations of this class the close contiguity of the facial nerve and the possibility of its sheath being injured must be borne in mind. Curetting, and all other maneuvers in the region of the foramen ovale, must be most carefully done and the face watched constantly for twitching. I was reminded of this more forcibly than pleasantly some two years ago. I had operated on an old long-standing suppurative, which I had labored a year or so to heal. Under ether I removed the remnant of malleus, incus and stapes, and as there was considerable granulation tissue I mildly curetted the inner surface of the tympanic cavity. No force was used at any time, and the operation was done under good illumination. I had not noted any particular twitching during the operation, nor was I expecting any; in fact, I do not recall having considered it as a possible complication. My surprise was, therefore, great to find the next morning that the face was drawn to one side and all the evidence of a very decided facial paralysis at hand. The patient laid the stiffness of the face, as he called it, to the ether, but I knew otherwise. The tympanic wound healed promptly with entire cessation of discharge, but the facial paralysis persisted for months in spite of the regular systematic use of faradism, galvanism, iodides and tonics; and even now, two years and more after, the recovery is not complete and he laughs somewhat one-sided and cannot whistle. Of all medical agents galvanism did this case the most good. Here the sheath, and perhaps some of the filaments were undoubtedly injured by the curette.

Bishop has had a somewhat similar case in which after an operation for excision of the ossicles through the meatus, paralysis affecting all the branches of the facial took place. Recovery took place after use of galvanic and faradic current for three or four months.

In chronic suppuration where ossiculectomy is done Burnett advises against the use of the curette even for the removal of granulation tissue. He reports 113 intra-tympanic operations without facial paralysis, in none of which was the curette used.

While I am not prepared to say, as does Burnett, that the sharp curette ought never to be used for the removal of granulation tissue in the tympanic cavity, it ought certainly to be used with care and with a due appreciation of the possibility of injuring the nerve. Whenever the dull wire curette will answer, it is preferable. It is also possible under certain circumstances to injure the nerve with the incus hook.

Even the use of the probe in these cases has its danger, though I do not know that paralysis has resulted. Strong medicinal measures used for the healing of suppurative and carious conditions may do some damage to, or excite some inflammation of the facial where there is any gap in the canal wall. There seems at present to be no way of accurate diagnosis as to the presence or absence of any gap.

Paralysis is common from pressure on the nerve at its outlet, due to purulent accumulations which break through the mastoid or glaserian fissure, or through the thin mastoid cortex of children. Such a case has recently come under my observation where a child during convalescence from scarlet fever developed an acute mastoiditis, with cerebral symptoms, and later acute meningitis. The pus worked its way to the surface in the neck near the facial nerve outlet before I saw the child, and produced a complete facial paralysis of that side of the face. After the evacuation of the pus and operation on the mastoid the paralysis rapidly disappeared, although no treatment whatever was directed to the nerve as such, the child being too sick from the cerebral and mastoid condition. By the time these were recovered from, the facial had fully regained its normal condition.

In the various operations which are done for the relief of acute and chronic mastoid inflammations, cholesteatomata and the like, the facial nerve is to be borne in mind. It is probably less likely to be overlooked than in the tympanic operations to which I have already referred. Here, again, its topographical relations to the parts operated upon must be borne in mind, not forgetting that the mastoid is a bone with many variations, and that the depth of the nerve and its relations to the posterior wall of the external canal are such that in a complete operation it may be absolutely impossible to avoid its injury.

Barr states:

"Below the level of the antro-tympanic passage the facial nerve takes a curve downward and outward, and may be injured by the operator. This is undoubtedly a risk in all cases, and may be inevitable when the nerve is denuded of bone and imbedded in granulation tissue. To avoid it we must keep well up and away from the lower portion of the posterior wall of the external canal."

It was the rule when I was a student in Schwartze's clinic to measure the depth from surface downward, and the rule was given that exceptionally the facial nerve was met at a depth of 18 millimeters, that at 20 millimeters the greatest care must be observed, and 25 millimeters never exceeded. In four specimens recently examined I found the depth to be, respectively, 16, 18, 20, 20 millimeters.

In an article on Schwartze's Clinic, published in 1895, I stated that injury to the facial in the course of mastoid operations was not uncommon, and that the nerve was affected "not only in that portion of the fallopian canal, which is above and behind the oval window and on the lateral wall of the antrum, but also further under and outward on toward the stylo-mastoid foramen, a portion of the canal that shows in its course many variations. The paralysis may not appear until some time after the operation. It is then the result of a peri neuritis extending from the point of injury, or caused by pressure from blood exudation in the canal. Whenever, during an operation, the region of the fallopian canal or the oval window is reached the greatest care is exercised and an assistant is told to watch for the slightest sign of spastic contraction, so as to avoid if possible any injury to the nerve." As an instance of the occurrence of paralysis some time after the mastoid operation, I have the following to report:

Miss S., aged 16 years, old purulent ear, with fever and threatening symptoms. Operation: Mastoid antrum and attic thrown into one cavity and remnants of the ossicles, with granulation tissue and cholesteatomatous material removed and bony parts curetted. The whole of the posterior bony canal was removed and the membranous wall slit posteriorly. Especial care was taken not to injure the facial nerve. The immediate convalescence from the operation was as speedy as could have been expected but the final recovery was delayed somewhat owing to the slow healing and final granulation of the extensive bony area involved in the operation. There was more or less purulent discharge for some time both through the canal and at the site of the mastoid wound. It finally healed, as I supposed, for good and the child was allowed in school, seeing me two or three times per week. One day, five months after the operation, her father brought her to me with unmistakable signs of facial paralysis. This had come on two days previously and was growing worse. The wound looked all right, and at this time I noted nothing special out of the way. I assured the father it would come all right and hoped it would. I took her out of school and began the use of iodides and strychnia with galvanism and later faradism every day. The paralysis gradually improved. At this time the original wound was all healed except a small fistulous tract at the upper portion of the mastoid wound. After the paralysis had lasted about two weeks I found some granulation tissue in the external canal, coming from what had been previous to operation the region of the attic. Removal of this seemed to hasten very rapidly the restoration of the power of the nerve. The paralysis was at no time absolute and the cervico-temporal group recov-

THE FACIAL NERVE IN ITS RELATIONS TO THE AURIST. 115

ered ahead of the cervico-facial ones. It was due to the too early closure of the external wound and consequent pressure from retained granulation products before complete healing had taken place in the deeper parts, combined with the fact that there must have been as a result of the old caries some point where the fallopian canal was bare. I have wondered why this did not occur in the earlier days when the amount of secretion was greater and the pressure of dressings and retained secretions would have favored pressure on the nerve. After the reopening of the wound the healing was speedy and no return of the paralysis has recurred, although several months have since elapsed.

In all operations where the attic and antrum are thrown into one cavity and the margo tympanicus with, it may be more or less of the posterior bony wall cut away, the fallopian eminence must be carefully watched and its injury avoided, for underneath this lies the nerve.

Occurring independently of any operative procedure it may be necessary to determine the exact seat of the lesion. This is not usually difficult though of course it may be. The history of the case will usually be sufficient. As an aid to this Ross gives the following as diagnostic points:

If the injury is external to the fallopian canal, the muscles of the face are alone paralyzed.

If, in the fallopian canal but below the point at which the chorda tympani leaves the facial, the muscles of the external ear are paralyzed in addition to those of the face.

If between the point where the chorda is given off and nerve of the stapedius there is in addition to preceding, abolition of taste on the lateral half of the anterior two-thirds of the tongue and diminution of the salivary secretion on the affected side.

If between the point of origin of the stapedius nerve and geniculate ganglion the same symptoms are present, together with abnormal acuteness of hearing.

If the geniculate ganglion is diseased all the preceding are present together with paralysis of the soft palate and distortion of the uvula.

If above the geniculate ganglion all the preceding are present except disorders of taste together with implication of the auditory nerve and dullness of hearing on the affected side.

If in the pons on a level with the facial one or more of the following—hemiplegia of the opposite side, paralysis of the sixth, auditory and branches of the fifth on the same side and a staggering gait with tendency to fall towards the side affected with facial paralysis.

The question of prognosis is one of the important ones in connection with facial paralysis. Will it get well? How long will it last? are questions sure to be asked over and over again. Fortunately most cases do sooner or later get well usually completely so, but not always. Occurring as a result of congestive processes or from pressure of inflammatory products it will pass off in reasonable time provided the pressure is removed. Even here some time may elapse before complete restoration of function takes place. I do not think that we can be certain of complete recovery if the condition has lasted any length of time as contraction or atrophy of the nerve or thickening of its neurilemma may take place. Politzer remarks that "occurring in connection with suppurative ear trouble it must always be regarded as a grave affection and significant as indicating a possible unfavorable extension of the disease toward brain abscess or sinus thrombosis." Should the lesion result in actual destruction of the nerve trunk the paralysis would appear suddenly, be complete and be likely to continue unimproved. M. Furet has sutured peripheral end of facial to trapezius. Some tonicity seemed to return but the results were not brilliant. The prognosis in connection with the operative cases depends, of course, entirely on how much the nerve is injured. If only the sheath has been injured the chances of eventual recovery are good. The reaction to the electric current is a valuable prognostic sign. "At first the irritability for both galvanic and faradic currents is diminished. In a week or two the galvanic is heightened while the faradic remains diminished. As case goes on the neuro-electric effect becomes less while the myoelectric effect becomes greater." If there is no response to faradic stimulation the paralysis may be looked upon as serious. All the fibres will not be found to be affected alike, some muscles will act much better than others and tone will come back faster to some groups than others. It has been my experience that the conditions are more troublesome to the patient when there is inability to close the eyes than when the mouth group are principally affected. The inability to close the eye produces a conjunctivitis and a degree of discomfort which is very great. As a rule I have found, fortunately, that this group were the first to regain their normal condition. It will occasionally happen that single branches remain permanently affected.

The treatment of these facial paralyses will depend primarily on their cause. If due to pressure or caries, so far as possible, the cause is to be removed. After the pressure is removed the remedies that seem to do the most good are the two forms of electricity combined with strychnia and the iodides. Of them all I regard the electric treatment as the most useful. Contrary to the experience of some I have found galvanism to do the most for me, although I frequently use both currents in the same case. Where possible I have the patient use a faradic battery several times daily at home, while I use the galvanic current myself. As faradism excites muscular contractility and galvanism acts directly upon the nerve endings, the advantage of using both currents is readily seen. The reaction of degeneration is not always a hopeless indication.

The question of mastoid operation will come up in many cases and must be decided according to the indications in the individual case. Urbantschitch regards facial paralysis as an indication for mastoid operation only when there is evidence that the mastoid is diseased.

The treatment is to be kept up for a long time if necessary, nor are we to become discouraged if at first the results of treatment are but slight. If the electrical reactions can be kept up and the nutrition of the individual muscles maintained at par or near to it, the chances of recovery, or at least of great improvement, are fair. Never say when recovery will take place. Say that it is probable that it will take place, but just when no one can tell. Even the cases that do not get completely well usually improve to a considerable extent: enough to justify all that is done in the way of treatment, since even a little improvement 'is worth a great deal to the individual concerned.

As a matter of precaution and of protection to the physician, it is wise in all operations in which there is any possibility that the facial nerve may be injured so to state to the patient or his friends in advance of the operation.

REFERENCES.

Politzer—Lehrbuch der Ohrenheilkunde. Stuttgart, 1893.
Barr—Diseases of Eye. Glasgow, 1895.
Bishop—Diseases of Ear. Nose and Throat, Philadelphia, 1897.
Dench—Diseases of the Ear, New York, 1894.
Sexton—The Ear and its Diseases, New York.
Ross—Diseases of the Nervous System, Philadelphia, 1895.
Burnett—System Diseases Ear, Nose and Throat, Philadelphia.
Burnett—Philadelphia Med. Journal, Feby. 26, 1898.
Weiss—Petersburg Med. Wocheschrift No. 39, 1897.
MM. Moure and Liaras—Journal of Laryngology, 1898, p. 623.
Panzer—Jour. Laryng., 1898, p. 395.
Cartaz—Reviewed Jour. Laryngology, 1897, p. 98.
Urbastschitch—Jour. Laryngol., 1898, p. 128.
Richards—Halle, The Aural Clinic of Prof. Herman Schwartze, Boston Med. and Surg. Jour., March 21, 1895.
Bacon—Diseases of Ear. Philadelphia, 1895.

ON THE USE OF RUBBER SPLINTS IN THE TREATMENT FOLLOWING INTRA-NASAL OPERATIONS.*

J. PRICE BROWN,

TORONTO.

In the August number, 1898, of the *Journal of Laryngology, Rhinology and Otology,* Richard Lake had a short article on the use of rubber splints in the intra-nasal work. I was impressed with his views at the time, as they seemed to supply a much needed want.

In my own experience, covering a period of more than ten years, devoted to special work upon diseases of the nose and throat, the evil effects of septal deformity could in the large majority of cases be removed by widening the narrow nasal passage, without resorting to fracturing or straightening the septum itself. Let a clear open chink be made, if only wide enough to prevent accumulations of mucus between the turbinates and the septum, and the catarrhal difficulties caused by the obstruction will, after healing of the mucous membrane, be in a great measure removed.

We rarely find, even in examination of healthy individuals, that the two nasal passages are approximately alike, the distance between the septum and middle and inferior turbinates on the right side differing from that on the left in the majority of instances. Still, provided the narrow passage is open, a considerable difference in the lateral dimensions of the two, will have little or no injurious effect upon the secretions of the mucous membrane.

Disease, however, arises when, from one cause or another, the septum touches the turbinate, or when the chink of the inferior meatus becomes so narrow that the mucous secretions accumulate in the passage, thereby inducing post-rhinal catarrh and preventing normal

---

*Read before the American Laryngological, Rhinological and Otological Society, June 3, 1899.

respiration on that side. In dealing with these cases, it is not the operative but the post-operative treatment that I have usually found the most troublesome. By saw or knife, drill or scissors or curette, single or combined, the projecting spur or ridge might be removed; synechiæ connecting the turbinate with the septum could be excised; or a partial turbinectomy, when necessary, might be performed; but to procure smooth equable pressure upon the incised tissues during the process of healing has been a much harder matter.

Some years ago a paper of mine on "Silver Tubage in Certain Cases of Septal Deformity" was read at the Laryngological Section of the American Medical Association in San Francisco, dealing to some extent with this subject. In many cases these silver tubes are useful, but in many others they are inapplicable: and in the latter class, in which the chink can only be a narrow one at best, I think that rubber splints, made as Lake advises, from thick rubber sheeting, do better work than anything else we have at our command. Their surfaces are smooth, compressible, and elastic: they can be readily cut to the required shape and they can be obtained of any thickness we desire.

After cocainizing the parts, and coating the plug with vaseline, it can readily be placed in position. Once in, it will retain its place and, by elastic pressure, give a smooth and even support to the raw surface to which it is applied, as well as prevent that profuse granulation which otherwise would sometimes occur. At the same time it does not retard the gradual extension of the new mucous membrane, while it moulds the tissues into a smooth and regular form.

The stiff pliable rubber, although not so hard on the surface, nor possessing the polish of vulcanite, is probably just as impervious to bacterial invasion. Sometimes, however, after prolonged use, it will acquire a peculiar unpleasant odor, in part arising from the rubber itself. In these cases new splints or tampons should be substituted for the old ones. As I have used these rubber plugs in a good number of instances, I might briefly quote the following ones from my case book.

CASE 1.—Oct., 1898. A boy, aged 6 years, was brought

by his mother to the Western Hospital for treatment on account of entire inability to breathe through the right nostril. This had been coming on gradually for several years, occasioned, the mother thought, by a fall on the face when two years old.

There was nothing striking about the external shape of the nose. There was, however, a marked curvature of the cartilaginous septum to the right, with a longitudinal ridge at its base. Chloroform being administered, the ridge was excised. Then to lessen the resistance, I cut into the convex surface of the curvature of the cartilage from behind forward. In one spot, although guarded by the little finger in the opposite nostril, the knife accidentally penetrated through the septum. Not heeding this, as it would probably unite by first intention, a rubber splint one-eighth of an inch thick, long enough to go beyond the triangular cartilage and as wide as the fossa would admit, was pressed into the nostril.

The child was kept under observation but the plug was not removed for two weeks. It was then found that the perforation had healed, and that the nasal passage was patulous. After cleansing, the splint was replaced and worn for several days more. The right passage was almost as large as the left and the patient was discharged, cured.

CASE 2.—Dec., 1898. A gentleman, aged 58, came for treatment for left nasal stenosis and "throat dropping." He stated that thirty years before, while at college, he went to a surgeon about his nose. The advice that he received was that there was a growth in the left nostril, but that it would be a difficult and delicate operation to remove it, and that unless it occasioned serious trouble he should leave it alone. He followed the advice given; and it was only during the last few years that it had given much inconvenience.

On examination I found a curved septum and a large round cartilaginous spur, filling up the anterior portion of the left nasal cavity. It was pointed, and impinged upon the opposite wall, just in front of the anterior end of the inferior turbinate. Behind it, the osseous septum was also curved for the greater part of its length toward the left side.

With a sharp curved knife, I excised the spur deeply, leaving a clean-cut surface. As this was followed by profuse hemorrhage, the naris was packed with absorbent cotton. On removing the tampon the following day, I found that the congested walls completely filled the cavity behind the site of operation; so after applying cocain and thus shrinking the parts, I at one slid in a rubber splint, the end being beveled to facilitate its entrance. It was made out of sheeting two-eighths of an inch in thickness. Slight irritation existed for a day or two but this soon passed away. At the end of the week it was removed. By this time congestion was over, the surface was smooth and it healed without further difficulty leaving a clear narrow chink.

CASE 3.—Feb., 1899. A carpenter, aged 23, had his nose broken when a child by a fall, producing a partial depression of bridge. For years has had almost complete stenosis on left side, resulting in pharyngeal catarrh and edema of uvula.

EXAMINATION.—Right nasal fossa enlarged, presenting concave, hook-notched septum on that side. Mucosa healthy and without catarrhal accumulation. On left side, large curvature with cartilaginous spur filling in the passage, together with osseous ridge extending to the posterior choana. In the centre a bony synechia connected inferior turbinated with septum.

The first operation was to remove the cartilaginous spur and put in a thick rubber splint. Four days later the osseous ridge with synechia was sawn out, and after hemorrhage had subsided, a long splint extending to the posterior naris was inserted. For a few days it was not disturbed. Then it was taken out daily, and after being cleansed, returned. The excisions in the case were very extensive. Still in six weeks the healing was very satisfactory, resulting in a clear chink from end to end of the passage with rapid reformation of mucous membrane.

CASE 4.—April, 1899. A boy, aged 7 years, was brought as a mouth breather for treatment. He had been stunned by a blow with a stick on the forehead when four years old. From that time, it was said, nasal breathing gradually became more difficult and finally ceased.

EXAMINATION.—Curvature of cartilaginous septum to left with ridge at base, columnar cartilage curved to right, also adenoids in naso-pharynx. Chloroform was administered. Ridge was first excised with knife. Then two longitudinal incisions from behind forward were made through the cartilage on the curved side, the finger in the right nostril acting as guide to protect the mucous membrane from perforation. A rubber splint two-eighths thick was at once inserted pressing the cartilage into central position. While still under chloroform a slip from the columnar cartilage on the right side was excised, and the adenoids removed.

Two weeks later the rubber tampon was taken out; the result being nasal breathing and good left nasal passage.

CASE 5.—April, 1899. Youth, aged 17 years. Nose externally twisted to right side. Says he was struck with a ball on the nose two years ago, since which time there has been increasing deformity and considerable nasal stenosis.

EXAMINATION.—Extensive ridge spur on left side with curve filling up the fossa, part of the cartilage being adherent to the middle turbinate. Under cocain I excised front part of ridge and after compressing septum to right with chisel, inserted one-eighth in rubber tampon. Four days later, under chloroform, I made two incisions from behind forward through septal cartilage, guiding, as in case 4, by finger in right nasal fossa and thus preventing perforation of mucous membrane. I then pressed out septal cartilage by passing a two-eighths inch rubber splint. The septum being straightened, the tampon was left in for two weeks. The front part of the fossa being now freely open, a bony ridge extending along the lower part of the vomer was removed by saws, and a long wide tampon one-eighth in thickness but extending from the anterior to the posterior naris was placed in position. After the first day it created no discomfort. As patient was returning home he was instructed to retain it in position for a month.

I might say with regard to the last two cases that the cosmetic improvement will be marked.

My own experience in the use of rubber splints has

so far been very satisfactory, and I earnestly recommend a trial of them to members who up to the present have not adopted Mr. Lake's advice in this matter.

In closing I might make one more remark. I have seen somewhere that it had been proposed to manufacture a species of perforated rubber quite distinct from the tubular splints already in use, in order to allow a certain amount of respiration and ventilation through it while in position. This I think would be a great mistake, as it would destroy all possibility of keeping the splints in an aseptic condition. Another thing, the perforations would be so quickly filled with nasal secretions of one sort or other, that the object for which the perforations were made would be nullified.

REMOVAL OF TONSIL AND ADENOID FOLLOWED BY FATAL RESULT.

J. A. STUCKY, M. D.,

LEXINGTON, KY.

LARYNGOLOGIST AND OTOLOGIST TO GOOD SAMARITAN HOSPITAL.

The history of a case requiring for its relief the removal of diseased and enlarged tonsils, and the operation followed by death within a few hours, is so unusual that I have thought it of sufficient interest to bring it before you for consideration and discussion.

J. A., aet 15, consulted me February 21, 1899, giving the following history: Had been in bad health for the past two months, caused by attack of "grippe," though not confined to bed or house; had been at school most of the time. He had been suffering with sore throat, tonsillitis and quinsy; being much worse for past two weeks, although he received the best medical attention. The throat trouble was aggravated by a "hacking cough." The attending physician referred him to me to have his tonsils and adenoid removed, as the only method of relief.

The appearance of the patient bore every evidence of genuine illness. For some days he had been having rigors and hot flashes at short intervals, accompanied frequently by profuse sweating. He was pale, except for the hectic flush; pulse quick and full. Complained of constant headache, and had the characteristic expression of a mouth breather from adenoid obstruction. Temperature 101° F., tongue coated, breath offensive.

Examination of the pharynx showed the left tonsil enormously enlarged, protruding beyond the median line, of soft, spongy and granular or fungus appearance. The tonsillar crypts and follicles were filled with offensive pus, evidently oozing from an old peri-tonsillar abscess. The anterior and posterior pillars were adhering to the tonsil in such a way that the tonsillar mass prevented free drainage of the abscess. Pushing the tonsil to one side gave freer exit to retained pus. Several small abscesses were noticed along the alveolar border, and the gums were soft, spongy, bled freely and were soaked with pus. The pharyngeal vault was filled with adenoid vegetation, covered with offensive discharge, similar in every respect to that in the tonsil.

There was entire absence of indications of active inflammation, no redness or marked induration being noted. A diagnosis of general septicemia due to auto-toxemia was made, and removal of diseased and suppurating tonsil and adenoid advised.

This was consented to, and the patient sent to St. Joseph's Hospital. Dr. John Scott, after examination of the patient, said there was no contra-indication to administration of an anesthetic, and gave him chloroform, after the parts had been thoroughly cleansed with antiseptic solution by means of atomizers. Very little anesthetic was needed, and was taken without an unpleasat symptom. The throat being large, every step of the operation was easily and quickly done. The tonsil was removed with tonsillotome, adenoid with Gottwein's curette. There was little more than the usual hemorrhage, and, after spraying the parts with iced hydrogen dioxide solution, the patient was put to bed in good condition. On account of the general septic condition and suspecting him to be a bleeder. I remained an hour and a half after he had recovered from the chloroform administration, and left him in good condition and quite cheerful. Instructions were left with the nurse to use iced spray (25 per cent. hydrogen dioxide in Seiler's solution) if there was any free oozing of blood.

Within thirty minutes after leaving him a hurried telephone message was received, saying patient had just vomited, and was bleeding profusely from nose and mouth. I was at his bedside within 10 or 15 minutes: the bleeding had checked considerably under use of the spray. Pulse was quick, expression anxious, great restlessness, and every indication of impending collapse. A hypodermic injection of ergotine 1/10, strychnia 1/30, and morphia 1/6 grain was ordered, while I proceeded to thoroughly remove all blood and clots. Examination revealed no special bleeding point, but a very general oozing of venous blood; very little arterial oozing was found. Most of the bleeding was from the tonsillar and post-pharyngeal surface. After drying the parts, an application of McKenzie's styptic solution, followed by ferri per-sulph., applied by means of cotton covered probe, effectively stopped all bleeding.

Before completing this treatment Drs. Scott, Kinniard and Patterson arrived, approved of the treatment pursued, and agreed with me that the patient would probably soon react and rally if there was no further bleeding. After waiting a few minutes, the pulse being fairly good, though weak and irregular, it was decided to use transfusion of hot normal salt solution. Within three hours three pints were used subcutaneously, and readily (apparently) absorbed. Whisky, strychnia and digitalis were also given hypodermically, as indicated. Efforts to sustain life by these means failed, and the patient died nine hours after the operation, and seven and a half hours after the secondary hemorrhage had been controlled. As to the imperative and immediate indication for the operation, there is in my mind no doubt. I am equally positive that death in this case was coincident with the operation, the latter being the exciting not the immediate cause.

It is well known that no operation (when indicated) gives such remarkable results as that for the removal of adenoid tissue; also their removal is accompanied by very free venous bleeding. In this case the loss of venous blood at the time of operation was little more than is usually the case, and all the bleeding stopped within a short time without the use of any styptic except iced spray of hydrogen dioxide and alkaline anesthetic solution. Within a few moments after the appearance of secondary hemorrhage I had thoroughly cleansed all bleeding surface; there was no special bleeding point discovered, and no evidence whatever of a vessel of any size being severed, but instead there was a very rapid, free oozing of venous blood, which was easily controlled by the application of the styptic referred to.

The most plausible theory, to my mind, as to the cause of the death is that the entire system, with all its recuperative force, had been so exhausted and undermined by sepsis that reaction was impossible, although every facility for promoting this was easily at hand and freely used. I know of no other way to account for the result, because I do not think enough blood was lost to cause death, and this did not occur for seven and a half hours after the bleeding was completely controlled. Admitting the hemorrhagic diathesis does not account for the unexpected and terrible result, I am forced to the conclusion in this case that I was dealing with a septic condition of affairs, that nothing short of what was done would have relieved the patient, complicated with a hemorrhagic diathesis, and that "something which passeth understanding."

DISCUSSION OF DR. STUCKY'S PAPER.

Dr. H. H. Curtis, New York:—Mr. President: I should like to say a word of sympathy for the doctor. We are fortunate if such a result has not come to us individually, for we know such cases will happen. I was going to operate upon a young man for trouble in the nose, in a prominent family in New York. I was about to operate on Monday. I felt his hand and said to his father, "that boy is not well." I took his temperature and found it 104.5°. The boy died on Thursday from pyemia. If I had operated upon him on Monday he would probably have died on Thursday anyway, and it would have been one of those cases such as the doctor spoke of as coincident. I tell all patients who come to me to operate on the tonsils, that there is danger of about one to forty thousand and if they are willing to take that chance I will operate. I think no criticism should be made on the case reported by Dr. Stucky.

Dr. Robert C. Myles. New York:—Dr. Stucky's case confirms something that I observed yesterday, that is that the statistics of the world belong to us. If his patient had been under our care we would have had the same misfortune. No power could have been brought there to change the result. It is very probable that the depression of the child and the extreme condition he was in from the septic poisoning, combined possibly with a reseasonable amount of hemorrhage, contributed to the shock. A great many surgeons think shock is due only to hemorrhage, but I think there is a great deal else contributed to shocks. It is my impression that the boy died of that, combined with his weakoned condition.

Dr. Charles W. Richardson, Washington:—I would like to ask whether the temperature was taken and whether there was any question of the presence or absence of fever. Was there any septicœmic temperature?

Dr. Stucky:—The temperature was taken four hours before the operation and it was 101, and after cleansing the parts thoroughly and giving the child a dose of salicylate of soda, to relieve his headache, his temperature dropped to 100.2° at the time of the operation.

Dr. Price-Brown, Toronto:—The gentleman could do nothing but use the best judgment he had, and I have no doubt if one of us had had the case we would have done the same as he did. It is an unfortunate result, but I do not think Dr. Stucky is in any way to blame for it. Physicians often go a lifetime without seeing such cases, yet sometimes we meet them and we are blameless of the results. I want to mention a case that resulted almost fatally in the practice of a gentleman I know. The operation was a tonsilotomy. There was slight hemorrhage for a short time and then relief. Later hemorrhage again came on more severely and the patient bled several quarts before it could be stopped. Every effort to relieve it was in vain and the patient went almost into syncope before it was finally stopped. I will also mention a case that I saw, in which there was septicemia. It was a death following the removal of adenoids by the same method used in this case. Notwithstanding all that could be done the patient died. Upon autopsy examination, it was proven that the internal carotid artery was out of its normal position and had been opened. It was impossible at the time of the operation to tell that the artery was not in its normal position. These unfortunate cases will occur, and I think Dr. Stucky is in no way to blame for the results in the case he reported, and he certainly needs our sympathy.

Dr. T. V. Fitzpatrick, Cincinnati:—I disagree a little with the last speaker in saying that Dr. Stucky needs our sympathy. I think he does not need our sympathy, for he discharged his duty just as he should have done. I recall a case of operation for adenoids, in which I operated in the face of a temperature of 103.5°. After operation and drainage the temperature fell to normal. The case was not complicated by any rise of temperature. I think the doctor managed his case as well as could be, and under such circumstances the physician must simply "take his medicine," which Dr. Stucky knows how to take very well.

Dr. Ewing W. Day, Pittsburg:—Cases like Dr. Stucky's are indeed unfortunate, and make a physician timid and liable to shirk his duty from fear of like results. In this case I do not believe the operation was the important factor, but an unfortunate coincident, that is liable to happen to any of us. The amount of hemorrhage was certainly not more than we often encounter. There is such a thing as cowardly surgery, if I may so call it, when a timid man fails to do the only thing that can give relief, from fear that his reputation may suffer if the operation is not successfull. If Dr. Stucky had been influenced by these considerations and allowed his patient to go on each day coming more deeply under the influence of the septic poison, he certainly would have shirked what was his plain duty.

Dr. J. E. Sheppard, Brooklyn:—I look upon it as purely a piece of luck that I have not had one of these cases. I have a sort of wholesome fear of it, for I am advancing in years and know that I must be nearing my fatal case.

I use the galvano-cautery in these cases. It seems to me the absorption of toxines or toxic poison must play an important role in these cases. I do not think this case would have died but for the toxic poisoning. Those of you who have not seen what the toxines will do, can hardly understand it. But those of you who have seen even a tenth of a minim of the toxin of erysipelas produce a very high temperature, will not believe it improbable that this case died from toxin poisoning. Possibly the hemorrhage had something to do with it, but that the patient died from hemorrhage I am very doubtful indeed.

Dr. J. A. Stucky:—I believe if I had not operated on this case I should have been guilty of criminal neglect. If I had a thousand cases to confront me, one right after the other, I should advise the same thing and do the same thing in the very same way, if the patient and the friends would consent. It is just one of those little cyclones that strike a fellow now and then. To have done this kind of work for twenty years and then have it occur, makes one feel "a little bilious." But I think you will agree with me, if I had done anything less than I did, I would have been guilty of neglect.

THE QUESTION OF POSTICUS PARALYSIS, PART II, AND THE INNERVATION OF THE LARYNX DURING BREATHING.*

A. KUTTNER AND J. KATZENSTEIN,

BERLIN.

Translated by CLARENCE LOEB, A. M., M. D., St. Louis.

I. Semon's law, its historical evolution and its foundation.
II. Grossman's hypothesis and its refutal.
III. The removal of the posticus in experiments on animals, formerly and at present.
IV. Semon's law in the light of recent experiments on animals.
V. The dilators of the glottis and the innervation of the larynx during breathing.

Grossmann's article in opposition to Semon's law, which appeared in Vol. VI of the *Archiv. für Laryngologie* we called forth a large number of replies, all† of which, with the exception of one by H. Krause,‡ have proved unfavorable to Grossmann. In the Vol. VII, of the *Arch. f. Lar.*, likewise took a position on Grossman's hypothesis and we explained that it appeared to us untenable. We then continued our experiments, in order to arrive at an opinion as to how far experiments on animals could settle the vexed question.

I. SEMON'S LAW, ITS HISTORICAL EVOLUTION AND ITS FOUNDATION.

Semon took the position, by reason of many facts which have been known for some time and whose interpretation then appeared free from objection, that, in a progressive disease of the recurrent, the fibres of the abductors suffered first while the adductors were affected later in the course

---

*Archives für Laryngologie, Vol. IX. p. 308.
†Vol. VI, VII, VIII Archives für Laryngologie.
‡Arch, f. Anat. u. Physiol., 1899, Physiol. Abtheil s. 77.

of the disease. The laryngeal picture which portrayed the three steps of progressive paralysis of the recurrent was already known; likewise the limitation of the outward movement, caused by the simple posticus paralysis (stage I), the median position, corresponding to paralysis of the posticus with secondary contraction of the abductors (stage II): and the cadaveric position, the sign of complete paralysis of the recurrent (stage III). The explanation, at first purely hypothetical and obtained by exclusion, that the first two stages were due to a lesion of the posticus, gained a strong support when Riegel proved in a post mortem that the clinical median position had a palpable degeneration of the abductors as an anatomic substratum. A further proof for the statement that the median position was to be explained by a paralysis of the posticus, was found in the results of v. Schechs*' and Schmidt's experiments on animals. They found that in dogs, when both postici had been severed, 48 hours after the operation at the latest, there appeared a median position of both vocal bands, which irrevocably, without artificial aid, resulted in death by suffocation. Furthermore, the observation had been made in the clinic that Stage I can gradually pass into Stage II, and further, that the median position can change into that which, since the time of Legallois, has been recognized as characteristic of paralysis of the whole recurrent.

Semon found these facts, which then were unopposed by any argument, already prepared for him. From these and from his own corroborating experiments, he formulated his law which, without giving any explanation whatever, was the simple statement of the conviction that in a gradually progressing disease of the recurrent. the succession of phenomena takes place in a fixed and regular manner. When on the contrary, Grossmann maintained that Semon's law is a purely theoretic hypothesis for which clinic and pathologic results gave only supplemental support, his assertions are completely contrary to the facts.

---

*Schech. Experimentelle Untersuchungen über die Nerven und Muskeln des Kehlkopfes. Zeitschr. f. Biolog. Bd. IX, 1873, p. 258.

Schmidt. Die Laryngoscopie an Thieren. Tübingen, 1873.

Archiv. für Laryngologie. Vol. VII. p. 376.

II. Grossmann's Hypothesis and its Refutal.

In part I of our paper we noted the objections raised by Grossmann against Semon's law, *i. e.*, not so much against this law itself, as against the interpretation of the clinical picture on which it is based. He maintained that Stage I (the simple posticus paralysis) has never yet been observed by anyone. In opposition to this we, as well as Semon, had furnished proof that this very laryngeal picture has been seen in practice and described by more than 20 observers though Grossmann pictured it theoretically. Regarding the Stage II, he claimed that in dogs, we obtain the picture corresponding to the clinical median position (adduction position, as he calls it) if a simple section of the recurrent is made, and that this position passes over into the actual cadaveric position if the section of the recurrent is followed by that of the superior laryngeal. From these experiments he drew the conclusion that the clinical median position has nothing whatever to do with a posticus paralysis. After checking over Grossmann's experiments, step by step, with an equal degree of care and with the same measuring instruments, we think we have proved that the position of the vocal bands in animal experiments, after section of the recurrent, is very different from the clinical median position to which, according to Grossmann, it should be analogous, and that every conclusion, therefore, which is based on the asserted congruence of both laryngeal pictures is fallacious.

III. The Former and Present Elimination of the Posticus in Animal Experiments.

After we had come to the conclusion that Grossmann's hypothesis was untenable in consequence of the considerations outlined above, we proceeded further in following up the above mentioned line of thought, namely, to imitate by means of removal of the postici muscles those conditions which we accept in the clinic as the basal facts of the laryngeal picture in question. These experiments, as we have indicated above, were carried on by Schech and Schmidt more than 25 years ago, and the results obtained thereby formed, at that time, a weighty support for the belief that the median position was caused by a posticus paralysis with secondary adductor-contraction.

Grabower, Klemperer and Grossmann took up these experiments and their results have been published co-incidentally.

Differing from other investigators, who use simply estimates, we carried on this part of our work in such a manner that, by means of Muschold's laryngometer telescope, we were able to measure exactly every separation and every oscillation.

In the beginning we conducted the operation in such a manner that we made a lateral passage to the postici muscles. Thus, we obtained on the whole the same results as Schech, Schmidt, Grabower and Klemperer. Later we changed our method of operation and made an entrance in the manner lately described by Grossmann.* We cut directly across the trachea about 2 cm. below the larynx, separated the upper part of the trachea and the larynx from esophagus, which was then turned upwards, whereby the posterior wall of the larynx with the postici was laid bare. After removal of one or both postici, both segments of the trachea were joined together by 4 to 6 sutures.

The results obtained by these proceedures were usually different from those by the methods which were previously used. Formerly, after bilateral section, at the latest 48 hours afterwards, a median position of both cords appeared, and in addition the highest degree of dyspnea which, in absence of artificial help, terminated in death by suffocation. This result was one of the chief supports for the theory of the relationship between median position and posticus paralysis.

This new method of experimentation taught, that median position and suffocation are by no means a necessary consequence of the removal of both postici. We can therefore not deny that the results abtained by the lateral method are influenced by accessory injuries and that the newer mode of operation, which with an equally careful removal of both postici, does not cause median position and suffocation, furnishes more unobjectionable results.

We removed the postici in about 50 dogs on one side or bilaterally and obtained the following results:

---

*Pflüger's Arch., Vol. 73, p. 184.

1. The vocal band robbed of its musculus posticus can not be abducted to the same degree as before; the maximum of abduction is now nearer the median line than before. Within the possible latitude of oscillations the rhythmic respiratory movements and the phonation movements are preserved.

2. During quiet respiration the respective vocal band is not abducted beyond the cadaveric position, as is the case after section of the superior laryngeal nerve. In forced respiration the outward movement goes beyond this measure.

3. Animals, in which both postici muscles are removed, do not show median position and do not die of suffocation. In quiet attitude the respiration is, it is true, audible, but not forced. During motion or psychic irritation dyspnea sets in, which, with complete median position, may cause suffocation.

4. The mode of movement of the vocal cords robbed of their postici muscles, suffers no variation at all, even if the animal is kept alive as long as a year after the operation. We cannot notice especially any approach of the vocal bands to the median line, any interference with the adduction or abduction movement, and any median position.

These results agree entirely with those obtained by Grossmann, but contradict in an essential point the theory lately advanced by Klemperer[*] on the ground of some new experiments. He never saw, after removal of the postici, an abduction greater than the width given by the cadaveric position. We have been very careful on this very point, and on the ground of correct measurements repeated time and again. We must decidedly hold the position that, after total removal of the posticus, the outward movement may excede the cadaveric position obtained by section of both laryngeal nerves.

IV. SEMON'S LAW IN THE LIGHT OF NEW ANIMAL EXPERIMENTS.

What, then, are the conclusions which may be drawn from the above facts affecting Semon's law and what are

---

[*] F. Klemperer: "The position of the vocal bands after removal of the crico-arytenoid. post. Pflüger's Arch., Vol. 74. p. 272."

the objections raised by Grossmann against the latter?

In regard to Stage I, the simple posticus paralysis, we have already stated the fact that numerous observers have described laryngeal pictures which are explained on the basis of a single posticus paralysis. Grossmann did not know of the cases, hence his erroneous statement that no one ever saw the simple, uncomplicated paralysis. These clinical pictures conform in the most complete way to those obtained by animal experiments. Here, as in each case, the outward movement is lessened; the affected band lies closer to the median line, and from this position its rythmic respiration and phonation movements continue. The whole deviation from the normal consists in the fact that abduction and adduction no longer takes place to so great a degree. The analogy of the clinical laryngeal picture and that obtained by animal experiments is a complete one and, if anywhere, we must here recognize the evidential strength of animal experiments, i. e. we must recognize that the clinical picture for which we claim that it corresponds to the Stage I of the progressing recurrent paralysis, is due to a disease of the abductors.

The observation had been made in the clinic that, in individual cases, the picture of a simple posticus paralysis suffers a change in the course of time. The vocal bands very gradually approach the middle line, the range of motion still remaining becomes smaller and smaller, and finally ceases entirely at the moment when the vocal bands attain the median position. Riegel, confirming this appearance, has expressed the opinion (Semon only accepted it) that there occurs after the paralysis of the abductors a gradual contraction of their antagonists, the adductors. The gradual transition from Stage I to II, seemed to speak for this opinion; then the results of the Schech-Schmidt's experiments, and, finally, the results of the microscopical experiments were not inconsistent therewith. The later experiments in section of the posticus destroy an important link. Since in this chain of argument apparently so well made, they show that the removal of the abductors, by no means as heretofore believed, must necessarily be followed by a contraction of the adductors. For one year after the operation, the vocal band deprived of its posticus was just as movable and just as far from the median line as immediately after the removal. We are unable to explain the contradictions between our observations and the later ones of Klemperer (l. c.) who obtained a gradual approach of the vocal bands to the median line. Our statements in regard to this point are in agreement with those of Grabower and Grossmann. We must, therefore, confess that we cannot experimentally obtain the median position by a simple removal of the posticus. Does this, then, oppose Semon's law? We think not. In the first place, it was already known, from clinical observation, that Stage I remains in some patients years at a time without the occurence of a contraction of the adductors, i. e., a median-position; we ourselves have observed such a case for about two years. In other patients, the change from position I to II occurs sometimes earlier and sometimes later. Does not this irregularity of cases argue with greatest probability for the fact that if this transition is to occur, some circumstance which exerts the determining influence must occur?

This belief obtains a still greater measure of probability if one examines the cases in which the vocal bands alternately assume one or the other position, as the history of the following case shows:

In the fall of last year, there appeared in the polyclinic of one of us (Kuttner) a railroad employee from Frankfort who suffered from asthmatic trouble. The thoracic viscera showed no anomalies; the examination of the larynx, however, showed that the glottis could not be opened more than 4 mm. Both vocal bands stood and moved symmetrically, made rhythmic respiration movements, and came together in the median line on phonation. The voice sounded somewhat hoarse; dyspnea was not present except when the patient exerted himself. After diligent questioning, we learned that the so-called asthmatic attacks were always accompanied by a cramp-like feeling in the throat. Further examination revealed a difference in the pupil. We decided upon a diagnosis of tabes incipiens, and concluded that the asthmatic attacks were nothing more than laryngeal crises. Our opinion was confirmed by Dr. Rothmann, a nerve specialist, whom we consulted.

For two weeks we treated the patient by the faradic current; no attack appeared during this time. The patient then returned home, but came back in about two weeks with the statement that he had suffered greatly while at home with attacks of suffocation. The laryngeal examination revealed that the left vocal band now stood just in the middle line and no longer made any movement; the right continued its excursions formerly observed, and moved from the median line as far back as 2 mm. The maximum of glottic width was no longer 4 mm., but was only 2 mm. During the examination, the patient was opportunely seized with an attack of suffocation. We could then see that even the right vocal band, in its ligamentous part, was drawn into the middle line and lay close to the left, while only in the cartilaginous part there remained a narrow space of about 1 mm. breadth, through which the air laborously was forced. For four days the right vocal band maintained its median position unchanged. On the fifth day after his return, it went back again to its former position. The glottis then again showed a breadth of about 4 mm. and the left vocal cord also once more commenced its movements. During this time a symptomatic treatment only was employed.

This observation teaches, in our judgment, with absolute certainty that position I, corresponding to simple posticus paralysis, can, under the influence of a new factor, change to position II, corresponding to the median position. This factor is lacking in our animal experiments, and that, in our opinion, is the reason that here Stage I does not pass over into Stage II.

But what sort of a thing is this factor? That it is not to be sought in Grossmann's hypothesis, is evident. The median position could never have yielded to that mobility of position I, if it had been caused by a total recurrent paralysis. We see no possibility of explaining the above case other than that an irritant of some kind caused temporary contraction of the adductors.

Can we imitate this condition in animal experiments? Experiments, which we have undertaken in this connectioon, have shown that we, as all our predecessors can not evoke the picture of median position exactly corresponding to reality, but can bring about something similar.

If, for example, a thread is loosely tied around a recurrent, the movements of the affected vocal band will cease and it takes a position nearer the middle line. If now the recurrent of the other side be severed, there is seen, if the thread be tied neither too loosely nor too tightly, an unmistakable asymmetry in the glottic width. It forms an approximate right-angled triangle; the right angle lies on the side of the tied recurrent. Faradic excitation of this nerve (the electrode must be placed about 2 cm. toward the center from the site of the thread) shows that the conduction is interrupted. After a longer or shorter period, in two cases about 20-30 min., this median position passes away and the vocal band lies symmetrical with that of the other side whose recurrent has been severed.

The result of this experiment agrees with similar observations of Krause,* Semon, Katzenstein and others. We think we may conclude that in all cases where the interruption of the conduction recurrent occurs in an irritated rather than normal condition, there will be found an irritation of the part lying peripheral to the point of irritation which finds its expression in an approach of the vocal band to the median line beyond the cadaveric position. Sooner or later this irritation ceases, and there appears the position corresponding to total recurrent paralysis.

We are confident that these experimentally prepared laryngeal pictures are indeed similar to the clinical median position and perhaps give an indication as to its meaning; an analogy to these does not, however, exist there, for if the experimental median position can and does last only a few days, the clinical, on the contrary, persists for months and years—through the kindness of B. Fränkel† and E. Meyer we were made acquainted with a case of double sided median-position lasting 23 years. We see a difference greater than one of degree, and we must insist that no one has as yet been fortunate enough to obtain experi-

---

*We mean here the observations which Krause related in his first work, Virch. Archiv., Vol. 98, 1894. The refutation in the actual observation of these and his last work (Arch. für Anat. u. Physiol., 1899. Physiol. communication) we cannot give.

†B. Fränkel—Laryn. stenosis following weakening of glottis abduction. Deutsche Zeitsch f. pract. Medicin., 1878. 6 and 7; A. Rosenberg, Vol. VIII, Archiv. für Laryng., page 13, case 10.

mentally a true analogy to the clinical median position. In the recognition of this fact, we see the great weakness of Grossmann's hypothesis.

From all these experiments and reflections thereon, we deduce the following conclusions regarding Semon's law and the grounds on which it is based:

I. Stage I of the simple posticus paralysis has been repeatedly observed and described. The removal of the posticus in dogs gives essentially the same conditions which we see in the clinic.

II. A condition analogous to the clinical Stage II, median position, does not occur in dogs after the removal of the posticus. This circumstance argues against the theory that every posticus paralysis necessarily involves a contraction of the adductors leading to a median position. We have not yet been lucky enough to produce experimentally the exact picture of a median position. That laryngeal picture, which in animal experiments, yields certain similarity, teaches, in agreement with clinical observation and autopsy records, that this Stage II, the median position, appears when there is added an irritation of the adductors to paralysis of the abductors.

That this explanation is the correct one for certain cases of median position seems to us without doubt. Whether it is the only one and fitting all cases must be tested by further experiments. Grossmann's hypothesis that the median position, after total paralysis of the recurrent, results from the influence of the cricothyroideus we consider false.

III. Animal experiments, in agreement with cinical observation, show that paralysis of the recurrent is necessary for bringing about the so-called cadaveric position. That the cricothyroideus does exert an influence on the paralysed band is not denied. But its influence on the conduct of the paralyzed vocal band, and, above all, the length of the influence are so minimal that by the clinical observation of individual cases, which ought to last more than a few days, a determining rule cannot be assigned to it.

We thus come to the conclusion that there actually is a simple, uncomplicated posticus paralysis and that only the question of the nature of the complicating factor requires further elucidation. The fundamental facts of Semon's law, against which Grossmann contends, have been proven correct, and thus falls every objection to the theory which Semon deduced from these premises and which of itself is not the subject of attack.

V. THE OPENING OF THE GLOTTIS AND THE INNERVATION OF THE LARYNX DURING BREATHING.

Not only for Semon's law put also for two other, formerly much discussed, questions does the experimental removal of the posticus seem to us to be of a determining meaning: (1) For the question whether besides the postici there is still some laryngeal muscle able to exert an abductor influence, and (2) for the question concerning the innervation of the larynx during breathing. In an earlier work* we have already taken a position on these points, but inasmuch as this publication, is concerned with the discussion of an important, vital principle and is found in a work little accessible to the laryngologist, it is fitting here to briefly review the trend of our communication.

Our experiments showed that, even after the removal of the posticus, the rhythmic respiration movements still remained, and by correct measurement we obtained the proof that the vocal band deprived of its posticus in energetic breathing can be carried in inspiration so far outward that it is separated further from the median line than the other vocal band, whose superior and inferior laryngeal have been severed. The outward movements can be either active or passive. If they be of a passive nature, they must be referred to a relaxation of the adductors, which is present during expiration. In this case, however, the outward movement of the vocal band could not possibly exceed that position which corresponds to a complete relaxation of the abductors which is the position assumed by the vocal band after section of the superior and inferior laryngeal. But since exact measurements have repeatedly shown that the outward movement of the band deprived of its posticus exceeds that of the other side in an appreciable degree, there remains only one

---

*Kuttner and Katzenstein—Experimental contributions to the Physiology of the Larynx.: Arch. f. Anat. und Physiol.: Communication. 1889, p. 274.

possibility, that even after the removal of the posticus there is still present some force which may cause an active abduction of the vocal cords.

Further experiments have made it probable that the cricothyroideus, the arytenoideus and, especially, the cricoaryienoideus lateralis may cause an outward movement of the vocal band of the same side, questionably of the other.

In regard to point 2, concerning the innervation of the larynx during breathing, we have determined that the Krause-Semon theory, until now valid, that in quiet breathing only the abductors are tonically innervated while the adductors, on the contrary, are in perfect rest, does not obtain for the dog. The permanent rhythmic adduction and abduction movement, as shown in the quietly breathing dog, speaks with a certainty against the theory of a tonic innervation of the abductors. For the explanation of this laryngeal picture, only two possibilities come into consideration. Either there is an alternate innervation of the adductors and abductors or a rise and fall in the innervation of the abductors alone—a third is not possible.'

The last mentioned possibility is excluded, however, by the condition of the dogs deprived of their postici. In these animals, after the operation, the vocal bands should maintain an equipoise if the adductors, during quiet breathing are not innervated and the abductors which alone should be innervated, have been removed; but this does not occur; more often there is found after the operation, exactly as before, even in quiet breathing a rhythmic adduction and abduction of both bands, only the range of movement is somewhat lessened.

If the postici were the only abductors, this fact would necessarily lead to the conclusion that the adductors as well as the abductors are innervated during quiet breathing. But—a fact overlooked by the adherents of the heretofore accepted belief—after the removal of the postici, other muscles have an abductive influence. This greatly complicates the above question, for it can be said that these muscles in taking up the function of the abductors also enter into their innervation. Yet, the proof can be shown that there cannot be simply a tonic or periodic innervation of the remaining abductors, but *de facto*, an innervation of the adductors must accompany that of the abductors. The continuation of the adduction and abduction movements opposes the theory of tonic innervation of the abductors alone; while against a theory of a periodic rise of the innervation of the abductors with complete inactivity of the adductors we have the fact that by laying bare the larynx there is seen at every expiration both before and after the posticus removal a plainly active contraction of the cricothyroideus. The rhythmic contraction of this muscle synchronous with the adduction of the vocal band, which, appearing in the form of a movement of the vocal band, can, not alone, but only in conjunction with the other adductors, cause the movement in question, shows with certainty that the expiratory adduction in the dog is not of a passive nature but is brought about only by an active participation of the muscles concerned.

We consider it correct to transfer to mankind the physiologic principle obtained from the dog, and these are our reasons:

(1) It is of highly probability that the vital conditions in dog and men are not essentially different.

(2) Krause and Semon constructed on the ground of laryngeal pictures this conclusion as to what was the rule in men in quiet breathing, viz.: rest of the vocal bands in the abduction position.

We admit the abduction position without argument. But the motionless state of the vocal bands in this position is not entirely opposed to our statement, for it is just as clear by our explanation as by the Krause-Semon, and this single phenomenon is not a chief one, according to numerous statements of Semon himself. The majority of cases, in which, despite all precautions, the vocal bands do not stand still in quiet breathing,—and every one of these cases is a proof for us and against Krause-Semon,—is so great that it is wrong to proclaim the motionless position of the vocal bands during quiet breathing as the usual appearance.

But in this way the only basis on which the Krause-Semon theory was founded falls. And of just as little weight are the arguments which Semon urged against other possibilities which come chiefly into consideration here are likewise of little weight. The following are Semon's arguments and our counter-arguments:

(1.) Semon, in contrast to O. Rosenbach, considers the totality of the adductors stronger than the abductors. If we took no position in opposition to Semon's argument he has gained nothing for his standpoint, for even the weakest muscle can do a markedly greater degree of work, as soon as its innervation is correspondingly increased.

(2) Semon insists, again in opposition to O. Rosenbach, that the adductors and abductors are as antagonistic to one another as the flexors and extensors of the extremities. In our opinion, this condition has no importance for the above question. Nevertheless the groups of muscles mentioned move the vocal bands in an opposing manner; whether or not there is as precise antagonism in the anatomic sense, seems to us immaterial.

(3) Semon claims that if the theory of the innervation of both groups of muscles is correct, there should be a widening of the vocal cleft on paralysis of the adductors. But as the widening does not appear in functional aphonia, where there is a paralysis of the adductors, he concludes that in quiet breathing the adductors are not innervated. We cannot accept this deduction, for the disease which affects the adductors in functional aphonia, affects only the voluntary vocalisers, while, as Semon himself explains, the possibility of a complete closing of the glottis on reflex stimulation is not vitiated. There is, then, no widening of the glottis in functional aphonia because no influence of the adductors is present in the respiration movements of the vocal cords, which indeed represent only a reflex action.

(4 and 5) Semon thinks that the greater vulnerability of the abductors and the central conditions of the innervation of both muscle groups speak against the physiologic preponderance of the adductors over the abductors. In our opinion, the physiologic preponderance of this or that group of muscles in the determination of this question comes as little into consideration as the anatomic, for not this but the momentary condition of innervation is that which gives the stronger force now to the abductors and now to the adductors.

(6) As his last and most potent argument against the

theory of the innervation of both groups of muscle, Semon says: "If the peripheral end of the severed recurrent is stimulated, the affected vocal band will be drawn towards the median line, i. e., the adductors are stronger than the abductors, although both groups of muscles were equally stimulated." These circumstances, which Semon considers, are not actually so. In breathing, the nerves which run to the adductors and to the abductors are not equally stimulated, but one stronger and the other weaker, and just from this alternation of force, from the alternating rise and fall of the stimulation, the different pictures which we observe in the larynx during the alternating phases of breathing result.

From all these observations we think the following conclusions may be drawn:

In men, as in dogs, both adductors and abductors are innervated during breathing (quiet as well as labored). During inspiration, the innervation energy of the abductors increases; during expiration that of the adductors. The movement, which is accomplished by the increase in the active force of a group of muscles is aided by the passive relaxation of the other group of muscles. All laryngeal pictures, which we observe in breathing from the inactivity of the vocal bands in the most quiet breathing, to the forcible, median position of the dead, rest on the same principle; the variability of laryngeal pictures is occasioned only by a variability of the energy with which one or the other group of muscles is brought into activity.

From the great number of our experiments, we give the following typical histories, which agree with the above in all essentials:

I. (Experiment 72). Black watch-dog, 7 k. g., about 6 months old; narcosis, morphine (subcutaneonsly), 0.075, ether.

Before the beginning of the operation, the glottis showed in quiet breathing, in very mild narcosis, a width of 10 mm. The vocal bands made a very small, short excursion, 1—1.5 mm. sideways.

The trachea was laid bare, and then cut across between the 4th and 5th cartilaginous ring. The upper end of the trachea and the larynx were separated from the esophagus, and dissected upwards. The right recurrent was then cut, at once the right vocal band stood motionless in a sloping position with a free, sharp border, 1.6 mm. distant from the median line. The left vocal band meanwhile continued its rythmical movements of respiration; on expiration it came within 1 mm. of the median line, and on inspiration moved 4-5 mm. away from it. The left posticus was then removed; the rythmical respiration movements were still continued, but the vocal band, even on deep inspiration was not carried more than 2.8 mm. from the median line. The right superior laryngeal was then cut; the ligamentous portion of the vocal band moved 0.4 mm. outward, the arytenoid cartilage remaining in the same place. The glottis now showed, on deep inspiration, a width of 4.7 mm. The right vocal band, whose connection with both laryngeal nerves was destroyed, stood 1.9 mm. from the middle line. the left on deep inspiration was carried as far as 2.8 mm. outward, and exceeded the cadaver position of the other side by full 1 mm.

Prof. H. Munk was kind enough to confirm these results. The dog showed on the 5th day after the operation the same picture. The breathing, at rest. was not dyspneic, but became so on the least exertion. On the 7th day the dog became sick, and on the 10th he died.

Necropsy gave as cause of death a bilateral pneumonia; the left posticus was entirely removed.

II. (Experiment 84). Brown spitz, ¾ yr. old, weighed 7½ kg; narcosis, morphine (subcutaneous), 0.12 g. ether.

Glottic width in quiet breathing 10 mm.; movement of vocal bands normal. The left superior and inferior laryngeal nerves were cut; the left vocal band became motionless in an inclined position, 2 mm. from the middle line.

The right posticus was laid bare from the side and removed. The rythmical respiration movements of this side continued but its abduction was clearly circumscribed. With quiet breathing, the right vocal band was carried, during expiration to 0.5 mm. from the median line; in inspiration it moved about 0.5 mm. outward, so that it reached quite, or nearly, the cadaveric position assumed by the other side, whereby the whole arytenoid was pushed like a coulisse from within outward. As soon as the breathing became easier, the right vocal band was plainly drawn further outward, and indeed full ½ mm., so that the distance from the right vocal band to the median line was greater than that of the other side. It gave the impression as though a particular torsion or turn in the direction of the crico-arytenoid was added to the earlier bend toward the side.

Prof. H. Munk was kind enough to confirm these statements by observation made with the graduated magnifying telescope.

III. (Exper. 70). Black-draught dog, 10-12 years old, 20 kg. in wt.; narcosis, morphine (subcutaneously), 0.12 ether.

The glottis showed at the beginning of narcosis a width of 12-14 mm. in quiet breathing; the vocal bands moved normally.

The trachea was laid bare, cut across, and dissected up. Thereupon the glottis showed, on quiet breathing, a width (inspiration) of 4-4.5 mm. On deep inspiration it increased in width to 5 mm. On every expiratory adduction, even with an entirely quiet breathing, a plain contraction of both crico-thyroidei was visible. The trachea which had been severed and dissected upward acted like a writing lever, showing every contraction by a movement upward. The dog was quite alert, and made efforts to break away from his fetters. Both ends of the trachea were approximated and the wound closed, but the laryngeal picture remained unaltered.

Eight hours later, the dog was still drowsy, but somewhat more lively and came when called into the room, etc. When quiet, the respiration was noiseless, 24 times a minute, becoming, on movement, quickly dyspneic and vehement. The dog was now tied down and slightly etherized. Breathing was quiet; glottic width 4.5 mm. The vocal bands showed only a small respiratory oscillation; the adduction movement appeared fairly energetic, slightly oscillating; the abduction movement appeared somewhat painful in 2 to 3 phases, giving the appearance as though the muscle, required for this work, was too weak for its task. In order to test whether the method of operating by making an entrance from the side to the postici had an influence on the movement of the vocal bands, the pharyngeal constrictors and fascia, on

POSTICUS PARALYSIS.   149

both sides, were cut through, so that the posterior face of the cartilaginous ring which had been deprived of its posticus was laid bare. The movements of the vocal bands and the glottic width remained unchanged.

The superior and inferior laryngeal on both sides were then encircled by loops; all four nerves show a typical muscle reaction with the faradic current (200 mm. distant). When all four nerves were cut, the glottis showed a width of 4 mm., i. e. it is 5 mm. narrower than the inspiration position in rapid breathing after removal of both postici. The dog was killed by a stab in the heart. The section showed that both postici with their respective nerves had been removed and, further, both superior and inferior were cut.

IV. (Experiment 81). Black watch dog, about 2 years old, 9 kg. in weight; narcosis, morphine (subcutaneously), 0.15. ether.

Glottic width with quiet breathing 12 mm. Movements of bands normal. Both superior laryngeal were laid bare and tied; then, from the side the right posticus was laid bare and removed. Thereupon the glottis, in quiet breathing, showed a width on inspiration of 6.3 mm.; the distance of the right band from the median line amounting to 2.2 mm., and that of the left to 4.1 mm. Section of both superior laryngeal nerves made the bands somewhat looser, but produced no change in the movements. The left recurrent was then cut; at once the left vocal band stood still (1.8 mm. from the median line) the right was abducted as before, 2.2 mm. so that at the moment of inspiration the glottis width was 4 mm. Now the crico-arytenoideus lateralis dexter was severed; at once the glottis was narrowed to 3.2 mm.; both bands were now approximately symmetrical. The rythmical movements of the right vocal band had become minimal; the difference amounted to about 0.5 mm. On stimulating the right recurrent, the vocal band of the same side pressed close on to the median line.

The dog was killed by a stab in the heart. On section, it was seen that the right posticus had been entirely removed, and the right lateralis had been cut through to a small fibre. Both superior laryngeal and the left recurrent had been cut.

V. (Experiment 92). Black poodle, about 10 years old, 22 kg. weight; narcosis, morphine (subcutaneously) 0.1. ether.

Glottic width, 8 mm. Movement of the vocal bands normal. Both recurrents were laid bare and encircled by a loop of thread; stimulation with the faradic current gives a typical reaction. The thread around the right recurrent was tied, the movement of affected vocal band became trembling and showed a decreased abduction. At the moment the knot touched the nerve without compressing it tightly, the right band approached close to the median line and stood still. The left recurrent was then severed by a sharp scissors, at once the left vocal band stood in a curved position, at the back 5.5 mm. from the right band. The glottis gave the appearance of a right-angle triangle, the right angle being adjacent to the right vocal band. When the right recurrent was stimulated distal from the knot or immediately behind this, a typical twitching of the vocal band occurred. When the electrode was placed 1-2 cm. to the central side of the knot, no reaction followed. About 30 minutes after the tying of the thread, the glottis again presented about the appearance of an isosceles triangle; the right vocal band also shows a curved position, corresponding to that of the left.

VI. (Experiment 71). Black watch dog 6 or 7 years old, 15 kg. weight; narcosis, morphine (subcuteneously) 0.1. ether.

Glottic width, 10 mm. Movement of vocal bands normal. Even in entirely normal breathing, at every inspiration a plain contraction of the crico-arytenoidei was visible.

Trachea cut through 1-2 cm. below the larynx. The upper end and the larynx were separated from the esophagus, and dissected upwards. Both postici were removed. Then the trachea was sutured together, The respiration movements were continued, the greatest glottic width measured being 4.5 mm. Thirty-eight days later, dog well, ate well. When quiet, breathing inaudible; under examination dyspnea quickly appears. The vocal bands move symmetrically and synchronously, and the greatest glottic width was 4.5 mm. Section shows that both postici had been completely removed.

ABSTRACTS FROM CURRENT OTOLOGICAL, RHINO-LOGICAL AND LARYNGOLOGICAL LITERATURE.

I.—EAR.

A Contribution to Diplacusis.

TEICHMANN, Berlin. (*Archives of Otology*, Vol. XXVIII, No. 1.) By diplacusis we understand a functional alteration in which an objective tone is heard double. The author in making tuning fork tests discovered that when the fork c^4 is dying away before his ears, a few seconds before it ceases vibrating, he heard a second tone, a minor third below. This minor third is less intense than the original, begins suddenly with a delicate buzzing and dies away gradually with the original note. The author agrees with Gradenigo, that diplacusis is an abnormal increase of physiological processes (transmissions of excitations, etc.,) and originating in the labyrinth and nerve centers. He learned, however, that diplacusis may depend on peripheral causes. Through exposure he felt a slight coldness in his right ear accompanied by ringing and roaring. There was moderate deafness, perception was reduced for c^1 and c^2 forks. Next day the tinnitus was less apparent, the c^2 fork sounding dull, but with a distinct overtone of f^2. Improvement came slowly, even after several days the fork c^2 was accompanied with a double tone on c^2 in the right ear, as the ground tone died away. A purely labyrinthine affection was excluded by normal perception of the c^2 fork by bone-conduction, false hearing by aerial conduction and alterations in perception after the air bag. So we are left to assume a combination of middle ear and labyrinthine disease. *Campbell.*

Movable Spongy Osteoma of the Cartilaginous Portion of the External Auditory Meatus.

EULENSTEIN, Frankfort-On-The-Main. (*Archives of Otology*, Vol. XXVIII, No. 1). The patient, a man, aged

36, complained of deafness, of one week's standing, in the right ear. Tuning fork test showed deafness by aerial conduction. The meatus was reduced to a mere slit owing to the presence of a tumor covered with normal epidermis. The contracted orifice was full of cerumen. After removal of the cerumen, by syringing, the hearing became nearly normal.

The tumor was movable with a rotary motion and the probe could be passed around it. With a snare the tumor was readily removed and its attachment was seen in the cartilaginous portion of the meatus. Macroscopically the mass, the size of a large pea, resembled medullated bone. Microscopically, it was diagnosticated as a typical exostosis with mucous medulla. The osseous trabeculae contained numerous osteoblasts. There was no actual pedicle, its locality being suggested by a spot without epithelium. *Campbell.*

Remarks on Mastoid Operations, with a Case of Bezold's Mastoiditis.

GRUENING. (*Archives of Otology*, Vol. XXVIII, No. 1.) The operation, which the author practices, is essentially that of Schwartz. It consists of the systematic removal of the external wall of the mastoid process, from the apex to the linea temporalis. Beginning below, he exposes the terminal pneumatic spaces, removes pus, granulations and diseased bone and converts the cells into one spacious cavity. The relation of the sigmoid sinus to the antrum is determined, and the antrum generally reached without difficulty.

The case reported is that of a man who suffered from pain in the right ear, and on whom an aurist did a paracentesis. The pain subsided and the ear began to discharge. In about two weeks the patient considered his ear trouble at an end. A few weeks later, he had violent chills, his right cheek and retro-maxillary region were swollen, and the tip of the mastoid and an extensive post-mastoid area were tender. The mt. was intact and with the right ear he could hear a watch at a distance of 3 inches.

On the mastoid being exposed, the cortex appeared sound, but when the tendon of the sterno-cleido-mastoid muscle was detached, a large quantity of pus welled up. Upon close inspection it was found that the pus occupied the diagnostic fossa, had burrowed under the parotid gland and also formed an abscess in the substance of the sterno-cleido-mastoid muscle.

No pus was found in the antrum, but its posterior wall showed a perforation, which led into the groove of the sigmoid sinus, where a large collection of pus surrounded the vessel. In the tip of the mastoid and in the posterior wall of the apex cell perforations could be traced, which must have communicated with the substance of the sterno-cleido-mastoid muscle and the digastric fossa respectively. *Campbell.*

A Case of Acute Mastoiditis (Bezold Variety) without Perforation of the Drum Membrane; Operation; Recovery.

KNAPP, ARNOLD H. (*Archives of Otology*, Vol. XXVIII, No. 1.) A man, aged 21, applied for treatment because of loss of hearing and tinnitus in the left ear. After measles the hearing in the right ear was permanently impaired, although there never had been any otorrhea. The present illness came on one week ago after a cold.

ON EXAMINATION:—Left ear: Conversational voice, $\frac{8}{60}$; Rinne, negative; the mt. is intact, of normal color and retracted; inflation, improved hearing; mastoid region, normal.

Right ear: Voice, $\frac{3}{60}$; no high tones; Rinne, negative; the mt. retracted and atrophic; no improvement in hearing after inflation.

The patient visited the clinic at intervals for six weeks. The hearing in the left ear, though at first improved by inflation, grew gradually worse. Suddenly severe pain in the ear, with mastoid redness and swelling, set in. Temperature, 101 F. mt. appearance unchanged. Paracentesis was followed by the escape of a small quantity of blood serum only.

The mastoid was opened and its cells found converted into one large cavity filled with pus, carious bone and granulations. A probe was passed through an opening beneath and in front of the antrum, which led into a cavity covered externally by the mastoid tip and the sterno-cleido-mastoid muscle. The tip and the entire inner wall of the mastoid were removed and the digastric fossa and styloid process exposed.

Recovery was uneventful. The hearing in the left ear quickly returned. Whisper, $\frac{10}{20}$. Rinne, positive. The mt. was thickened and depressed. *Campbell.*

A Plea for the More Accurate Definition of Tuning-Forks.

GREEN, J. ORNE. *(Archives of Otology*, Vol. XXVIII, No. 1.) The author refers to the difficulties encountered in comparing tuning-fork tests of different observers on account of the different systems of notation used in different countries.

In France the semivibration (vibration single, or v. s.) is used; while in Germany, England and America the full vibration (vibration double, or v. d.) is used.

In the interest of both authors and readers it is suggested that any note to which reference is made should be described in full, giving the number of vibrations and whether they are double or single; for example, fork 512 v. d., or C 1024 v. s., or F 682.4 v. d. *Campbell.*

Dry Air in the Treatment of Suppuration of the Middle Ear.

ANDREWS, New York. *(Archives of Otology*, Vol. XXVIII, No. 1.) Knowing that dry air is unfavorable to the growth of bacteria, the author has successfully applied this treatment to the middle ear suppurations. He has modified his powder blowers by adding a wooden or glass handle, so that it can be held by the operator while heat is being applied to the metallic cylinder. *Campbell.*

Otitis Media Purulenta; Cleansing the Ear Therein.

JACKSON, CHEVALIER. *(Journ. Amer. Med. Assn.*, Jan. 28, 1899.) In addition to the usual method of thoroughly cleansing and sterilizing the infected field, the author advises the use of a solution of carica papaya for cleansing purposes on account of its digestive properties, and in his hands it has proved more valuable than any other preparation for this purpose. *Scheppegrell*

Suppuration of the Middle Ear, Complications and Consequences, with Report of Illustrative Cases.

MCCONACHIE, A. D. *(Maryland Med. Journal*, Jan. 28, 1899.) Drugs dropped into the ear for relief of pain should be dissolved in oil rather than in water. A five to 10 per cent. solution of cocain in lanolin is preferred.

Scheppegrell.

The Grippe Ear.

EWING, F. C. *(Tri-State Med. Journ.*, Jan., 1899.) A characteristic of the grippe ear is persistent pain after rupture of the tympanic membrane. It is not less likely, however, with suitable treatment to end in a perfect cure. The treatment does not differ from that of simple inflammatory affections. *Scheppegrell.*

Hypertrophic Catarrh of the Middle Ear.

LAFORCE, B. D. *(Journ. Amer. Med. Assn.*, Feb. 25, 1899.) In many cases, the cure of the nasopharyngeal catarrh, associated with the majority of these cases, will prevent recurrence of the disease in the tympanum. In addition to local treatment, the hygienic and sanitary surroundings should be carefully looked after.

Scheppegrell.

Progress in Otology.

HARDIE, T. MELVILLE. *(Journ. Amer. Med. Assn.*, June 3, 1899. An interesting review of the progress made in otology, with a careful bibliography of the subject.

Scheppegrell.

Earache in Children.

GLEASON, E. B. *(Medical Council*, Jan., 1899.) In many cases, gentle inflation of the middle ear is more efficacious for earache than either heat or cocain. Instead of the Politzer bag, the author prefers for this purpose two feet of rubber tubing provided with end-pieces of glass or other material, one for insertion into the patient's nostril and the other to be placed in the operator's mouth. The child is asked to puff out his cheeks, when the operator blows gently into the tube, and, the nostril being otherwise closed, the air enters both middle ears. Extreme gentleness should be used in this method.

As a prophylatic of earache, the child should be taught to blow the nose properly. The best method is by means of the Politzer bag, which is inserted in one nostril and the secretion blown from the other, this procedure being practiced two or three times daily. Local abnormalities should be corrected, and attention paid to the genera health of the patient. *Scheppegrell.*

A Case of Phlebitis and Thrombosis of the Sigmoid Sinus and the Jugular Vein of Aural Origin.

KOLLER, CARL. *(Medical Record*, Feb. 11, 1899.) The patient, a girl of 13 years, had a septic appearance, was conscious but very hard of hearing. Soon after admission to the hospital she had a chill of five minutes duration, temperature 104 F. This was soon followed by another chill and a temperature of 106 degrees, the symptoms becoming worse.

The patient had had scarlet fever when three years of age and since then had repeated suppuration from the ear. A careful examination showed a thrombo-phlebitis of the right sigmoid sinus and jugular vein. Surgical interference was decided upon. The usual operation was made, the patient, however, being already unconscious before the anesthesia was commenced. The thrombosed vein was incised and a sharp spoon passed up toward the bulb. The thrombus was found to be half solid. Incision was then made into the sinus and several drams of foul smelling pus evacuated. Immediately after the operation the patient looked cyanotic. but soon reacted. The next day pain developed in the chest and crepitant rales over the left lung. Respiration became shallow and finally ceased entirely. No autopsy was allowed.

Scheppegrell.

Ear Diseases Co-existent with Adenoids of the Nasopharynx; an Analysis of 110 Cases.

BRAISLIN, W. C. *(Phil. Med. Journal*, Feb. 25, 1899.) A great many of the ear diseases of childhood depend for their etiology upon adenoids of the nasopharynx, The demonstration of the presence of adenoids should in every case lead to investigation of the state of the ear. Ear diseases in some degree, will almost invariably be found to accompany the adenoid growth. The treatment of ear diseases should always be continued for a variable time after the removal of the adenoids. Removal of the growth checks, to a great extent, the onward progress of the ear disease, but the operation does not eliminate the requirement for subsequent treatment of the ears in suppuration, tubal obstruction or other well established pathologic conditions. The presence of adenoids has a continuing de-

generating influence on the ears, while under the influence of colds or attacks of acute contagious disease of childhood, adenoids are prone to produce active disorders.
Scheppegrell.

Diagnosis and Treatment of Suppuration of the Middle Ear.

TOEPLITZ, MAX. (*Medical Record*, December 31, 1899.) A review of the diagnosis and various forms of treatment of suppuration of the middle ear, commencing with simple cases, in which mild irrigations are recommended, to the most severe forms in which Stacke's operation and its various modifications are recommended. With the aid of these methods under favorable conditions, aural suppuration may be cured. *Scheppegrell.*

Electricity in Deafness and Strictures of the Eustachian Tube.

NEWMAN, R. (*Medical Record*, Dec. 17, 1898.) The method advocated is that described by Dr. A. B. Duel, in which small copper bougies varying from numbers three to six (French scale) mounted on No. 5 piano wire passed through insulated catheteters into the Eustachian tube, the bougies being connected with the negative pole of a battery, and the circuit completed by the patient holding the positive pole. Two of five milliamperes are applied.

In the cases reported it is demonstrated that electrolysis is useful in cases of deafness due to stricture of the Eustachian tube. *Scheppegrell.*

The Use of Nosophen and Antinosin in Purulent Disease of the Ear.

MILLENER, F. H. (*Buffalo Med. Journal*, Dec., 1898.) In purulent suppurative otitis media, the antinosin solution, two and a half to three per cent., is allowed to remain in the ear for five minutes. Nosophen is then applied by means of a powder blower. In the cases reported by the author, this treatment gave good results. *Scheppegrell.*

A Case of Acute Mastoiditis; Perforation of the Medial Plate of the Process and Consecutive Abscess of the Neck.

BURNETT, CHAS. H. (*University Med. Magazine*, Feb., 1899.) Three ways of propagation of otitic and mastoid suppuration of the neck are recognized, viz., by way of the veins, of the lymphatics, and by direct escape of the pus through a spontaneous opening in the medial plate of the mastoid process, beneath the insertion of the sterno-mastoid muscle, the latter being known as Bezold's mastoiditis.

In a case reported, there was a copious discharge from the right ear. The mastoid region behind the auricle appeared perfectly normal, but behind and in front of the insertion of the sterno-mastoid muscle there was a swelling extending downward about three inches, most prominent however behind the sterno-mastoid muscle and toward the nucha. The appearance of this swelling was coincident with the cessation of the mastoid pain. Firm pressure on this swelling forced pus from the middle ear through the perforated membrana tympani.

An incision two inches long and three deep was made into the abscess and a fluid dram of odorless pus escaped. A grooved director was passed down to the bone, which entered an opening apparently in the base of the mastoid process and passed with little pressure into the cells. Sterilized water syringed into the incision passed directly into the middle ear and out of the external auditory meatus. The case healed without further operation on the mastoid.

The author calls attention to the efficacy of this treatment, and concludes with some excellent advice on the treatment of acute otitis media and consecutive mastoiditis.

Scheppegrell.

II.—NOSE AND NASO-PHARYNX.

On Congenital Closure of the Choanae.

JOEL, Gotha. *(Archives of Otology,* Vol. XXVIII, No. 1.) The author describes these closures as unilateral or bilaterial, complete or incomplete, bony or cuticular. He then relates the history of a case, in which he entitles the obstruction unilateral, complete and mixed. *Campbell.*

The Toilet of the Nose.

RICHARDS, G. L. *(Medical Progress,* Feb., 1899.) To prevent the development of catarrhal conditions of the nose, the author believes that we should pay the same attention to the toilet of the nose that we do to that of the teeth, and that the nostrils should be washed as frequently as the teeth. *Scheppegrell.*

A New Method for the Operative Correction of Exaggerated Roman Nose.

GOODALE, J. L. (*Bost. Med. and Surg. Journal*, Feb. 2, 1899.) After the patient, a girl of 13 years, had been etherized and placed in the Rose position, a pair of short curved scissors with the convexity uppermost was introduced into the left nasal vestibule. One blade was made to penetrate the triangular cartilage at its anterior extremity immediately beneath the integument, and a cut made along the superior margin of both cartilaginous and bony septum, terminating at the junction of the perpendicular plate of the ethmoid with the cribriform plate. The superior margin of the septum was thus separated from the integument and from the nasal bones by this incision, the outline of which was essentially parallel with the angular outline of the bridge of the nose. The extremities of this angular incision were next connected by a straight cut made through the septum with straight scissors, and the portion of the septum included between the two incisions was removed with forceps. A septum with a straight superior outline was thus produced.

The next step consisted in depressing the bony bridge of the nose so that it should rest upon the now straight septum. A small nasal saw was introduced with the teeth uppermost into the left nasal passage and the articulation of the nasal and maxillary bones sawn through from below upward, this procedure being repeated on the right side. A few comparatively gentle taps upon the nasal bones sufficed to break the frontal articulation and depress them, still firmly united with each other, until they came into contact with the upper margin of the septum. With the depression of the nasal bones, the bridge of the nose assumed a straight line from tip to forehead, but a ridge at the same time appeared on either side, formed by the maxillary bone along the line of the nasal articulation. This was corrected by two or three light blows, with a protected mallet, upon this ridge fracturing the maxillary bone which is here very thin.

The operation occupied about 40 minutes and was

attended by comparative slight hemorrhage. An external splint was applied to hold the nasal bones and fragments of maxillary bone in proper position. The result was satisfactory. *Scheppegrell.*

Some Remarks Upon the Use Suprarenal Gland of the Sheep in Nasal Surgery.

VANSANT, E. L. *(Phil. Med. Journal,* Feb. 25, 1899.) After a preliminary cleansing, the parts to be operated upon are painted with a five per cent. solution of cocain. Five to 10 minutes later the prepared solution of suprarenal gland is thoroughly applied by means of cotton wool. Both the anesthetic and ischemic effects of the cocain are increased and prolonged. In acute and subacute inflammatory conditions of the nasal chambers, such as coryza, hay fever, etc., this method gives much greater and more prolonged relief from the congestion than when cocain alone is used.

The solution is prepared by adding the contents of one capsule of dessicated extract to one-half ounce of camphorated and distilled water, to which 11 grains of boracic acid have been added. *Scheppegrell.*

Sarcoma of the Nose.

EVANS, THOS. C. *(Medical Times,* Feb., 1899.) A post mortem examination of a man of 54 years, who died from sarcoma of the nose, showed complete absorption of the basilar process of the occipital bone as well as the body of the sphenoid. It has also grown into the middle fossa of the scull, the tumor in this region being about the size of a partridge egg, and also into the orbit, causing exophthalmos which destroyed the eye.

Scheppegrell.

Indications for Operation in Adenoid Disease of the Nasopharynx.

BALL, J. G. *(Clinical Journal,* Dec. 28, 1898.) Among the indications enumerated are habitual mouth-breathing or labored nasal breathing or suffocative attacks at night. If ear symptoms, or bronchitis develop, or reflex disturbances such as nocturnal enuresis, chorea, etc., the author also recommends operation. *Scheppegrell.*

The Etiology and Diagnosis of Empyema of the Accessory Sinuses of the Nose.

SCHADLE, J. E. *(St. Paul Med. Journal*, Jan., 1899.) A review of the etiology and diagnosis of empyema of the accessory sinuses of the nose illustrated by eight excellent half-tone cuts of preparations from the author's collection.

Scheppegrell.

Chronic Perichondritis of the Nasal Septum.

SOMMERS, L. S. *(Medical Record*, Jan. 14, 1899.) Localized perichondritis forming spurs on the nasal septum is a frequent occurrence, while general inflammatory thickening of the soft tissues and cartilages occurs in but a small proportion of nasal diseases. In the development of these cases there is first congestion, increased glandular activity and dilated blood vessels, followed by hyperplasia with increase in the number of cells.

Falls and blows on the nose are most common causes, although indirect traumatism as when due to deflected air-currents is responsible for a certain number of cases.

Scheppegrell.

Occlusion of the Posterior Nares; with Report of Two Cases.

FRITTS, J. R. *(Medical Times*, Feb., 1899.) A girl of 10 years, while being treated for a polypus of each naris, was found to have complete occlusion of the posterior nares, the palatal bone passing upward and backward, attaching to the sphenoid and forming a union between the center and the vomer, which caused a large ridge or thickened center of the abnormal bony part closing the posterior nares. The sense of smell was absent but the hearing was excellent. An opening was made by means of a dental burr, and was kept pervious by a hard rubber catheter.

The second case was that of a man 24 years, in which the obstruction was similar. The author collates 17 other cases which he has succeeded in finding in medical literature, making a total of 19 cases on record.

[The author has omitted a case, reported by Dr. W. Scheppegrell, of congenital occlusion of the posterior nares, published in the *Annals of Ophthalmology and Otology*, Apr., 1894. The reporter has also operated on another case of congenital occlusion which has not yet been published.]

Scheppegrell.

III.—MOUTH AND PHARYNX.

Foreign Bodies in the Pharynx and Esophagus.

JONES, R., in a paper *(Lancet,* May 6) describes his experiences with foreign bodies in the pharynx and esophagus. Often, a small body may set up alarming symptoms. He has removed fish bones, orange peel, pins, apple, match, date-stone, plum-stone, camel's hair brush, slate pencil, set of false teeth. coins, etc. He advocates prompt energetic treatment; first explore, after anesthetizing locally the sulci between the fauces and the tonsil. Then examine with the laryngoscope. Examination with the head lowermost is favorable to coughing up the body and there is not so much danger of dislodging the particle into the larynx or trachea. Extract by means of the double coin-catcher or the sponge probang. Do not push rough. angular bodies into the stomach. Perform laryngotomy or tracheotomy if there is great obstruction to the breathing. Esophagotomy or gastrotomy is the last resort, the latter if the body is more than 30 centimeters from the teeth. As practical rules he gives (1) bodies which have remained some time and given rise to symptoms of obstruction, irritation or dyspnœa should be operated upon at once. (2) Forcible attempts at extraction *per os* are to be condemned. (3) Sharp, irregular impacted bodies especially require esophagotomy. (4) Sometimes gastrotomy. and sometimes a combination of gastrotomy and esophagotomy is required. (5) No stitches should be used where the wound in the esophagus is jagged, or its walls inflamed. (6) Otherwise stitch with continuous suture. not piercing mucous coat. (7) Only when no danger of infection or suppuration exist, should the external wound be closed. (8) Liquid food may be given *per os* in 24 hours after the operation. *Loeb.*

Acute Diffuse Cellulitis of the Sub-Maxillary Region.

WEBBER, H. W. *(Lancet,* Sept. 17, 1898.) CASE I. Woman, aet 43, was admitted, suffering from inflammatory swelling of the neck with dyspnea. When observed by the writer she was moribund from asphyxia. A deep incision was at once made from the chin to the sternum and the trachea, which was five inches from the skin surface, was opened and a sinus forceps inserted. A tracheotomy tube not being at hand a Symonds' esophageal tube was inserted and the cavity around this was powdered with iodoform and tightly packed with strips of sal alembroth gauze to avert if possible the entrance of septic material into the trachea. On the next day a dozen free incisions were made into the neck and drainage tubes inserted into two of them. Patient recovered.

CASE II. Woman, aet 20, was found sitting up in bed, face of a dusky purplish, pulse 120, temperature 102°, with a brawny inflammatory swelling extending from the lobule of the right ear to the hyoid bone, chin and lower jaw. The floor of the mouth was involved, dyspnœa was present, and speech impossible. Deep incisions were made and tracheotomy later performed. Patient died.

Loeb.

Leucoplakia.

MARSHALL, J. H. *(Journ. Amer. Med. Assn.*, Feb. 25, 1899.) An interesting description of the varieties, etiology, predisposing causes, symptoms, diagnosis, prognosis and local and surgical treatment of this affection.

Scheppegrell.

Pseudo-Membranous Angina Due to the Pneumocococcus.

VIDEL, A. AND V. *(Medical Record,* Dec. 31, 1898. *Bull. Med.*, No. 64, 1898.) A report of five cases of severe angina in children due to the pneumocococcus. In three cases the erythematous type existed, the children recovering in from seven to eight days. The other two cases were of the pseudomembranous type, the children from the outset giving evidence of severe infection. All therapeutic efforts were without effect. The injection of diptheria antitoxin produced no result and the children died. *Scheppegrell.*

A Contribution to the Pathologic Histology of Acute Tonsillitis.

GOODALE, J. L. *(Bost. Med. and Surg. Journal,* Jan., 1899.) The article embodies the result of the histologic examination of 16 cases of acute tonsillitis, illustrated by three excellent plates. Two distinct types of histologic lesions were encountered. In all cases a diffuse proliferation of the lymphoid and tissue cells was present. Four cases showed also local foci of suppuration in the follicles. Acute tonsillitis due to infection by the streptococcus pyogenes and the staphylococcus pyogenes albus and aureus is characterized histologically by a diffused inflammation of the parenchyma of the organ, appearing in the form of an increased proliferation of the lymphoid cells and of the endothelial cells of the reticulum, due probably to the absorption of a toxin formed in the crypts. While bacteria are rarely demonstrable in the tonsillar tissue in cases characterized by purely proliferative lesions, yet at times infection of the interior of the follicle occurs, giving rise to circumscribed suppuration and the formation of abscesses which eventually discharge into the crypts.

Scheppegrell.

IV.—LARYNX.

Eversion of the Ventricle of the Larynx and Cyst Involving the Larynx and Side of the Neck.

INGALS, E. F. (*Journal Amer. Med. Assn.*, Feb. 18, 1899.) A man of 39 years suffered from difficulty in speaking. A cystic tumor about one and a half inches in diameter was located just below the angle of the jaw upon the right side. By pressure, it could be almost disseminated, but it immediately returned when the pressure was removed, and when pressure was made the patient was unable to speak.

Examination showed a bulging of the right side of the base of the tongue crowding somewhat against the epiglottis. The swelling was about five-eighths of an inch in diameter, having a cystic appearance. It extended downward, at the right side of the larynx crowding the epiglottis inward and causing a bulging of the ventricular band so as completely to hide the right vocal cord and ventricle upon inspiration, and the left cord and a part of the left side of the larynx on attempted phonation. An exploratory syringe demonstrated the cystic character of the tumor.

About two ounces of a mucilagenous semi-transparent fluid was removed from the cyst by means of an aspirating needle. Three-fourths of a dram of equal parts of carbolic acid and glycerin were injected by the same needle and allowed to remain in contact for several minutes. Upon examination of the larynx it was found that the cystic tumor at the base of the tongue and at the side of the larynx had disappeared. There was a smooth reddish tumor about five-eighths of an inch in length and three-eighths in diameter projecting from the right side of the larynx just above the vocal cord, which was demonstrated to be a prolapsed ventricle.

Fearing acute swelling, a number of incisions were made into the mucous membrane with forceps. The everted ventricle was removed by means of the polypus snare. Some days later, the voice was still hoarse in consequence of the thickening of the vocal cords, but otherwise the patient appeared well. *Scheppegrell.*

V—DIPHTHERIA.

Membranous Croup.

SAWYER, J. L. *(Iowa Med. Journal,* Vol. IV, No. 8.) Membranous croup is contagious and identical with diphtheria. Antitoxin has a very favorable influence, proving almost a specific, and combined with intubation, the mortality is very small. Peroxide of hydrogen does not influence the diphtheritic poison, but it prevents mixed infection. A powder containing iodoform and sulphur has proved a useful local remedy. *Scheppegrell..*

Remarks on Antitoxin, Diphtheria, the Practitioner and History.
A Practical View of Antitoxin and Diphtneria in Private Practice.
Antitoxin, Diphtheria and Statistics.

RUPP, DR. ADOLPH. *(Medical Record,* Nov. 5, 1898; Dec. 31, 1898; Jan. 28, 1899.) Although many able practitioners and scientists claim that antitoxin for diphtheria has ceased to be a question, and that all argument concerning the fact of its utility is futile labor, other equally able observers and equally well equipped practitioners claim that the remedy is useless and at times harmful.

From a careful analysis of 24 cases treated, the author concludes that the remedy has no well marked favorable effect on the general and clinical course of the disease; it neither shortens nor lessens its severity. In the croupous cases it exerts no beneficial or inhibitory influence on the progress of the croup. The operation in these cases and not the antitoxin saved the children. In none of the 24 cases, nor in others seen by the author, did the antitoxin seem (beyond a reasonable doubt) to cause the pseudomembranes in the throat to disappear sooner than would have been the case had no antitoxin been used.

In all his cases, the initial dose of antitoxin was given not earlier than the third day, and not later than the fourth. None of the cases were markedly "septic" or "mixed," nor were they severe cases in any sense.

Antitoxin is a substance and a remedy of a variable and irregular unit strength. The same make of antitoxin may be highly praised at one place and condemned at another on account of its high mortality. It is an organic substance which is easily impaired by age and unfavorable temperatures. It has no power to check or to mitigate diphtheritic laryngeal processes. The differences in mortality rates vary as much in different places and at different times among antitoxin cases as they do among cases that have not been antitoxinized. *Scheppegrell.*

The Health Department and Diphtheria.

JONES, C. H. (*Maryland Medical Journal*, Jan. 28, 1899.) A report of the work of the health department of Baltimore against diphtheria during the last six months of 1898. Instead of using yellow flags, yellow cards were substituted which gave good results. The rule established in 1897 that children from an infected house should not attend school until the house had been thoroughly disinfected and until after all the throats of the children had been examined was enforced. Antitoxin is strongly advocated as a curative as well as a preventive agent.

Scheppegrell.

The Death Rate of Diphtheria in the Large Cities of the United States.

JORDAN, J. E. (*Phil. Med. Journal*, Feb., 1899.) The following table shows the decrease in the average death-rate from diphtheria since the introduction of antitoxin:

City.	Average death-rate from diphtheria and croup, per 10,000 population. 1886-1894.	Average death-rate from diphtheria and croup per 10,000 population. 1895-1897.	
266 German towns	10.6	4.4	
Berlin	9.3	4.0	
Paris	6.1	1.5 (2.1)*	
London	6.0	5.8	
New York	15.1	8.0†	(7.1)‡
Chicago	12.0	6.4†	(5.2)‡
Brooklyn	14.7	10.1†	(8.8)‡
Philadelphia	10.1	11.0†	(10.7)‡
St. Louis	11.7	7.5	
Boston	11.8	10.9	
Baltimore	7.0	6.2	
Milwaukee	13.5	7.3	

This table shows considerable divergence during the antitoxin period as well as during the pre-antitoxin period. The lowest rate for the pre-antitoxin period was in London, 6.0, the highest in New York, 15.1; the lowest for the antitoxin period was in Paris, 1.5, the highest in Philadelphia, 11.0. The greatest actual diminution occurred in New York, the largest percentage decrease in Paris.

In explanation of the fact that the decrease in London has been very small, Dr. Corbett has suggested that "there was more room for improvement on the Continent than in London," and that "the prevailing type of diphtheria during the past three years has been very severe in London." That the results have been better than those obtained in the United States he claims to be due to the fact that the serum used in this country is inferior to that used abroad, and that the remedy is more extensively used there than here. *Scheppegrell.*

The Management of Diphtheria from a Public-Health Standpoint.

STOKES, W. R. *(Maryland Med. Journal,* Jan. 28, 1898.) For disinfection, formaldehyd gas has been used, but the tests have not yet been completed. Anti-

---

*1894-1897. †1895-1898. ‡1896-1898.

toxin in the treatment of diphtheria gave gratifying results. Isolation hospitals for treatment is recommended.
Scheppegrell.

The Use of Antitoxin in the Treatment of Diphtheria.

WILLIAMS, J. J. (*Medical Review*, Feb., 1899.) A detailed report of 14 cases of diphtheria treated with antitoxin, the result being favorable in all the cases except the first in which infection had progressed too far.

The author states that the accepted idea of the maximun period of incubation being 72 hours is erroneous, as his experience warrants the belief that the germ of the disease may be deposited in the fauces and there retained for a much longer period. *Scheppegrell.*

MISCELLANEOUS.

Arrest of Hiccough by Depressing the Tongue.

KOLIPINSKI, L. (*Maryland Med. Journal*, Feb. 25, 1899.) In a case of singultus, in which the symptoms had continued until they were alarming, complete relief was afforded by firm pressure on the tongue by means of a large spoon handle. *Scheppegrell.*

Surgical Treatment of Exophthalmic Goitre.

SCHWARTZ, M. (*Medical Times*. Feb., 1899.) In a case of marked symptoms, a resection of the right sympathetic and two-thirds of the superior cervical ganglion was made, and at the same time that of the left sympathetic involving also a section of the spinal. The operation was followed by marked amelioration of the symptoms observed.
Scheppegrell.

Pulmonory and Laryngeal Tuberculosis Treated with Antiphthisic Serum T. R., with Remarks on the Etiology of Tuberculosis.

FREUDENTHAL, W. (*Medical News*, Feb. 18, 1899.) The antiphthisic serum was used in four cases. Although the results were not as gratifying as those obtained by Drs. Carl Fisch, A. M. Holmes and F. B. Waxham, they were sufficiently good to encourage the author to continue his investigations. *Scheppegrell.*

Removal of Foreign Bodies from the Respiratory Tract.

TAULBEE, J. B. *(Medical Progress*, Feb., 1899.) The foreign body was the escutcheon of a keyhole of a chest, measuring one-half by one and a half inches, which had been retained in the trachea of a five-year-old boy for three days. After the administration of chloroform, it was removed by means of a long curved Leonard forceps *per vias naturales*.

In the experience of the author, thyrotomy, laryngotomy and tracheotomy for the removal of foreign bodies, although simple in technique, result unfavorably in the majority of cases. In six operations by the author, all save two proved fatal. *Scheppegrell*

WESTERN SECTION, AMERICAN LARYNGOLOGICAL, RHINOLOGICAL AND OTOLOGICAL SOCIETY.*

ANNUAL MEETING, SAN FRANCISCO, CAL., MARCH 31, 1899.

DR. H. L. WAGNER, the Vice-president, delivered the address of welcome.

DR. M. A. MARTIN, of San Francisco, opened the regular scientific portion of the business by the presentation of two short papers:

1. Hemorrhages Following Adenoid Operations.

DR. MARTIN stated that there is much more danger in adenoid operations than we are accustomed to believe. He had had no trouble for some years, but afterwards found that his early experience was nothing but good luck. Illustrative of the sort of complication referred to he cited three cases occuring during the past year.

DR. ADOLPH BARKAN considered the operation for the removal of adenoids a very serious one. The pharyngeal tonsil was the one of the three most liable to give trouble, and he makes it a rule never to operate on the three at once, and always insisted on having the patient in the hospital for at least 24 hours for adenoid operations. He does not operate unless the patients go to the hospital. As a routine of procedure, he first introduces a silk thread through the soft catheter, leaving it in place through the nose and mouth until several hours have past and no trouble is evident. It is easier to pass the thread before the operation, when the patient is quiet, than in the face of a severe hemorrhage, when the patient is excited and weakened by severe bleeding.

DR. SAMPSON TRASK in discussing hemorrhage after adenoid operations, stated there was always to be remembered the danger of hemorrhage following operations on the throat. Especially is this true after operation on the third tonsil. He considered it unwise to operate in these

---

*[Journ. Amer. Med. Assn., Apr. 15, 1899.]

affections during the existence of any acute inflammatory process.

Dr. Martin stated that undoubtedly the hospital was the best place in which to operate for the removal of adenoid growths, but unfortunately all patients do not have the necessary means to go to a hospital and also pay for the operation. The placing of a silk thread, before the operation, as suggested by Dr. Fehleisen was also an excellent idea, but in two of the three cases which he had just reported, the thread would have been removed even by Dr. Barkan long before the hemorrhage occurred. In one there was no hemorrhage to amount to anything within 24 hours, and no considerable hemorrhage until the fifth day. In the last case the alarming hemorrhage did not appear until the ninth day after the operation. He had made use of extract of suprarenal capsule in two or three cases, and so far thought well of it. In one case in which he had operated to divide a band between the middle turbinate and the septum, he met with excessive hemorrhage after the first cut, in spite of the fact that he had made liberal use of the solution of the suprarenal gland. What it would really be worth remained to be seen.

II. A Case of Empyema of the Sinus Frontalis.

Dr. Martin saw this case first in May, 1898, when there was a large swelling at the inner angle of the orbit of the left eye; the root of the nose on that side was also swollen. Some polypi had been removed. The orbital trouble had been increasing for some time past; the man had been afflicted with the disease for several years. The anterior portion of the left middle turbinate was snared away and an attempt was made to probe the cavity without beneficial result. The collection of pus over the orbit was producing so much discomfort that this was first removed by making an incision into the pus-sac just below the orbital ridge. About two ounces of pus discharged. Three days later the frontal sinus was opened by means of a bone flap just to the left of the inner angle. The frontal sinus was found to be greatly involved; this was curetted carefully. The ethmoidal cells were involved somewhat; drainage tube was introduced and retained for three months; wound then closed. Shortly afterward pus was found to have gathered again, and the wound was reopened. It was now washed and irrigated and a solution of iodoform in absolute alcohol injected. The patient eventually made a good recovery with slight deformity.

Empyema of the Sinus Frontalis.

DR. F. FEHLEISEN, of San Francisco, read a paper on this subject and said that he had operated on some thirty cases, some acute, with fever, etc., but most were old chronic cases which had persisted for years. He formerly trepined just above the eyebrow, near the nose, and cleaned out the sinus as well as he could through this small opening. This method he found not satisfactory and he now performed a more radical operation. In the old method all the diseased tissue was rarely removed, and as a result there was long continued discharge from the sinus and poor healing of the wound. Now he removes the entire interior of the sinuses, scraping out not only the mucous membrane but also the diseased bone which he finds present in nearly all old cases. The anterior table over the entire affected region is removed and no diseased tissue allowed to remain. The wounds heal readily and the condition is cured. He presented two cases.

In the author's opinion, it is not enough to remove the mucous membrane alone, but the softened bone, and where there is a fistula, the floor of the sinus should be removed. The cases where exhibited, and though the disease was undoubtedly cured, the resulting scars were very unsightly. As the anterior wall had been removed, there was a marked depression over the eyes, of irregular outline and ragged appearance.

DR. KASPER PISCHELL said that in both cases the eyes had been pushed down and out; there was hypophoria and esophoria, but no binocular vision. In both cases the vision was very good. Transillumination had been tried on these cases, but it was of but little value in affections of the frontal sinus. He thought it useless to try syringing in these cases, as there was always a great degenerated mucous membrane and often much granulation tissue, which must be removed before suppuration will cease and the wound heal. A more radical operation is much to be preferred.

In most cases there is no direct effect on the sight due to optic nerve involvement, but this does occasionally occur. He cited a case on which he had recently operated, in which the optic nerve had become involved through spread of the inflammatory condition. The disease was caused by repeated severe colds following an attack of grippe. He could nearly, but not quite, probe the frontal sinus through the natural opening, and found the customary pain when the probe came near the inner angle of the orbit. Operation relieved the trouble with the sinus, but not the nerve affection.

DR. F. B. EATON thought it rather too heroic to make such a radical operation as was done in the cases presented by Dr. Fehleisen, The disfigurement was very great. He thought it much more advisable to treat it by packing through the sinus or a small opening, and most if not all cases could be cured. The treatment is tedious but efficacious. He called attention to the method advocated by Gulovein, in which hot steam is injected into the cavity through a small tube first introduced either through the fistula, if such be present, through the natural opening, if it be possible, or through a small incision.

DR. SAMPSON TRASK stated that the cases presented by Dr. Fehleisen were instructive as to the result of the more radical operation, but he himself was more conservative, thinking that simpler means, though they consumed more time, yielded cures and did not leave unsightly scars. He called attention to the normal opening into the frontal sinus, which he thought might be more extensively used. Removal of the anterior portion of the middle turbinate generally made it possible to probe the sinus in this manner.

DR. FEHLEISEN was much interested in the proposed use of hot steam for dealing with these affections, but had not yet been able to make use of it. He thought that if there had been, for some time, a fistula in the floor of the frontal sinus only, the removal of the diseased bone would effect a cure.

DR. MARTIN thought the specialist gave more thought to the ultimate cosmetic result of an operation than did the general surgeon. Specialists therefore were rather loath to undertake the extremely radical operation for removing the anterior table over the frontal sinuses, and were willing to spend more time in the slower method of treatment when there is any chance of its being eventually successful. He had quite cured a severe case by the injection of a solution of iodoform and absolute alcohol through a canula introduced by means of the ostium frontalis. The treatment covered a period of one and a half years, but the patient was more than willing to put up with the long treatment rather than be disfigured by the radical operation.

Unilateral Hypertrophy of the Face.

Dr. Douglass W. Montgomery, of San Francisco, made a further report of a case of unilateral hypertrophy of the face, with unilateral hypertrophy of the hard palate and gums. The history of this exceedingly rare condition, which was so unusually well-marked in the case exhibited, was as follows: The patient, a man of about 45 years, had been under observation by the doctor since 1893. His parents were German and so far as could be ascertained perfectly normal. The man's birth and life up to the fifth year certainly were also normal. A photograph taken at the fifth year shows no peculiarity. A photograph taken at the tenth year, on the contrary, shows the condition commencing. A slight enlargement of the left side of the face could be noticed in the photograph taken when the patient was 10 years old. The hypertrophy was very well marked and was absolutely unilateral, the median line very exactly marking the line of hypertrophy. Along the head and forehead was a large roll of bone; the malar was very prominently enlarged; the skin as well as the bones on that side of the face was the seat of hypertropic changes. Indeed, in 1893 a very large roll of skin had been removed by operation, from just beneath the left eye, where it acted mechanically in disturbing the man's vision. This had not returned since the time of operation. The lips, gums and hard palate were greatly enlarged, and a cast of the mouth, which was shown, exhibited this peculiar condition most strikingly. The skin on the left side of the face was markedly red, but Dr. Montgomery stated

that it was not as red as it had been three years ago: he
thought the condition was receding to a certain extent.
Certainly it seems to be at a standstill. There was no enlargement of any other portion of the body, either on that
side or the other. There was no enlargement of the
hypophysis, and the case was surely not one of acromegaly.
Dr. Montgomery called attention to the excessive rarity of
the condition, and mentioned all the cases thus far reported, the first being by Beck in 1836. Almost all of the
cases reported were, however, congenital; this one was
not, but a clear case of acquired, progressive unilateral
hypertrophy of the face. No explanation of the phenomenon was attempted.

Otologic Experience During the Naval Battle of Santiago, on Board the Iowa.

Dr. M. M. Simons, U. S. N., first considered the two
kinds of powder used on board, the brown prismatic and
the so-called smokeless powder. In the brown, some of
of the grains are unburned, and by the explosion are
finally powdered. This dust is often blown back on the
decks of the ship and is somewhat irritating to the mucous
membranes. It causes slight congestion which passes
rapidly away. The smokeless powder does produce some
slight amount of smoke, or rather haziness in the air,
after a discharge of a large amount. Carbonic oxide gas
forms in the breech, and when the latter is opened it is
changed to carbon dioxide; no ill effects were noted
from this gas. After a number of discharges, the decks
become hazy with the fumes from this powder, and there
is noticed a slight though decided acid smell. It is
extremely irritating to all mucous membranes, though no
serious trouble results. When the decks are washed down
after the firing has ceased, all this passes away. After
the battle of Santiago there were several cases of nasal,
tonsillar and ocular inflammation. These were simple and
yielded at once to treatment. A few were deaf, some for
from two to four days, but they all recovered by the use of
inhalations and Politzerization. Only two cases were
observed in which there was rupture of the tympanum.
Here there was no pain, but the patients complained of

tinnitus. He stated that he was himself slightly deaf as a result of that engagement. When a gun is fired there is a feeling of a sudden blow, something like the blow from a bar of iron. With the small guns this is quite sharp, but with the large guns it is more heavy. Some complained of general muscular soreness after the battle. The two cases of perforation occurred in the vicinity of the eight-inch gun. In his opinion, the deafness was the result of the irritation of the throat, primarily, this producing a closure of the Eustachian tube; a subsequent heavy discharge would force in the tympanum, thus forcing out a small amount of air which could not return and hence the drum would be somewhat retracted and slight deafness result.

Dr. M. A. Martin had seen some cases of rupture of the drum membrane from explosions, and thought the explanation was not that suggested by Dr. Simons, but more in the nature of an outward thrust of air in the middle ear, occurring at the time when the wave of rarefaction, or partial vacuum, struck the ear; in other words, that the rupture is due to pressure within.

Dr. F. B. Eaton agreed with Dr. Simons, and artillerymen had told him to keep his mouth open when near a large gun about to be fired, saying that this prevented injury to the ear. He thought the impact of the wave could do more damage than the pressure from within.

Dr. Simons had not noticed any cases of deafness until the guns had been firing for at least 15 minutes, during which time the gases had accumulated and produced their irritating effect. He spoke of the difference which relative positions had in the effect, but said that one could not choose his attitude, for guns were going off above, below and on all sides. The irritation of the fumes was much like the irritation produced by holding the head over a vessel of weak formalin solution. *Scheppegrell.*

AMERICAN LARYNGOLOGICAL ASSOCIATION.*

TWENTY-FIRST ANNUAL CONGRESS, HELD AT CHICAGO, MAY 22-24, 1899.

FIRST DAY.

President's Address, delivered by Dr. William E. Casselberry, of Chicago. I think all are conscious of the change in the drift of laryngological practice wrought in recent years; first by the evolution of nasal pathology, and secondly by the appropriation of the ear. From a physician treating affections of the throat and chest, the laryngologist is fast becoming a surgeon with a routine limited to local measures as applicable to the upper respiratory tract alone. While freely conceding that practice has been realized along local surgical lines, I deprecate the tendency of the day to deal with the throat and nose exclusively in a mechanical way as if they were organs detached. I believe it engenders narrowness of thought and that through habitual disuse there is gradually lost to the physician much of that fundamental knowledge of pathology and thereapeutics which is so essential to the welfare of the patient. There are now nasal enthusiasts whose idea of the entire subject seems to consist in the establishment of a wide patency of the nostrils. The intimacy of the ear with the nose and throat is conceded. The laryngologist has the treatment of aural affections actually thrust upon him. Even the patient's discernment convinces him that the laryngologist's skill and equipment are the means best adapted to the end, and having appropriated the ear, one should cultivate an exhaustive knowledge of the organ, but as an addition to laryngologic lore, not as substitute for a part thereof. It is not claimed that the laryngologist must of necessity embrace in his practice all pulmonary diseases, but it is urged that he be ready to apply all the arts of diagnosis and that he be conversant with every resource known to medicine in the treatment of pulmonary condi-

*From the Philadelphia Medical Journal.

tions in their relation to the throat. Nor will it answer to omit attention to the heart, aorta, or mediastinal contents. Paralysis of a vocal cord through pressure on the recurrent nerve by an aneurysm is a simple proposition, but more complicated ones which require a high degree of diagnostic skill are continually encountered. Above all, one should study the conditions of natural immunity and susceptibility in order that having made an early diagnosis one may direct the mode of life and place of abode best adapted to arrest the disease and overcome the susceptibility. This implies a ready familiarity with sanitation, hydrotherapy, climatology and sanitarium resources and methods. All laryngologists are familiar with the many throat conditions which appear as salient features of underlying systemic states, and yet in the overswing of the movement toward localism general pathologic data, systemic therapeutics and hygienic aids are not always utilized to the utmost. The laryngologist should continue to be first of all a good physician and after that, something more—a specialist.

Is the So-called American Voice Due to Catarrhal or Other Pathological Conditions of the Upper Air-passages? By Dr. John W. Farlow, of Boston. By American voice is meant the charistic nasal twang so commonly noticed. Conditions of the larynx and fauces, which causes their own distinctive vocal impairment are not here considered, as we do not mean by American voice the thick voice, and that of low carrying power. Any condition which hinders the passage of air into the nose tends to diminish the nasal resonance, and does not contribute to the production of the nasal voice. When anterior septal deviations and spurs exist the air during vocalization vibrates abnormally in the nasal chambers, so also when the nose is narrowed at the tip which, perhaps, may be bent downward and when there are anterior polyps and anterior turbinal enlargements, chronic catarrhal inflammation of the mucosa often occurs without hypertrophy. From all these causes there result anesthesia and paresis of the soft palate, and some of the tones which should be formed in the mouth are formed higher in the nasopharyngeal space. This explains the occurrence and persistance for a while after operation of the peculiar voice in adenoids until the plate has regained the tones lost from disuse. Atrophic rhinitis does not seem to cause a nasal voice. In those under 12 years old the nasal voice is common, but the lesions enumerated above which might seem to stand in a causative relation thereto are not common. In those between 12 and 30, they are more common than in the first class, but the occurrence of the nasal voice is not more common and does not appear, therefore, to have any constant relation with the apparrent formative causes. In those over 30 the nasal voice is less common than earlier in life. It is thus obvious that the nose is not the determining factor in the formation of the nasal voice, because the nasal lesions do not determine the classification into nasal and non-nasal voices. The voice can often be improved by training without any medical or surgical intervention, and pathologic states have more to do with the range and power of endurance of the singing voice than with the quality in the middle register. The nasal voice is more often merely a matter of imitation. Good public speakers and actors with good voices have, from a purely anatomic point of view, very poor vocal organs. Anterior nasal obstructions are common in all civilized races. Some of the most pronounced nasal voices are heard in our country villages. The great influx of foreigners has introduced all sorts of vocal sounds and methods of speaking English into common speech, and we are quite indifferent as to the best methods of speaking. Hence children should be carefully trained in this respect, for in them, imitation of the vicious vocal methods they constantly hear is the most potent factor in the production of nasal voice.

In discussion, Dr. G. Hudson Makuen, of Philadelphia, said, that far more attention should be given to proper vocal methods. As a rule, physicians knew little about the subject. Laryngologists should instruct singing teachers. The American voice is due to the excessive tension of modern American life. The active anatomic factor in nasal voice is probably a low-hanging palate during speech. The sound-vibrations are centered in the postnasal space. We must train the levator palati muscles, which can be done by having the patient practice several times daily with *open* mouth before a mirror. Dr. I. Amory DeBlois, of Boston, thought that racial condition had much to do with the character of the voice. We are all familiar with the guttural of the German, the vibratory nasal French voice, and the high-pitched voice of the Yankee, while the English voice, no matter what the social rank, is always of an agreeable low pitch. Dr. John O. Roe said that the nasal sinuses had much to do with voice-production, as they acted as resonators. The voice depends on the shape and size of the cavities, which vary greatly. Anterior nasal obstruction does not cut off the influence of the resonators, while posterior obstruction does. So, also, the peculiarities of the various languages are a factor. Those who use many consonants, as Germans and Russians, have low-pitched voices, while an excessive use of the vowels conduces to the opposite result. Dr. A. W. de Roaldes, of New Orleans, spoke of American conversational habits, and of the fact that in the South the conversational tone is much lower than in the North. This may be due to a mixing of various race-bloods in the former, and to the lazier mode of life. The negro voice was not much addicted to the tonalities. Dr. Thomas Hubbard, of Toledo, thought that much of our high-pitched speaking was due to the noise of our American cities, where it was often necessary to speak in a loud high tone in order to be heard. So, also, from the noisy environment the ear becomes obtunded to the appreciation of such misuse of the voice. Dr. Farlow said that the low palate was often due to a lack of use, but was not the original cause of the nasal voice. Its causative relation was secondary not primary. He had noted that nasal voice was common in children, in whom the sinuses were scarcely at all developed. English is spoken in many countries, in none of which does the voice have the same qualities as in this country. Voice is not, therefore, primarily a matter of language. He had often noticed the most pronounced cases of nasal twang in country villages.

Adeno-carcinoma of Nose, with Report of Case. By Dr. James E. Newcomb, of New York. Up to last year there were on record the notes of 23 similar cases in which the clinical diagnosis had been verified by microscopic examination. This case was that of a widow, aged 61 years, of American parentage, who had suffered since June, 1898, with obstruction in the left naris and almost daily nosebleed, severe on one or two occasions, but never requiring operative interference. At times there had been a slight watery discharge with a slightly offensive odor. She had lost some flesh and strength. In September last, she had blown from that side of the nose what she described as a "fleshy bean" and which was probably an ordinary polyp. The left middle turbinate was considerably enlarged and the mucosa over it seemed to be in a condition of polypoid degeneration. About it were some fleshy proliferations which bled easily upon manipulation with the probe. There were no glandular enlargements and no pressure-symptoms. The masses were removed under cocain-anesthesia without incident. The report of the microscopist was adeno-carcinoma. Radical interference was refused by the patient. Cancer of the nose is relatively rare. It is claimed that there is nothing more than coincidence between the association of the ordinary polyp with carcinoma. Tissier does not believe that the epitheliomatous degeneration of simple polyps has ever been definitely proven. What we know, he says, about the etiology of polyps explains their occurrence in a cancerous nasal fossa. On the other hand, Plique states that it is pretty frequent after the ablation of numerous benign polyps, to find new polypi appearing, composed this time of epitheliomatous tissue. Dr. Newcomb could hardly accept the truth of the latter statement, for polyps were very numerous and were often removed by very crude means, while the occurrence of cancer here was so rare. Dr. G. V. Woolen, of Indianapolis, related the history of a case of malignant disease of the nose occurring in a girl of 8 years, following an injury received two years before. Nasal polypi, so called, were removed and pronounced by the microscopist to be nonmalignant and only simply mucoid in character. Pressure-symptoms were prominent. The growth was found to recur over an increasing area, and the case was pronounced to be nonoperable. Later clinical history clearly demonstrated the fact of malignancy. Dr. J. L. Goodale, of Boston, mentioned a case of adeno-carcinoma of the nose occurring in a man of 51 years. The left naris was filled with a soft bleeding mass showing after removal and examination a fibrous stroma with epithelial cell-nests. There was bulging of the eye, much pain and nasal obstruction. Removal was followed by recurrence. Dr. Newcomb urged more hearty cooperation and association of the clinician with the microscopist in the early management of cases of doubtful nature in their incipiency.

Removal of Foreign Body from Bronchus Through Tracheal Opening was a paper by Dr. A. Coolidge, Jr., of Boston. A teamster, aged 23, was admitted to hospital a year ago, having worn a tracheal tube for 20 years. Twelve hours before admission the tube had become detached from the shield and had been drawn into the bronchus. On admission, severe cough, dyspnea and noisy breathing were present. Examination by x-ray was negative. Patient was etherized on the back with the shoulders over the end of the table with the head held downward and rotated to the right side. The tracheal wound was enlarged downward. A urethroscope one-half inch in diameter and three inches long was passed down the trachea with the stiletto in place, and when the stiletto was withdrawn, a speculum was pushed down the trachea without any difficulty to within about one inch of the bifurcation. Under sunlight illumination with the hand-mirror, the tube was seen in the right bronchus with its upper end half an inch below the bifurcation. Alligator forceps introduced through the speculum extracted it without difficulty. No ill after-effects were noted. There was some inconvenience at first from coughing, but none from secretion. The mucosa of the trachea was cocainized. In case a body is large enough to make it probable that it has gone no lower than a primary bronchus, immediately tracheotomize and explore by means of straight tubes as large as possible. If the body is in a secondary bronchus operative interference is justifiable if there is a good chance of reaching it after illuminating the primary bronchus, as would be the case, especially on the right side. A body loose in the trachea is much less dangerous than the same body impacted in a bronchial tube, and hence any violent respiratory excitement should be avoided. It is better to tracheotomize under cocain than under ether. In special cases tracheotomy may be avoided by introducing straight tubes through the glottis from above, but if the body is being rattled to and fro in the trachea, it would not be justifiable to run any farther risk of its being inhaled. Even when it is in the larynx in such a position that attempts at removal are likely to push it farther down, it is safer to do a tracheotomy and reach it from below. That absolute surgical cleanliness is necessary to present septic pneumonia goes without saying. A favoring circumstance in the use of the straight tubes is the flexibility of the lower part of the trachea and bronchial tubes, which permit considerable straightening or bringing into line the bronchus under observation. Dr. H. L. Swain, of New Haven, said that statistics show that foreign bodies are frequently expelled, sometimes months after they have entered the air-tubes. Ulceration is set up and they become dislodged. Dr. Roe stated that in thousands of instances the body had been expelled. The best results came from leaving it alone if it is not producing active symptoms. Dr. de Roaldes believes that we should not hesitate to operate. In 7 out of his 8 cases, the body had been removed through the tracheal wound. A low tracheotomy should be done and the bronchus titillated so as to excite cough, the edges of the tracheal wound being held open. We ought not to advise non-interference. Dr. Woolen thought that if there were no immediate symptoms and if we knew the body to be of such a nature that it could be easily expelled it was advisable to defer operation.

Exhibition of a Case of Stammering with Demonstration of Methods Employed in the Treatment, was considered by Dr. G. Hudson Makuen, of Philadelphia. The patient was a civil engineer, aged 29, who had stammered since the period of voice-formation. There was no assignable cause for the affection, which seemed to be the outcome of a congenital neurosis. The chief characteristic of the defect was spasmodic contraction of the muscles of the soft palate resulting in sudden closure, during attempts at vocalization and articulation, of what Dr. Makuen called the posterior palatolingual chink. These spasms were of variable frequency and duration and came on at most unexpected times. They gave the speech a peculiar jerky character and sometimes blocked it entirely. The defect was more pronounced in reading than in speaking. There was also a sort of mental hesitation and he could not always think correctly. In normal speeh, the action of the muscles is entirely automatic, and when any of the mechanisms employed fails to perform its functions this automatic action becomes impaired and it is the effort to control the lagging mechanism and to bring it into harmony with the other mechanisms that constitute the chief difficulty of the stammerer. In this particular instance the respiratory mechanism was at fault. Resort was had to direct nerve-muscle training, that is the singling out of the muscles with faulty action and by training them by means of voluntary exercises to properly functionate. This is superior to the indirect method which leads the patient unconsciously by means of approximately correct speech to use the muscles properly. The former methods develops the nerves as well as the muscles and establishes a volitional control over the faulty mechanism.

Report of Cases of Chronic Empyema of the Antrum of Highmore Operated upon by the Caldwell-Luc Method was made by Dr. A. W. de Roaldes, of New Orleans. Five cases were reported, in all a radical and speedy cure being obtained. It is surprising that this plan of operation, originally devised in this country, has not been more generally followed. It is believed to be superior to the older plans. The various steps of the operation can be summarized as follows: (1) A buccal incision is made parallel with and near enough to the upper gingivolabial fold in order to allow of the subsequent easy union of the muco-periosteal flaps. (2) The anterior wall of the antrum is opened in the canine fossa, the opening being ovoid in shape. Its extremeties give easy access to the tuberosity on one side and to the nasal wall on the other. (3) The cavity is thoroughly curetted and all diseased tissue removed. (4) A portion of the anterior extremity of the inferior turbinate is removed. (5) A large artificial opening is made in the nasal wall of the antrum as close as possible to the angle formed by the floor and anterior wall.

(6) The cavity is finally inspected, cleansed, dried, and lightly dusted with iodoform, followed by suture of the mucoperiostal flaps. Iodoform gauze is gently packed into the antrum and also into the nasal fossa, changed on the third to fifth day and afterward on alternate days until the twelfth day. The patient is then allowed to irrigate the cavity with a syringe and cannula, using boric acid solution. In one case a little secretion could sometimes be found at the entrance of the sinus, but this was ascribed to an old ethmoidal trouble, the pus leaking into the sinus through an opening in the nasal wall from old necrosis. Dr. de Roaldes dwelt especially upon the importance of locating any other possible focus of suppuration, as the latter may prove a serious complication or materially retard healing. Dr. Rose said that we ought to consider the level of the antral floor with reference to that of the nose, and also that the existence of pockets or septums in the antrum often prevents proper drainage of the latter even after incision has been made. Dr. E. L. Shurly, of Detroit, did not believe that operation was always necessary. Acute cases (as after grip) often get well of themselves, but in chronic cases curetting will often be necessary. He did not believe there was any positive evidence of infection through the mouth, but the opening into the nasal fossa would in his judgment offer far greater chance of such infection. Dr. Makuen said that the law of gravity did not seem to hold good in normal drainage of the antrum, as the opening was nearer the top than the bottom. Drainage seemed to be regulated by some sort of capillary attraction. Dr. G. A. Leland, of Boston, remarked that man was not always an upright animal and that when lying down, the antrum was drained by gravity. If the natural opening could be kept patent, many of the cases could be cured without operation. The Caldwell-Luc method was really a combination of those of Mikulicz and Jansen.

Septic Phlebitis with Thrombus as a Complication of Peritonsillar Abscess. Report of two Cases, was the title of a paper by Dr. M. R. Ward, of Pittsburg. His first personal case was that of a woman, aged 30. Three weeks previously she had had pain and soreness in the left tonsil, which subsided. Three days later there was a right peritonsillitis, with a tumor in the right side of the neck, and enlargement of the postcervical glands. Palpation did not detect fluctuation. The muscles of the neck swelled, fever developed, with evidence of pneumonia in the lower lobe of the right lung, and of pus beneath the superficial fascia of the neck. Incision was made, but pyemia proved fatal on ninth day. Autopsy showed a thrombus in the internal jugular, with septic phlebitis running to the tonsillar plexus. His second case was that of a German male, aged 42, who had an ordinary quinsy, which was incised. Two days later there was a chill with renewal of all throat symptoms. Death occurred on the sixth day from pyema. Autopsy as in first case, with, in addition, multiple abscesses in the kidneys.

Acute Suppurative Process in the Faucial Tonsils was the subject of Dr. J. L. Goodale, of Boston, and was based upon a study of 8 cases of intrafollicular abscess, calling special attention to the etiologic relation of special bacteria to these cases, of the latter's relation to peritonsillar abscess, of their prognostic significance, and their clinical recognition. Streptococci were more numerous than staphylococci. Of the 8 cases, 2 were followed by peritonsillar abscess. In all there was severe infection, as evidenced by the swelling in the neck, temperature curve, and general course of the symptoms. In most cases the tonsil presented no gross appearance, which would suggest intrafollicular abscess, which varied in size and number. The fibrinous exudation was more marked than in simple proliferative tonsillitis. The pyogenic infection of the follicles seemed secondary to that of the crypts, and in the two peritonsillar cases there seemed to be a discharge of the abscess into the efferent lymph-channels.

Peritonsillar Abscess was considered by Dr. G. A. Leland, of Boston. The best method of relieving these cases is to make a long incision vertically through the tonsil, and then with the sterilized finger to thoroughly explore through this incision the pus-pockets. The method is not dangerous, but is very painful, and a few whiffs of an anesthetic may be necessary. Its advantages are that the abscess is drained from the bottom, recovery is prompt, and there are no relapses. The patient is able to swallow liquids in 6 hours, and solids in 12. In case the incision is slow in healing, applications of iodin and glycerin are made to the bottom. This operation is really a century old, but has fallen into disuse.

Peritonsillar Abscess Associated with Diphtheria: Report of Cases, was a paper by Dr. Thomas Hubbard, of Toledo. A farmer aged 30 had acute tonsillitis. Five days later a right peritonsillar abscess developed. Löffler bacilli were present on tonsil. Antitoxin was given, but the next day there was evidence of membrane in the trachea. Dyspnea soon became so marked that tracheotomy was done. The patient stopped breathing, but was brought around after an hour of artificial respiration. Death from pulmonary edema followed in 18 hours. A second similar case was described. In the latter instance different members of the same family were variously affected, ranging from ordinary simple tonsilitis to fatal diphtheria. In incising such cases care should be taken to definitely locate the pus so that the knife should pass through tissues devitalized by softening. Dr. F. C. Cobb, of Boston, exhibited some photographs illustrating wax injections through the tonsillar crypts into the pharyngo-maxillary space, the injections showing the same direction of extension as did the pus in the course of the actual abscess cases. Dr. Newcomb mentioned a case (recently reported by Sendziak) of diphtheria associated with multiple abscesses in the various tonsils and empyema of both antral cavities. The phlegmon about the lingual tonsil opened spontaneously and a profuse hemorrhage ensued.

A case of **fibro-lipoma of tonsil** was reported by Dr. T. Amory De Blois, of Boston. Patient, male of 30 years, presented a growth in left tonsil, pedicled and of the size of a peanut-kernel. Removal was done under cocain and with the cautery snare. A section of the growth was shown under the microscope.

SECOND DAY.

Discussion: The Relation of Pathological Conditions in the Ethmoid Region of the Nose and Asthma. Dr. Henry

L. Swain, of New Haven, in considering the pathologic side, said asthma means a hyperesthesia of the bronchial lining with a resulting explosion of energy. In addition, some other structure is diseased, irritations of which sets up the bronchial spasm, a condition called the neurotic habit. The nose is only one of many diseased sites which can produce an asthmatic attack. It may arise from some intranasal irritation. Certain irritations applied to certain nerve-fibers will produce conjection and chronic inflammation with swelling and watery discharge. Then come soaking of the tissues, edematous changes and polypi. In many of these cases of hyperesthetic mucosa there seems to be all over the body a thinness or flabbiness of the blood-vessel walls and a vasomotor responsiveness, which make possible the explosions which are the bane of the existence of these afflicted mortals. This character of vessel wall may be the inherent peculiarity of the neurotic subject. Such a theory would explain the headache, asthma, neuralgia, etc., which at the start are only vasomotor spasms. Later, by continued distension of the vessels, their walls become permanently stretched and flabby, and thus organic lesions are possible. In general the ethmoid or middle turbinate changes in all these cases are hypertrophic and confined to the mucosa. Septal spurs and bends tend to keep up the middle turbinate disease and increase the possibility of pressure. Repeated irritations gradually lead to vascular relaxation and the occurrence of edematous tissue. In turn, come contact areas and areas of pressure. With the latter present, the initiative of the asthmatic attack is easy if at the same time the bronchial apparatus is also diseased or susceptible. The clinical phase of the subject was discussed by Dr. E. Fletcher Ingals, of Chicago. In 3 cases the patients had asserted that they had had spasm in only one lung which corresponded to the side in which intranasal lesions were present. One patient was subject to attacks when on the ground, but was relieved by going to the sixth story of a character only a few blocks away has cured others. Even a change from the brick portion to the wooden portion of the same building has sufficed to relieve some cases. Dr. Ingals had found that a spray containing 3 per cent. of cocain and 5 per cent. of soda nitrate would often relieve aparently by a reduction of the sensitiveness of the bronchial mucosa. Treatment was discussed by Dr. F. H. Bosworth, of New York. He believed that the whole question turned upon the integrity of the respiratory function of the nose. The bronchi were only air-conductors, but there was an intimate relation between their mucosa and that of the nares. In asthma the condition was one of vasomotor paresis, not of muscular spasm. Behind the polypi, the polypoid degeneration and the edematous hypertrophy, is an ethmoiditis. The great indication for treatment is to relieve the ethmoiditis, to relieve the intracellular pressure by breaking down the honeycombed mass. A radical operation is necessary. It is not enough merely to remove the polyps. We must uncap the eggshell-like ethmoid. Points of contact should also be eliminated, not so much because of the pressure there exerted but because they encroach upon the lumen of the nares. In his own experience cutting forceps were unsatisfactory. He used rounded and oval burrs from one sixteenth to one quarter inch in diameter driven by power. We should burr down, stop, use burr as probe, revolve it again until all the thin trabeculæ are broken down and free drainage is established. In nearly every case treatment confined to the anterior cells is sufficient. Dr. Bosworth stated that he had never had any bad results in following the above rules. He formerly held that if cocain applied to the nose did not relieve an asthmatic attack it was useless to do any intranasal operation, but later experience had led him to modify his view. Purulent ethmoiditis does not cause asthma, but inflammatory disease does. Many colds in the head are doubtless acute ethmoiditis. In the general discussion Dr. E. L. Shurly remarked that the question was a difficult one on account of the complex physiology of the vasomotor system. We should look further back than the ethmoiditis and the edematous rhinitis. Recent experiments have shown that the filaments from the cranial nerves and the spinal cord are really conducting cables and that in the same one there may be filaments for various functions. In different animals and in different individuals of the same group, there is a great difference in the communicating branches between the nerve-trunks. These anatomic variations explain the peculiarities in individual cases. Moreover, it must not be forgotten that the nose has an olfactory as well as a respiratory function, though in man the former is very rudimentary. Impulses may travel along the sensory filaments and owing to defective insulation, as it were, may produce hyperesthesia. Even asthma cases should be divided into those due to local disease and those due to psychic influences. He must take exception to the statement that the bronchial tubes were mere air-conduits. There is a peculiar arrangement of the adenoid tissue in them, the function of which we do not know. Dr. Mackenzie said that the primal cause of asthma was not confined to any special peripheral organ, but that it resides in the individual as a whole. The area of nerve-explosion depends on the seat of the local pathologic process. Irritation may come from a peripheral organ as the nose, a distant organ as the uterus or from a general dyscrasia as gout. On this theory mere contract-areas and pressure-points are of no value. Nasal cough may arise in atrophic states of the organ. Final explanation is not to be sought in nasal or bronchial abnormities, for all the theories thus far advanced fail to conform to the requirements of a logical hypothesis. All polypoid degeneration is not due to an ethmoiditis, as is evidenced by both clinical and pathological observation. As to treatment he had no difficulty with the forceps. It is useless to temporize and we must curette, burr, and gouge freely. Dr. Makuen said that asthma depended on faulty nervimuscular action which might be due to any one of the thousand causes. Dr. Hubbard called attention to the autotoxemia theory. Such autopoisoning may be due to gastroenteric absorption or faulty elimination. Such a possible factor must be borne in mind in the study of each individual case.

Recurrence of the Tonsil after Excision: a Case of Hysterical Larynx, was the title of a paper by Dr. F. E. Hopkins, of Springfield, Mass. In discussing the subject of the recurrence of tonsils after excision, Dr. Farlow remarked that partial removal left behind diseased tissue, especially in the lower prolongation of the organ where the irritation of the lingual movements against the hardest part was apt to produce recurrence. Guillotines were usually employed for removal, but the ideal instrument was one which would actually get in between the anterior and posterior pillars. All tonsillar tissue should be removed and not merely the part projecting beyond the pillars. Dr. Farlow strongly advocated the use of scissors or forceps. He also called attention to the development, especially in young adults, of the plica triangularis, or fold of tissue running down diagonally over the anterior surface of the tonsil, which was often mistaken for a portion of the anterior pillar. It should be included in the tissue to be removed. Dr. Newcomb believed that recurrence often happened in tenement-house children who, after operation, were compelled to go back to the same general bad surroundings. Dr. Woolen said that a tonsil once removed can never return. He believed that what we call a tonsil is really a pathologic product which he viewed as he would a wart or a pipilloma, not indeed always requiring removal. He preferred the term "enucleation" to "excision." He used a guillotine without a fork—lifting the tonsil from its bed with a vulsellum. He tested the organ removed with a probe passed down through the crypts, and if it reached a solid bottom, he knew that he had removed the follicles entire, but if it easily passed through and came out on the other side, he was convinced that the operation had been incomplete. These roots of tissue left behind might easily provoke a quinsy. Dr. D. Braden Kyle, of Philadelphia, would look upon a recurring tonsil as a purely pathologic product, benign, it is true, and of a hyperplastic nature. It was the soft and spongy tonsil which was apt to recur. The recurring mass was more of a tumor than a tonsil, and somewhat on the order of an adenoma. Dr. Makuen called attention to the necessity of properly separating adherent faucial pillars before any instrument for removal was applied, and had devised for that purpose a special set of cutting blades.

Third Day.

Fibro-lipoma of the Epiglottis and Base of the Tongue was the subject of Dr. E. Fletcher Ingals, of Chicago. His patient was a farmer, aged 28, who, for 3 or 4 years, had had difficult breathing, swallowing, and speaking. In

1896, the cautery had been used on the base of the tongue, along with the snare and scissors, and a mass removed. He came under Dr. Ingals' observation in February of the present year. Symptoms had all increased in severity during the two preceding months, especially dyspnea on lying down. On examination, a smooth mass, with a congested surface, could be seen in the laryngopharynx, apparently attached to the right side of the pharynx and base of the tongue, being apparently of a fibrous nature. On attempting its removal with a cold snare, carrying a No. 5 wire, the latter broke three times. Later, a properly bent uterine ecraseur, carrying a No. 8 wire, was used, and removal effected without difficulty. Several sittings were necessary to secure entire removal. Some of the masses were purely fibrous, some fatty, and some of the mixed type. Attachment was, as above noted, with, in addition, the right side of the epiglottis. The patient was recently seen, and it was noted that the right side of the epiglottis had become adherent to the right side of the pharynx and the base of the tongue. Such adhesions would prevent the epiglottis from closing down over the larynx during deglutition, but in this case there was no difficulty with this function. Dr. Woolen mentioned a somewhat similar case. The patient had been exhibited several times to students previous to operation, and the wire slipped over the growth, in order to show them the mode of removal. When the latter was actually attempted, the patient suddenly stopped breathing. Artificial respiration restored him, but upon a second attempt the same accident recurred, and this time proved fatal. No anesthetic, local or general, had been used. If such had been employed, the sudden death would probably have been ascribed to its use.

Confined Suppuration of the Frontal Sinus with Spontaneous Rupture, Including Report of a Case, was the title of a paper by Dr. D. Braden Kyle, of Philadelphia. The patient was a woman, aged 60, who was seen in January, 1898. Her trouble had begun with an initial fullness at the inner angle of the left orbit and a profuse nasal discharge from the corresponding naris. Her face was swollen and the frontal region tender. Two months later the orbital swelling began to increase in size, and finally a rupture with the escape of pus occurred in the forehead a little to the left of the median line. It took two months after rupture to heal the wound. Dr. Kyle had been unable to find another case identical in all particulars with his own, though several resembled it in many particulars.

The Presence of Partitions and Diverticula as a Cause of Retarded Recovery in the Treatment of Sinuses of the Maxillary Antrum was a paper by Dr. John O. Roe, of Rochester, who remarked that in the treatment of antral diseases we should take into account the position of the sinus, its size, shape, and conformation, the thickness of the walls, and the relation of the tooth-roots to its interior. The above points were demonstrated upon a series of beautifully prepared skulls. Dr. Roe also showed an "antrum searcher," consisting of a probe-pointed flexible steel spring, running in a cannula, and capable of being projected from the latter after it was in the antrum. With this device the cavity could be thoroughly explored through a very small opening. Dr. Roe used a fine saw to enlarge the opening, rather than the rongeur forceps, which were apt to splinter the edges of the bone. Dr. Mackenzie did not think it always necessary to interfere with septums in the antrum. It was important to remember that in some antrums the ostium maxillare was high up and posterior to its normal site. In such cases the discharge would be more apt to drain posteriorly and high up in the nasopharynx, and such location might determine failure in operation. In cases of painful sinusitis, Politzerization of the cavity through the natural opening would often afford almost instant relief, and, in cases of pain extending over a wide area, often enable the determining of the exact point of origin. He was inclined to regard drainage-tubes as pus producers.

"Taking Cold" was considered by Dr. G. V. Woolen, of Indianapolis, who summarized the current views upon this topic. He believed that external influences were reflected upon internal surfaces, thereby causing disturbances of nutrition and the latter produced deficient calorification.

By a series of observations he had found that patients who habitually take cold have a subnormal temperature, even as low at times as 95 in people who appear perfectly well-nourished, and this subnormal temperature may explain much of their indefinite malaise. There is deficient hematosis, and this deficiency appears in the circulatory apparatus through vasomotor agencies. In some persons this condition dates back to improper care as to bathing, etc., during the first week of life. This may be the critical period. If a child passes into adolescence without getting into the "taking-cold" habit, he regarded it as safe in this respect for life. This habit may in some instances be the reflection of hereditary syphilis in the third or fourth generation. This same deficient hematosis may also arise from nasal stenosis and precisely the same train of symptoms be set up. Dr. Goodale said that we must not forget the role played by microorganisms which appear to affect the lymphoid tissue. The symptoms of an ordinary cold are in a mild degree those of an ordinary infection. In moderately severe cases the staphylococcus, and in very severe, the streptococcus predominates. Dr. Bosworth could not quite agree with the preceding speaker, for the rapid onset of a coryza does not always allow bacterial agencies to come into play. We take cold in many ways other than in the head. Acute rhinitis is only a local expression of a general disturbance. Prophylaxis may be summed up in two words, clothing and cold bathing.

During the executive sessions the following were elected to active fellowship: Dr. F. C. Cobb, Boston, thesis, "Peritonsillar Abscess;" Dr. F. J. McKernon, New York, thesis, "A Contribution to the Technic of Modern Uranoplasty;" Dr. Max Thorner, of Cincinnati, thesis, "Direct Examination of the Larynx in Children."

Election of officers for the ensuing year resulted as follows: President, Dr. Samuel Johnston, of Baltimore; first vice-president, Dr. T. Amory De Blois, of Boston; second vice-president, Dr. Moreau Brown, of Chicago; secretary and treasurer, Dr. Henry L. Swain, of New Haven; librarian, Dr. Joseph H. Bryan, of Washington; member of council, Dr. William E. Casselberry, of Chicago. Committee of arrangements, Dr. T. Morris Murray, of Washington.

PROGRAM OF THE SIXTH OTOLOGICAL CONGRESS, TO BE HELD IN LONDON, AUGUST 8-12, 1899.

I.—NORMAL AND PATHOLOGICAL ANATOMY.

1. BIRMINGHAM, Dr. A. (Dublin).—"The topography of the Facial nerve in its relation to mastoid operations, with specimens."
2. CHATLE, Mr. A. H. (London).—"The Petro-Squamosal Sinus."
3. COSTINIU, Dr. (Bucharest).—"L'état des oreilles, du larynx, et du nez observé chez les vieillards."
4. COZZOLINO, Prof. Vincenzo (Naples).—"Contribution a l'histologie du squelette des cornets pour la pathogénèse de l'ozène." (avec demonstration).
5. DENKER, Dr. (Hagen).—"Zur Anatomie des Gehörorgans der Säugethiere, mit Demonstration von Präparaten und Zeichnungen."
6. RUTTEN, Dr. (Namur).—"Présentation d'une exostose du conduit auditif droit."

II.—PHYSIOLOGY AND METHODS OF EXAMINATION.

7. BARATOUX, Dr. J. (Paris).—" L'unification et la mesure de l'ouïe."
8. BONNIER, Dr. Pierre (Paris).—"Un procédé d'acoumetrie."
9. GRADENIGO, Prof. G. (Turin).—"Sur l'examen fonctionnel de l'organe de l'ouïe, et sur la notation uniforme des résultats."
10. KAYSER, Dr. Richard (Breslau).—"Experimentlele Untersuchungen über acustische Phaenomene in flüssigen Medien." (mit Demonstration).
11. SCHMIEGELOW, Dr. E. (Copenhagen).—"On a new method of measuring the quantitative hearing power by means of tuning forks."

III.—PATHOLOGY AND THERAPEUTICS.

12. AVOLEDO, Prof. (Milan).—"Due casi di complicazioni patologiche della faccia in seguito a propagazione di un processus suppurativo acuto dell 'Orecchio Medio e Esterno."
13. AVOLEDO, Prof. (Milan).—"Risultati della chirurgia intratimpanica nei riguardi della funzione acustica, ma solo per la forma suppurativa."

14. BABER, Mr. Cresswell (Brighton).—"Turbinotomy in nasal obstruction."

15. BAR, De Louis (Nice).—"Abcès antérieurs de la mastoide et furonculose du conduit auditif externe."

16. BOBONE, Dr. T. (San Remo).—"L'involution précoce du tissu adenoidien sur la Riviera."

17. BRIEGER, Dr. O. (Breslau).—"Uber Tuberculose des Mittelohrs."

18. CHEATLE, Mr. A. H. (London).—"A case of Adenoma of the meatus in a patient suffering with chronic middle ear suppuration."

19. COSTINIU, Dr. (Bucharest).—"Résultats des exercises-acoustiques chez les sourds-muets."

20. COZZOLINO, Prof. Vincenzo (Naples).—"Statistiques des mastoidotomies simples et radicales, et des opérations de chirurgie oto-endocranienne pratiquées dans ma Clinique universitaire depuis l'an 1883."

21. COZZOLINO, Prof. Vincenzo (Naples).—"Pseudo-actinomycosis auriculaire externe avec ostéomielite diffuse a la zone mastoidienne, causée par un nouveau bacille filamenteux pyogenique" (avec demonstration.)

22. CURSETJEE, Dr. J. J. (Bombay).—"Some aspects of Aural practice in India, with special reference to Bombay."

23. DADYSETT, Dr. H. J. (Bombay).—"A paper on various domestic remedies with their effects, used by the people of India for certain diseases of the ear."

24. DELIE, Dr. (Ypres).—"Panotite avec complication cérébrale—opération—mort—autopsie."

25. DENCH, Dr. E. B. (New York).—"The Operative Treatment of Mastoid Inflammation."

25. DE SANTI, Dr. P. (London).—"The Radical Cure of chronic suppurative Otitis Media by Antrectomy and Attico-Antrectomy, with notes of thirty cases."

27. DE SANTI, Dr. P. (London).—"Some cases illustrating the Intracranial complications of neglected Otorrhœa."

28. EEMAN, Prof. (Ghent).—"La sclérose de la caisse tympanique."

29. FARACI, Prof. Guiseppe (Palermo).—"Sulla possibilita d riaprire la finestra ovale nei casi di anchilosi ossea della articolazione stapedo-vestibolare."

30. FARACI, Prof. Guiseppe (Palermo).—"Importanza acustica e funzionale della mobilizzazione della staffa."

31. FARACI, Prof. Guiseppe (Palermo).—"Utilita della miringectomia temporanea e consecutiva mobilizzazione di tutta la catena degli ossicini nel periodo sub-acuto di un otite catarrale decorsa senza perforazione timpanica."

32. FISCHENICH, Dr. Fr. (Wiesbaden).—"Die Behandlung der katarrhalischen Adhaesivprocesse im Mittelohre, durch intratympanale Pilocarpininjectionen."

33. GARNAULT, Dr. Paul (Paris).—"Mobilisation (two years ago, and extraction (one year ago) of the Stapes, in the same patient) with great improvement in hearing and typical phenomena."

34. GARNAULT, Dr. Paul (Paris).—"Mobilisation (three years ago) of the Stapes, in a man seventy-two years of age, deaf for forty years, absolutely so for fifteen years, with great and permanent improvement in hearing."

35. GARZIA, Dr. Vincenzo (Naples).—"Experimental study of the influence of malaria in diseases of the ear."

36. GOLDSTEIN, Dr. M. A. (St. Louis, U. S. A.)—"Therapy of the Nasal Mucous Membrane."

37. GRANT, Dr. Dundas (London).—"Diminished 'Bone-conduction' as a Contra-indication of Ossiculectomy."

38. GRAY, Dr. Albert (Glasgow).—"A case of unilateral Deafness caused by a Tumour of the Medulla, producing other remarkable symptoms—Post Mortem" (with microscope slides of the medulla).

39. GRAZZI, Prof. V. (Florence).—"Nuova cura della faringiti catarrali croniche in rapporto specialmente alle malattie dell 'orecchio."

40. HAIGHT, Dr. Allen T. (Chicago). "Naso-Pharyngeal Adenoids as a causative factor in Ear Diseases."

41. HEIMAN, Dr. Th. (Warsaw). "De l'inflammation primaire de l'apophyse mastoide."

42. KEIPER, Dr. Geo. F. (La Fayette, Ind.)—"A Description of a set of Mastoid Gouges."

43. LACROIX, Dr. P. (Paris).—"Complications otiques de l'ozéne."

44. LAURENS, Dr (Paris).—"Otite moyenne chronique supurée avec thrombose du sinus latéral et abcès du cervelet."

45. LERMOYEZ, Dr. Marcel (Paris).—"La Contagiosité des Otites moyennes aigues."

46. LUBET-BARBON, Dr. (Paris).—"Note sur les abcès aigus de l'apophyse mastoïde sans abcès de la caisse."

47. LUCAE, Prof. W. (Berlin).—"Zur Radicaloperation bei chronischer purulenter Mittelohrentzündung."

58. MALHERBE, Dr. (Paris).—"Traitement chirurgical de l'Otite moyenne chronique sèche par l'évidement pétro-mastoïdien, avec et sans tubage."

49. MELZI, Dr. Urbano (Milan).—"A case of retropharyngeal abscess of auricular origin."

50. MELZI, Dr. Urbano (Milan).—"A case of nasal hydrorrhea."

51. MELZI, Dr. Urbano (Milan).—"A case of endothelioid fibroangioma of the external auricular canal."

52. MENIERE, Dr. E. (Paris).—"Traitement des suppurations chroniques de l'attique."

53. MILLIGAN, Dr. W. (Manchester).—"Some observations upon the diagnosis and treatment of tuberculosis disease of the middle ear and adjoining mastoid cells."

54. MINK, Dr. P. J. (Zwolle).—"Pneumamassage unter höherem Drucke."

55. MOURE, Dr. E. J. (Bordeaux).—"Sur quelques cas de complications endocraniennes d'origine otique."

56. MOURE, Dr. E. J. (Bordeaux).—"Sur quelques points de technique à propos de la trépanation de l'apophyse mastoïde."

57. NUVOLI, Dr. G. (Rome).—"Sula cura pneumatica nelle malattie dell 'orecchio."

58. OSTMANN, Prof. (Marburg).—"Uber die Heilbarkeit bisher unheilbarer Schwerhörigkeit durch Vibrationsmassage des Schallleitungsapparates."

59. PASSOW, Prof. (Heidelberg).—"Chirurgische Eingriffe bei Sklerose und bei Ménièreschen Symptomen."

60. POLITZER, Prof. Adam (Vienna).—"On the extraction of the stapes, with demonstration of histological preparations."

61. ROHRER, Dr. F. (Zurich).—"On blue ear drums, 'tympanum coeruleum.'"

62. ROHRER, Dr. F. (Rurich).—"The appearance of varices on the ear drums."

63. RUDLOFF, Dr. P. (Wiesbaden).—"The operation of the removal of adenoid growths with the head hanging over the table, while the patient is under the influence of chloroform"

64. RYERSON, Dr. G. Sterling (Toronto).—"Objective noises in the ears."

65. SNOW, Dr. S. F. (Syracuse, N. Y.)—"Twentieth Century Prognosis in Chronic Catarrhal Deafness."

66. SZENES, Dr. (Budapest).—"Zur primaerer Erkraukung des Warzenfortsatzes."

67. TANSLEY, D. J. Oscroft (New York).—"Shall we use cold in acute Middle Ear or Mastoid Affections—if so, how long?"

68. TANSLEY, Dr. J. Oscroft (New York).—"Additional Remarks upon Ear Diseases caused by Deflected Septa."

69. TERVAERT, Dr. G. D. Cohen (The Hague).—"A case of thrombosis of both sinus cavernosi as a complication of chronic mastoditis ex otorrhoea, which ended in recovery."

70. TURNBULL, Dr. Laurence (Philadelphia).—"Some of the most important Discoveries in Otology; many of which have stood the Test of 35 years."

71. UCHERMANN, Prof. V. (Christiana).—" Rheumatic diseases of the ear."

72. VEYRAT, Dr. Ernest (Chambéry).—"Des améliorations de l'ouïe obtenues par le tympan artificiel, dans l'otite moyenne chronique sèche, ou sclérose tympanique."

73. VEYRAT, Dr. Ernest (Chambéry).—"Des Injections interstisielles de sublimé dans le traitement des lupus du nez."

74. WHITE, Mr. F. Faulder (Coventry).—"The Curability of Suppurative Otitis Media, without operation."

IV.—DEMONSTRATIONS.

75. HARTMANN, Dr. Arthur (Berlin).—"Lantern-slide demonstration on the Anatomy of the Frontal Sinus."

76. KATZ, Dr. L. (Berlin).—"Demonstration microscopischer und macroscopischer Präparate des Gehörorgans."

77. SZENES, Dr. S. (Budapest).—"Demonstration pathologisch-anatomischer Praprate: (a) 'Melanosarcoma auriculae et meatus.' (b) 'Osteoma liberum meatus auditorii externi.'"

78. TURNER, Dr. Aldren (London).—"Lantern-slide demonstration on the course and connections, of the Central Auditory Tract."

ANNALS
OF
OTOLOGY, RHINOLOGY
AND
LARYNGOLOGY.

THE BACTERIOLOGY AND HISTOLOGY OF OZENA.*

PROF. V. COZZOLINO,

NAPLES.

Translated by R. B. H. GRADWOHL, M. D., Bacteriologist to the St. Louis City Hospital.

Much has been written during the past two years upon the etiology of this form of atrophic rhinitis from a bacteriologic point of view. Some have urged that the pseudo-diphtheria bacillus is the exciting agent concerned; others contend that it is the bacillus mucosus which is at work, while still others affirm that the disease is due to the combined action of the two micro-organisms. Opposed to all these radical views we have the theory of a fourth class, comprising those who claim that there must first be a local predisposition for the disease; in other words, that the microbic factor is but secondary and not capable of provoking the disease in the absence of the local predisposition. Having observed then the diversity of opinion as to the etiology of the disease and likewise as to its treatment, based, of course in most instances on erroneous ideas, I determined to pursue an investigation along these lines, taking my observations from a close study of 42 cases of true ozena which presented types of the disease from the

---

*Annales des Maladies de l'Oreille, du Larynx, du Nez et du Pharynx, July, 1899.

mildest to the most severe and occurring in patients ranging in age from 5 to 50 years. I have endeavored to trace out the role played by bacteria in the causation of the disease in relation to these two factors of the ozenous triad, viz., fetidity and crusts. I have thus established the value of a bacteriologic examination of the ozenous crusts and the importance which should be attached to their constant presence.

Faithful to my old opinion about this phase of rhinopathy, I will say here in the beginning of my dissertation what perhaps should be said in conclusion, that I hold the micro-organisms concerned as only secondary in their effects to the primary factor of favorable soil for their development. I consider that there are other accessory causes for the development of ozena to which the presence and action of bacteria are secondary. If this disease were due entirely to the activity of a specific micro-organism and not to a local predisposition, how could we explain the fact that the disease is a strictly local one and of a markedly chronic character? I contend that the bacillus (bacillus mucosus) is not the prime etiologic factor in ozena, but that this micro-organism finds a favorable soil for its development in the nasal fossae which are endowed, by virtue of their embryologic and, also their teratologic natures, with features highly commendable for a microbic habitat; moreover, in nasal fossae which are normal and not so endowed, there will be no growth of this micro-organism. I shall not attempt to explain just how the micro-organism begins its work in this affection, nor what the nature of this work is, but shall leave that for other investigators.

I subjoin a list of the patients, arranged in classes according to age:

From	5 to 10 years,	-	-	4 cases.
"	10 to 20 "	-	-	10 "
"	20 to 25 "	-	-	17 "
"	25 to 35 "	-	-	6 "
"	35 to 50 "	-	-	5 "
			Total,	42

In these 42 cases of ozena I made bacteriologic examinations of the crusts from the nose and found great numbers of the encapsulated bacilli of Loewenberg, or the *bacillus mucosus ozena* of Abel. In 28 instances I isolated a bent and granular form of the pseudo-diphtheria bacillus which according to the general rule did not stain by Gram's method. I can confirm the statements of Abel and Loewenberg that the bacillus mucosus is not found on the outer surface of the crusts where we find the common saprophytes, but that it seems to thrive and, in fact, is found only on the inner surface next its point of attachment to the nasal mucous membrane, and also in the submucosa after the crusts have become detached.

I have been able to isolate the bacillus mucosus in all the 42 cases: the pseudo-diphtheria bacillus was found only 8 times; the staphylococcus pyogenes aureus 9 times; the staphylococcus pyogenes albus 7 times; the sarcina aurantiaca once; the bacillus subtilis twice; the megaterium once; streptococcus pyogenes, in long chains, 4 times; a filamentous bacillus once; bacillus prodigiosus once; non-classified cocci, twice; bacillus pyocyaneus, green variety, 7 times; the tubercle bacillus once in a patient who had been in contact with a phthisical patient. It is evident from these findings that the bacillus mucosus is present in every case of this form of atrophic rhinitis—a fact already established by Loewenberg (1888–94), and by Abel (1894), in a close study of 100 cases of ozena. According to the classification of Kruse, the bacillus mucosus belongs to the fourth group of schyzomycetes, between the bacillus aerogenes and that of rhinoscleroma. Morphologically it is to be placed with the bacillus of Friedlaender, the bacillus of Frisch, the bacillus of Passet, the encapsulated bacillus of Bordoni-Uffreduzzi, the bacillus coli immobilis and the bacillus lactis aerogenes of Escherich. It has been confounded by many experimenters with the bacillus of Friedlaender, but it differs from that micro-organism in many particulars.

In the crusts, the bacillus mucosus is a short rod, measuring from 1 to 1.25 μ in width, of variable length, with rounded ends; often it is grouped as a diplo-bacillus and surrounded by a capsule visible by means of all the methods of staining, and especially by Ribbert's method.

At times it assumes a size twice or thrice the normal dimensions.

In different cultures of recent date, the bacillus mucosus may take on a protean form: it sometimes takes the form of a short diplo-bacillus and sometimes appears as longer rods. In certain culture media, notably glycerin-agar, blood serum, etc., a capsule can be observed. In old cultures various involution forms can be observed—from chains of cocci and slightly colored filamentous forms to irregularly swollen and distorted bacilli. Inoculations afterwards from these old cultures will eliminate the suspicion of contamination. When the micro-organism is passed into the organisms of inoculated animals, viz., guinea-pigs, rabbits, mice, cultures present the original form of a short rod or an encapsulated diplo-bacillus. The capsule can be made to disappear by successive transplantations.

The bacillus mucosus is not stained according to Gram, but stains easily with the aniline stains. It develops readily upon all the ordinary culture media provided they have a slightly alkaline reaction. In nutrient broth, at the end of 24 hours, at 37°C. we see a slight cloudiness or a deposit of a whitish or greyish white color at the bottom of the tube. In most cases the culture has the nauseous odor of stagnant mucus.

In plate cultures upon agar-agar, three kinds of colonies are to be observed:

1. Superficial colonies: (a) small, round, closely-crowded colonies (with minimum dilution); (b) pearly colonies, round as pin-heads, convex, resembling dew-drops, transmitting light poorly, and with a tendency to spread and to form irregular groups (with medium dilution); (c) large discoid or convex colonies, of a "mucus-like" or whitish mucoid color, having a diameter of 1 2 to 3 4 centimeters, and having a tendency to coalesce in the form of a mucus-like deposit on the plate (with maximum dilution, 5, 6 and 10 colonies to the plate).

2. Deeper colonies, arranged perpendicularly to the surface of the plate, of small size, always lozenge-shaped and of an ivory-white color.

3. Deep colonies of a form resembling the round superfi-

cial colonies; they are seen scattered through the depths of the medium.

Under the low power of the microscope, the round superficial colonies and the lozenge-shaped bodies appear as colonies with an opaque, dark-colored, rhomboidal center and yellowish, granular border.

Upon agar-agar, as streak cultures, they appear as a thick, elevated, slimy band, with more or less sharply-defined edges; at the edges separate round, opaque colonies develop. By degrees, in the course of development, the whole slimy growth falls to the bottom of the tube.

Upon glycerin-agar, the growth occurs more quickly so that at the end of 12 hours at 37°C. there is an abundant colony. Bacilli grown on this culture medium are encapsulated.

Gelatin is not liquified by this micro-organism. Plate cultures upon gelatin are like those upon agar. In stab-cultures upon gelatin, during the first days of the growth we generally see a hob-nail growth like that of Friedlaender's bacillus with more of a tendency to spread over the surface of the medium. Along the stroke we see the development of a small series of round colonies or a break here and there along the stroke. Gas-production was observed in 30 per cent. of the cases.

Upon liquid blood serum, growth is faint, but it becomes more abundant if 20 per cent. nutrient broth is added to it.

Upon coagulated blood serum, the growth is the same as that upon glycerin-agar, even to the encapsulation of the bacilli.

Upon potato, the bacilli grow as a whitish grey colony, sometimes brownish and this coloration is given to the substratum of the potato.

Coagulation is produced in milk at the end of 48 hours. at 37°C. · This finding is different from that of Loewenberg. The fact that the bacillus is capable of provoking fermentation, however, was proved by the following researches upon sugar.

In fresh eggs, after from 48 to 72 hours at 37°C, we see slimy flakes appear, and can detect a very disagreeable odor.

In baked eggs, the yolks seem to be rendered mealy, and

appeared to be disintegrated around the periphery. Development is vigorous.

In fluids deprived of albuminous substances (Arnauld and Charrin, Hutschinsky) the bacillus mucosus develops very feebly. and it is only at the end of 15 days that a culture can be obtained.

The power of decolorization possessed by these micro-organisms is seen in liquids colored with fuchsin or methylene blue.

The bacillus mucosus has fermentation powers in different saccharine media.

Lactose broth colored with red litmus is not changed for 24 hours. but at the end of that time it begins to lose its color.

If melted lactose agar is inoculated with the bacillus mucosus and is then solidified. gas is produced in the substratum and bottom of the tube to some extent, but the micro-organism very quickly dies out.

In saccharose bouillon (2 per cent.) the reaction produced is strongly acid.

Glucose and levulose are fermented by this bacillus just as it ferments lactose.

It does not produce indol, but produces creatinin.

In bouillon containing urea. the bacillus grows feebly without the development of an ammoniacal odor, ammoniacal vapors not being produced in the presence of hydrochloric acid.

The action of light which. in the majority of cases, tends to retard the development of micro-organisms, seems to favor that of the bacillus mucosus. and at the same time does not tend to lower its pathogenic powers. After an exposure of a culture for 24 hours to the direct rays of the sun, at 35 C, it was kept for several months. and at the end of that time it was found to be possessed of marked pathogenic powers for guinea-pigs. The only change produced by the action of the sun's rays upon the bacilli was a marked swelling at one end. looking quite like a spore-formation.

The bacillus mucosus can live in temperatures as high as 56° to 60°C. At the end of a year, the cultures of the bacillus mucosos retain their vegetating powers.

Following out the thought of the difficulty in the therapeusis of this disease, on account of the anatomy of the nasal fossae, I made some experimental therapeutic observations *in vitro* with this micro-organism. The same thing was done with the pseudo-diphtheria bacillus, and I will speak of that later on. The following list comprises the antiseptics used in these experiments:

Trichloracetic acid, 3, 4, 5, 6, 9 per cent.   solutions.
Chromic acid. 1.05 per ct. 1 500 and 1 1,000   "
Corrosive sublimate, 1 to 1000, 2000 and 5000   "
Trichloride of iodine, 3, 1½ per cent. 0.75 per cent.   "
Silver nitrate, 2. 1, 1 2 per cent   -   -   -   "
Actol, 2 per cent. 1 per cent. and 1 2 per cent. -   "
Microcidine. 3. 2, 1 1 2 per cent.   -   -   "
    Drops of formalin—1 drop to 120 ccm. of water.
                      1 " " 60 " " "
                      1 " " 30 " " "
Salicylic acid, 1, 1 2, 1 4 per cent. solutions.
Water charged with oxygen in 80. 75, 50. 25 per cent solutions.
Potassium carbonate, 4, 2. 1 per cent. solutions.
Tincture of iodine, 20, 4, 2, 1 per cent. solutions.
Balsam of Peru, 100, 50, 25 per cent. solutions.
Creosote, 30, 25 per cent. solutions.
A pinch of tobacco in 5 ccm. of a culture.

Inoculations were made after 10 minutes. one-half hour, two hours and four hours contact with the antiseptic substance. After from ten minutes to a quarter of an hour's contact with the substances which are best tolerated by the nasal mucous membrane. I can verify the bactericidal action of only trichloracetic acid, 100 per cent. creosote, 4 per cent. trichloride of iodine and corrosive sublimate in from 1 per cent. to 1 2 per cent. solutions: a 1 to 5,000 solution is useless.

Formalin is only bactericidal in the proportion of 66 drops to 1,000 of water—a solution insupportable by the nasal mucous membrane. I wish to make a note here of that fact that several drops of formalin manufactured by the house of Schering of Berlin, were put into an agar culture of the bacillus mucosus; after several months another tube inoculated from this culture gave a growth of the bacillus mucosus. Salicylic acid, oxygenated water, potassium carbonate, nitrate of silver, microcidine, tincture of iodine, balsam of Peru and a pinch of tobacco, in solutions which could be tolerated by the nasal mucous membrane, show an absolute innocuousness for the bacillus mucosus. As we shall see later on, the results obtained with the pseudo-diphtheria bacillus were quite different.

An explanation for the resisting powers of this micro-organism for both physical and chemical agents (light and antiseptics) may be found in the presence of the capsule which is composed of a slimy substance secreted by the bacilli, making a slimy mass upon agar and gelatin plates, and in the bottom of bouillon cultures. This characteristic does not seem to be secondary to the viscid nature of the ozenous crusts.

62 animals were inoculated with the bacillus mucosus. In making the inoculations, the animals were divided into 9 series, each series representing a particular case of ozena. The bacillus was found to be slightly pathogenic for guinea-pigs, rabbits and white mice, by either intra-peritoneal or subcutaneous inoculation. The most pathogenic effects were seen when the inoculation was made into the pleural cavity, but owing to the technical difficulties involved in this method, it is impossible to say that the pathogenic action of the bacilli per se was responsible for the results obtained. All of the inoculations were made with broth cultures which had been grown for 24 hours at 37 C. With 1 to 1 1-2 ccm. of the broth cultures, intra-peritoneal inoculations in guinea-pigs produced death in from 7 to 12 hours. By successive passage of the micro-organism through the bodies of the animals, the virulence was increased to such a degree that death of a guinea-pig was induced in 4 1-2 hours, of a rabbit in 6 hours, and, with a dose of 0.1 ccm., of a white mouse in 9 hours. When subcutaneous inoculations were made into guinea-pigs with cultures obtained directly from the crusts, after the lapse of two or three days, we noticed a marked painful infiltration around the site of inoculation which was succeeded by a pouring out of a fibrino-purulent exudate. On the contrary, death always ensued when cultures of the bacillus mucosus were used, in from 15 to 19 hours. The white mice were most susceptible to subcutaneous inoculation. Post mortem examination of these animals showed in all cases an acute or purulent peritonitis, with fibrinous

flakes covering the different abdominal organs which were all markedly congested. The only change to be noted in the thoracic cavity was an occasional congestion of the lungs, unless, of course, the injection was made into the pleural cavity. Microscopic examination of the blood. the peritoneal exudate and the fibrinous flakes revealed the presence of numerous diplo-bacilli surrounded by a capsule. Bouillon cultures from the blood gave off the same disagreeable odor which emanated from the ozenous crusts.

By successive inoculations with small quantities of broth cultures, after a local infiltration which disappeared. the condition of the animal became refractory to larger doses of the cultures. Inoculations with serum from these animals mixed with a virulent culture of the bacillus mucosus were negative. while animals which were inoculated with the virulent culture alone quickly succumbed. The same phenomenon was observed when cultures were exposed to the antiseptic agents already recorded.

Attempts were made to inoculate animals with the bacillus by inoculating the mucous membrane of the nasal fossæ after it had been thoroughly scraped. but without avail. This is not to be wondered at when we remember that in veterinary pathology we have no mention of animals being affected with ozena or any fetid disease resembling it.

I will now report the results obtained in a parallel series of experiments with the pseudo-diphtheria bacillus which had been isolated in seven instances from ozenous crusts.

These experiments were carried on with the view of establishing or refuting the pathogensis of the pseudo-diphtheria bacillus and also. of noting the effect of the antiseptics upon it in the same proportions that we used in experimenting with the bacillus mucosus. In six guinea-pigs inoculated both by the intra-peritoneal and subcutaneous methods, with from 1 to 3 ccm. of a glycerin-bouillon culture, no effects could be seen after from 24 hours to 7 days and what is more, no local edema ensued such as has been described by Belfanti and Della Vedova. This result confirms the observations of Spronck, of Utrecht, who concluded. after a patient series of experiments, that the pseudo-diphtheria bacillus has nothing in common with the Klebs-Loeffler bacillus from a pathogenic point of view, although this observation is not in accord with the experience of A. Peters with seven cultures of the Klebs-Loeffler bacillus and eleven of the pseudo-diphtheria bacillus. I tried several times to inoculate animals by introducing crusts into pockets under the skin but failed signally, not even edema being produced. I found that the antiseptic solutions which were but relatively bactericidal for the cultures of the bacillus mucosus were strongly inimical to the growth of the pseudo-diphtheria bacillus, even after as short an exposure as ten minutes. Cultures exposed ten minutes to the action of such antiseptics as trichloracetic acid, etc., were barren of pathogenic action when injected into guinea-pigs. I have arrived at the conclusion that the pseudo-diphtheria bacillus is only a simple, innocent saprophyte as seen in the nasal fossæ of ozenous patients. This saprophytism has been known for some time: Hoffman, in 1887, isolated it in twenty-six cases from the buccal mucous membrane. The fact is worthy of note that the bacillus mucosus was found by me in that part of the ozenous crusts which were adherent to the mucous membrane, while the pseudo-diphtheria bacillus and other micro-organisms, was found on the outer surface.

I have been accused of having a groundless prejudice against the use of diphtheria antitoxin in cases of ozena; I will say that I have always opposed the many illusory therapeutic agents which have been advocated during my twenty-five years of rhinologic practice and will not admit their efficacy unless their administration is rational; certainly, the use of the diphtheria antitoxin is not so in cases of ozena. It is inconceivable to me how the serum of a bacillus which is *not* the etiologic factor in the causation of ozena can be efficacious in its cure; it is a fact well established by my own observation and that of others that the pseudo-diphtheria bacillus is not concerned in the production of ozena but it is only an incident. The error of using the antitoxin of diphtheria in such cases becomes even more apparent when we consider that, while in true diphtheria, 1,000 to 3,000 units have been sufficient to obtain certain immunizing effects, the observers, Belfanti, and Della Vedova, have used from 20,000 to 50,000 units as immunizing agents in cases of ozena. The use of the antitoxin of diphtheria in ozena is not only without curative effects but it is expensive and is not without its harmful effects upon the patients. In order to substantiate the above statements I will detail the use of antitoxin of diphtheria in twelve cases of true ozena in patients varying from very tender years up to adult life. I injected from 20,000 to 25,000 units in all these patients at short intervals. In the course of this treatment I observed the following effects; I may add that these effects disappeared with the cessation of the antitoxin treatment:

1. Hyperemia with engorgement of the nasal mucous membrane.

2. Excitation of the glandular secretion of the nasal mucous membrane, with diminution of the mucus stagnation and lessening of the fetidity.

3. Upon microscopic examination of the crusts, I noticed a decided decrease in the number of the bacillus mucosus and a corresponding increase in the number of the pseudo-diphtheria bacillus.

These results, at first glance, seem to justify the tremendous enthusiasm of the novice in the use of the antitoxin treatment; we shall see, however, that the glandular excitation is responsible for the diminution in the number of the bacillus mucosus and consequently of the decrease in the fetidity that may be present. The same results can be obtained by the internal use of the iodide of potassium or sodium. The use of the artificial serum has only one redeeming feature and that is its temporary effect of reducing the fetidity by the simple physiologic effect of increasing glandular activity.

I have had the best results in the treatment of ozena by thoroughly curetting the nasal mucous membrane, followed by the use of successive applications of antiseptics in the hope that repair of the ozenous patches will ensue. If we take into consideration the fact that the deeper layers of the nasal mucous membranes which are exposed by such a curettage harbor no micro-organisms, we become sensible of the fact that this is the most rational treatment for this rebellious affection. Although I was possessed of the idea in my researches along this line that the above detailed surgical treatment was of the most efficacious, I still clung to the hope of finding a useful antitoxin derived from the cultures of the bacillus mucosus, the supposed pathogenic agent at fault in this affection. I may say, however, after a conscientious trial that the use of such a serum was without good effects. After establishing the fact that the bacillus mucosus produced toxins, I inoculated a Fernbach flask containing glycerin boullion, 2 per cent., with a culture obtained from the heart of a guinea-pig which had succumbed nine hours after inoculation. A centimeter of the culture was used and kept in the thermostat at 37 C for 12 days. I then added a ½ per cent. solution of carbolic acid to destroy the micro-organisms which might be present. I injected ½ ccm. into four guinea-pigs by the subcutaneous method: one weighed 380 grammes, another weighed 375 grammes and I inoculated this one by the peritoneal method. The third and fourth pigs had been previously inoculated with small doses of the bacillus mucosus; these received each 1 ccm. and ½ ccm. of the supposed toxin. The four pigs gave not the slightest sign of a reaction. To eliminate the question of a possible neutralization of the toxins by the use of carbolic acid, two cultures were subjected to 60 degrees C for ten minutes and sterilized by control; two pigs were then inoculated and neither showed any reaction with the exception of a local infiltration of such a slight degree that it disappeared in a few days.

I tried to filter cultures with the Chamberland filter but without success on account of the viscid nature of the material. From this it may be fair to conclude that the pathogenic property lies not in the toxin elaborated but in the bacillary vitality.

I probably would have been able, if I had tried, to find a bactericidal serum in lieu of an antitoxin by slight immunization of animals with tentative doses of the bacillus directly inoculated into animals, but the results which I obtained with the serum discouraged me from making any such attempt. Prof. Paltauf, of Vienna, told me that he had tried an extract of the toxin of the bacillus of Friedlaender without success; the near parentage of the two micro-organisms explains the similitude of the experimental results already obtained. The difference in morphology and cultural characteristics between the bacillus mucosus, the bacillus of Friedlaender and that of rhinoscleroma is hardly perceptible. The culture of the bacillus mucosus is more viscid and fluid than that of the pneumococcus, but resembles very much that of the bacillus of Frisch. It may be well to add here that I made a comparative study of the three micro-organisms and that I was assisted in this through the courtesy of Prof. Pane who sent me a culture of the bacillus of Friendlaender and of Prof. Oro, of the dermatologic clinic of Naples, who sent me a culture of the bacillus of Frisch. The bacillus of rhinoscleroma is stained according to Gram differently from the bacillus of Friedlaender and the bacillus mucosus. Wilde has published a note contradictory to this, saying that the absence or presence of Gram's stain of the pneumococcus is due to relative hardness or softness of the micro-organism, while Ducrey, of Pisa, affirms that he never succeeded in staining the bacillus of Frisch according to Gram. Paltauf has said that in organs hardened in Müller's fluid the bacillus of Friedlaender loses the Gram stain; Schmidt too has proved experimentally that the ordinary colon bacillus retains the Gram stain provided it has grown in greasy media for a long time. As for the differential test proposed by Loewenberg between the pneumococcus and the bacillus mucosus, that the former causes a coagulation of milk while the latter does not, I can say that in seventeen cases I have seen a decided coagulation of milk by the bacillus mucosus. On the other hand, Abel, Kruse and Wilde maintain that the pneumococcus does not cause milk coagulation; Paltauf and Loewenberg say that the pneumococcus causes milk coagulation while Denys and Martin observed this coagulation of milk by the pneumococcus in a few instances. Ducrey has established the fact, by comparative experimental studies upon several encapsulated bacilli found in the naso-pharynx, that it is quite difficult to isolate them as individual species and that there are many varieties of encapsulated bacilli found in the rhino-pharyngeal mucous membrane which bear such a close relationship morphologically to the bacillus of Frisch that it is impossible to differentiate them. This is particularly true of several varieties found in cases of rhinoscleroma which Ducrey could not differentiate from the bacillus of Frisch. Close study of the chemic functions of these micro-organisms give us but a faint clue to their true classification. For instance, Schmidt has cultivated a colon bacillus which will not coagulate milk: Vincent has found a colon bacillus in water which will not produce gas in lactose bouillon; Sanarelli has isolated a colon bacillus which will not give the indol reaction and has found a cholera vibrio which will not produce nitrates. Kruse and Pansini have demonstrated that there is a pneumococcus which will not coagulate milk.

As for a possibility of differentiating these micro-organisms by their pathogenesis, I can say that the results of different observers have differed in that regard also: for instance, while I succeeded in clearly proving the pathogenesis of the bacillus mucosus for guinea-pigs, rabbits and white mice, Abel could not always do so. I succeeded in producing a passing slight edema of the tissues by direct inoculation with the ozenous crusts; Abel could not do so. Yet we were both working with the same micro-organism, i. e., an encapsulated bacillus which we had invariably found present in ozenous crusts. Finally, my experience has not been in accord with that of Abel and Loewenberg who write of an *agreeable* odor that arises from their cultures of the bacillus mucosus; on the contrary, I noted a *disagreeable* odor about these cultures—an odor reminding me of the disagreeable odor of stagnant mucus which is to be noted in connection with cases of true ozena.

I would add that I isolated the bacillus pyocyaneus in seven cases of ozena, the typical green color being produced in the media. I attribute the green color of the ozenous crusts to the presence and action of this micro-organism.

II.—HISTOLOGY.

I shall now give an account of the histologic examination of the inferior turbinate bodies in cases of true ozena, with all the symptomatology in many different phases of the disease. It may be well to speak of the different theories in regard to the histology of the affection. From a perusal of the literature upon the subject one is struck with the apparent mystery surrounding the true nature of the disease, as explained by the many different eminent authorities in this field. Krieg has well said that "this affection has been an enigma to us ever since it was first observed." According to Gottstein, E. Fraenkel, and Zuckerkandl, it consists of an inflammation of the mucous membrane which terminates by gradual atrophy; Habermann, Krause and Rethi say that ozena is a process of fatty degeneration of the glandular epithelial cells and of the recently formed connective tissue; Schuchardt considers that it is a metaplasia of cylindrical epithelium. During the past year, Drs. Cholewa and Cordes, of Berlin, have published an exhaustive and valuable contribution to the pathology of this affection. These authors attempt to prove that the initial atrophy resides in the bony framework of the turbinate bodies while the mucous membrane is only indirectly affected. They say that an energetic and rapid lacunar resorption of the osseous substance takes place, that the interior as well as the exterior layers of the bone are affected, under the influence of the periosteal alteration by numerous osteoclasts or myeloplacques. In the physiologic state, the osseous resorption is rapidly compensated for by the process of apposition; from the very first year of life the bone is formed from the periphery by numerous layers of osteoblasts which constitute a new lamellary system, causing the increased growth and thickening of the bone. In ozena, this process of restitution or apposition is at fault and is in a state of abeyance, while resorption continues to grow on as rapidly as before. This lack of restitution or repair is not only limited in ozena to the inferior turbinate bones but it is also true of the whole nasal skeleton in many cases; thus, these authors explain the relative frequency in ozenous patients of "flat-nose" or of the embryonic nose, and in general of the diminution in the length of the partition (Hopmann), resulting in an anomoly of development of different parts of the nose. The beginning of the process in the bone explains the fact that in cases where the mucous membrane is but slightly changed, the bone has already become thinner than normal and presents numerous lacunæ of Howslip and osteoclasts both upon the periosteal border and in the medullary spaces. The hypothesis then to be gathered from the above is that ozena is a consequence of an arrest of development either of the turbinate alone or of the whole nasal skeleton: this is not a new idea nor is the observation concerning the histologic examination of the lacunæ of Howslip and of the osteoclasts recent, but the merit belongs to these authors, Cholewa and Cordes, of combining the etiologic relation of the disease ozena with the histologic researches already made. Again consulting the literature upon this disease, we see that Zaufal says that ozena begins, perhaps, as an arrest of development of the turbinate; Potiquet has demonstrated that ozena occurs by preference in patients with a poorly developed nasal skeleton, especially when the transverse dimension is increased, which Broca attributes to an arrest in development; Chatellier is of the opinion that ozena is a form of rarifying osteitis; Schestakow considers it to be a primary atrophy of the turbinate bones caused by an alteration in development upon which is grafted a rarifying osteitis. In short, all of these authors conclude in a rambling sort of fashion that the faulty development of bone will favor the beginning of ozena, but not a single one can base their assertions upon delicate histologic researches. On the contrary, Krause, Habermann and Zuckerkandl have noted histologically that there is a form of progressive atrophy of bone terminating in the reduction of the bone to a thin sheet; they cite also a form of lacunar resorption with lacunæ of Howslip and osteoclasts, but the two authors first mentioned, i. e., Krause and Habermann, attach an etiologic importance only to the fatty glandular degeneration, while Zuckerkandl affirms that simple ozena is a chronic catarrh terminating either in an atrophy of the mucous membrane or of the turbinate bone. However, he says that perhaps the atrophy of the bone induces that of the mucous membrane and admits that even though the atrophy of the entire turbinate body attacks all parts of it with equal intensity, it is the *osseous* atrophy which predominates. Cholewa and Cordes demonstrated conclusively that this osseous atrophy predominates and precedes the atrophy of the mucous membrane. According to these two authors, this process of resorption is congenital or continues from fetal life; they have no belief in an inflammatory causation. Upon this point my researches do not agree with theirs; they also call attention to a periosteal proliferation, a fibrous alteration of the cells and the presence of the granular cells of Ehrlich in the periosteum and in the medullary spaces.

Cholewa and Cordes affirm that the ozenous turbinate bones are so soft that they can be cut without the previous process of decalcification so commonly resorted to: I do not agree with them in this regard, for I tried to imbed in parafin and cut sections without previous decalcification and was unable to obtain satisfactory results in that way. All sections which I made were kept for from five to ten days in picric acid for decalcification. I do not believe that a resorption of the calcareous salts of the bone takes place in ozena as I was able to make very thin sections and obtained almost as good results as if I had been working with any osteoid substance.

The comparison made by these authors of ozena with osteomalacia on account of the resorption of calcareous salts which takes place in osteomalacia and which they pretend takes place in ozena, is faulty by reason of the fact that the process of osteomalacia attacks by predeliction spongy bones—for instance, the ethmoid bone (Pommer)—and likewise the fact that not only one portion but the whole nasal skeleton may be attacked. Osteomalacia is an alteration of osseous tissue already formed, with an absorption of calcareous salts, while for some time the fundamental organic substance is unchanged. Moreover, in osteomalacia, the resorption progresses from the center toward the periphery; it begins in the medullary spaces and the Haversian canals and extends toward the exterior of the bone. This is never true in the case of the ozenous process. Cholewa and Cordes say that the alteration in the soft parts is a consequence of the change in the blood due to the ozenous process. They observe that the fatty degeneration of the glandular epithelium considered by some to be specific of ozena and the cause of every case, is met with in other affections of the turbinate bodies, for instance, in hyperplasia. They consider that under certain conditions the nasal mucous membrane reacts in the manner already demonstrated. I now give you the results of my histologic examination of ozenous tissues taken from cases in different stages of the disease. I have divided the disease into three stages, viz., first, slightly advanced, second, advanced, and third, very advanced cases.

FIRST STAGE (BUT SLIGHTLY ADVANCED).

The epithelium is of variable thickness: at some points it has three or four layers of cells and even ten layers in some places. At other times it exhibits a tendency to hypertrophy with penetration into the tissue, when the layers of cells run up to twenty in number. In some spots the papillæ can be seen against the thickened epithelium. Most of the layers consist of normal epithelium, of cylindrical epithelial cells, either distinctly cylindrical or of a lengthened oval form and superficially placed; we see ciliated epithelial cells among which are scattered calciform cells and mucus.

The cells are arranged with their principal axis of their nuclei perpendicular to the surface. White blood corpuscles are to be seen through these layers on their way to the surface of the epithelium. In traversing the layer of epithelium longitudinally, we can see in places the gradual metamorphosis of cylindrical cells into flat cells. The layers increase in spots; the cilia disappear; the form of the cell changes from oval to circular; the cellular layers are no longer separated but are closely crowded together as if under the influence of a reciprocal pressure.

The cells in the deeper layers where the epithelium is thicker (where there are from fifteen to twenty layers of cells) are cylindrical with their nuclear axis perpendicular to the surface. In the layers midway between the surface and the internal border, the cells take on an ovoid appearance with their nuclear axis oblique to the surface, while in the cells most superficially placed we see irregular forms—flat cells, fusiform cells and vesicular cells, all but slightly colored. At these points, the epithelium assumes an epidermoid character, with a tendency to penetrate into the stroma. The surface is covered with granular detritus in the form of amorphous plates arranged in a haphazzard way and containing chromatin residue. Between the cells we see many leucocytes in various degrees of degeneration—some even seeming to be proliferating.

The stroma of the mucous membrane consists of formed connective tissue with numerous fibers which in the deeper layers are thicker and arranged like connective tissue fascia. The fixed corpuscles are not very abundant, are large, flat and of a vesicular outline and consist of homogeneous protoplasm. Some have one diameter longer than

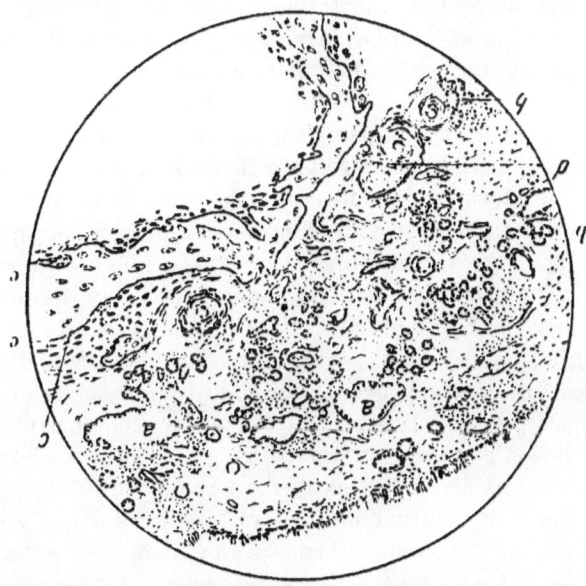

Fig. 1.—Section of right inferior turbinate body in a case of ozena (second stage, Koristka, obj. 2, oc. 3, 16).
   a. Dilated glands in the sub-epithelial layers.
   b. Acinous glands in the deep layers.
   c. Thickened periosteum.
   d. Venous cavities in the atrophic cavernous tissue.
   e. Osseous substance.

the other and seem to be composed of a rare chromatic tissue. In the adenoid layer, we see a very diffuse infiltration which is very pronounced around the vessels and glands. The infiltration extends along the glandular tubes and is also to be seen in the intertubular tissue especially in the vicinity of the glands adjacent to the periosteum.

The glands present themselves in a variety of forms. In the sub-epithelial tissue, one sees but a trace of the glands. In spots an extremely dilated gland is seen (Fig. 1. a) which is formed as a result of the fusion of different septa. This fusion is very evident at certain points. These spots are placed upon two or three flattened layers of cells.

They are strongly colored and consist of homogeneous protoplasm. Among these dilated glands we see some which are nearly obliterated and others occluded by the proliferation of the infiltrated glandular epithelial cells. The deep glands which, traversing the cavernous tissue, come in contact with the periosteum, are arranged in clusters (Fig. 1, b) but their orifices are markedly contracted. Their protoplasm is quite uniform, very compact and the plasmosomes are conspicuous by their absence. The protoplasm seems to cover the opening of the gland while on the periphery there is a proliferation of epithelial cells. The intertubular spaces are filled with round cells. The whole cavernous tissue is notably diminished in quantity. The venous sinuses which normally extend through the tissue in great numbers, are diminished in number and size and are surrounded by fibrous bands and bundles of elastic fibers which can be demonstrated by Van Gieson's method of staining. The venous channels are pushed aside by this exuberant growth of fibrous tissue while the walls of the nasal mucous membrane are greatly thickened by this dense fibrous tissue composed of mixed cells and leucocytes. At some places, these venous lacunæ are completely obliterated (Fig. 1, d). The orifice may be contracted in size and filled with blood: some of the orifices are open and some contain many crystals of hematoidin.

The bone is slightly thinner than normal. The periosteum is very much thickened (Fig. 1, c): at certain spots, the layer of osteoblasts is represented by three or four rows of large, flat, mononuclear cells. At other points, on the contrary, we see next to the bone a very thick layer of large fusiform cells (Fig. 2, d). Upon the surface of the bone we see several lacunae of Howslip and osteoclasts. At other points the medullary spaces are very rich in vessels whose walls are greatly infiltrated with white blood corpuscles (Fig. 2, a); at such points also may be seen numerous products of organization of blood. The osseous substance is but slightly altered. An infiltration can be made out around the blood vessels which burrow down in the depths of the bone.

Middle Turbinate.—The epithelium is only slightly thickened; the superficial parts are broken here and there. The middle layer of the epithelium is more highly colored than the deeper layers; the superficial parts consists of

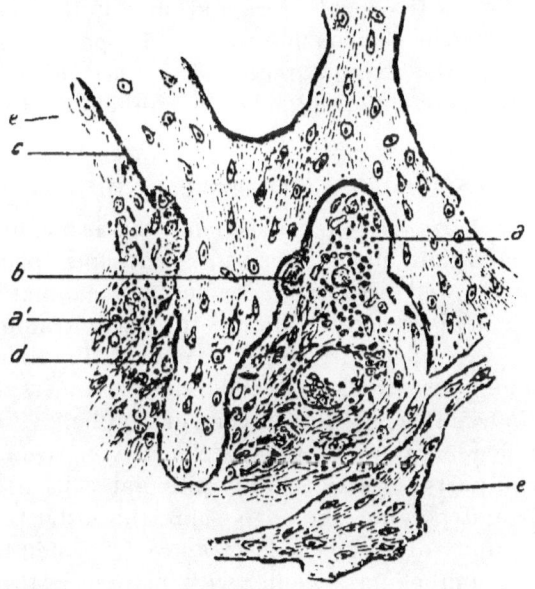

Fig. 2.—Section of osseous trabeculum, from preceding turbinate body (see Fig. 1).
    a. a'. Parvicellular and perivascular infiltration in a medullary space and in the periosteum.
    b. Osteoblast in a lacuna of Howslip.
    c. Osteoid substance.
    d. Fusiform periosteal cells.
    e. Osseous substance.

swollen and altered cells. In the deepest layers, near the connective tissue stroma there is a rich blood supply; these blood vessels are dilated and tortuous because they contain so much blood: their walls are infiltrated with small round cells. There is also quite an infiltration of white cells in the sub-epithelial tissues, amounting almost to the formation of peri-vascular or peri-grandular vessels. Ordinarily the cells are small, irregular and polynuclear. By appropriate staining we can make out a quantity of granular cells of immense size. In some preparations the glands are normal but for the most part they are dilated, with flat epithelial cells, no longer cylindrical but cubical, with faintly colored protoplasm and filled with multiple drops of fat. The peripheric outline of the cells is serrated or broken here and there, the connective tissue filling up the gaps in the continuity of this border; this connective tissue is yellow in color. Here and there in the glands we see dense mucus flakes more or less atrophied under the influence of the fixing agents. There are also acinous glands and polylobular glands, the epithelium of which is well preserved.

SECOND STAGE (ADVANCED PERIOD OF THE DISEASE).

Inferior Turbinate.—*Tout ensemble*, this is a counterpart of the condition just described. In some places the epithelium is well preserved, composed of elongated, uniform cells in close juxtaposition, with finely granular protoplasm. The nuclei are round or ovoid. In other places the epithelium is thickened and presents a metaplasia of the cylindrical cells into pavement epithelial cells; a marked degeneration of these cells can be seen in the superficial parts. Between the epithelial cells are mononuclear and polynuclear corpuscles, the latter predominating. Many of these corpuscles are free upon the surface of the epithelium together with more or less modified red blood cells. The adenoid layer is freely infiltrated with round cells, polynuclear for the most part, well colored and having a homogeneous protoplasm; the nuclei have a chromatic reticulated formation. These masses of cells are gathered about the blood vessels in the superficial portion of the epithelium, while in the deeper portion the connective tissue is dominant, with round cells and fibrous tissue; these round cells follow the course of the grandular tubes. In the stroma of the connective tissue we see many dilated glands (Fig. 3, a) as in the first stage of the disease; in the interior we see here and there granular detritus with round cells. These glands have thin walls, are infiltrated with small cells and lined with cylindrical, nonstratified epithelium having the character of superficial epithelium. Alongside of these dilated glands we see others (Fig. 3, b) in which the epithelium has proliferated to such a degree that the orifices of the glands are partially or completely occluded. Neither in the deeper nor in the superficial layers can we see glands arranged in clusters. The blood vessels, though not numerous, are filled with blood and have their walls infiltrated with small cells, but the perivascular cells are few and far between. The

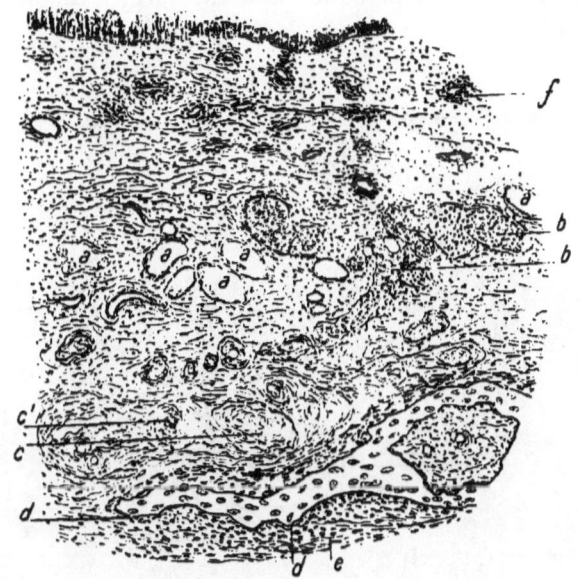

Fig. 3.—Section of the left inferior turbinate body (advanced stage of ozena).
  a. Dilated glands.
  b. Dilated glands filled up with proliferated epithelial glandular cells, thus occluding glands.
  c, c'. Atrophic cavernous tissue; the venous sinuses are partly obstructed, partly filled with proliferating endothelium.
  d. Osteoblasts.
  e. Thickened periosteum.
  f. Parvicellular infiltration under the epithelium (principally perivascular).

cavernous tissue is still more atrophic than that in the first stage, and there is a very pronounced occlusion of the venous sinuses which are contracted and warped by the growth of elastic fibrous tissue and endothelial proliferation. (Fig. 3, cc).

In comparing a section of the turbinate bone in this condition with a healthy bone, we are struck by the enormous dilatation of the medullary spaces and the thinning out of the osseous trabeculae which occurs in places to such a degree that a rupture of the partitions actually occurs. The tissue found in the medullary spaces is not like that found in the normal state, but it is filled with a formed fibrous connective tissue with numerous fibroblasts, (Fig. 1) while the myeloplacques and the large cells normally

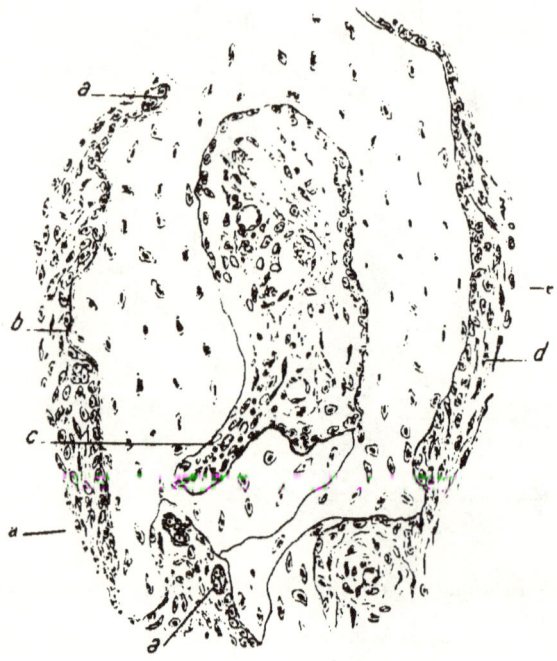

Fig. 4.—Section of osseous trabeculum, from preceding figure (3).
  a. Osteoblasts.
  b. Osteoid substance.
  c. Granular cells (*Mastzellen*).
  d. Fusiform cells in the thickened periosteum.
  e. Fusiform cells in medullary spaces.

present are rare. In places the layer of osteoblasts is well preserved, uniform and continuous, but in other places it shows a decided proliferation.

The cells are no longer cubical, but are elongated and even fusiform in spots (Fig. 4. d). Near the edge of the bone we see that here and there the continuity is broken by deep cavities of a semielliptic form filled with large polynuclear cells (Fig. 4. a), or with cells resembling osteoblasts. Small capillaries surround these places. The margin of the bone is serrated and eroded, and it is more highly colored than the rest of the osseous mass (Fig. 3. c, Fig. 4. b). This probably bears some relationship to the resorption of calcareous salts which ordinarily precedes the erosion and disappearance of bone. In the beginning there is a slight proliferation of cells, and again this proliferation of osseous cells is a mark of an attempt on the part of bone to regenerate: this regeneration is never complete, so that in a disease of bone of this kind there is never *restitutio ad integrum*. The vessels in the medullary spaces and in the periosteum are thinner than normal in most cases but may be thickened; the adventitia especially presents a fibrous proliferation. In the midst of the periosteal tissue we see granulations or crystals of hematoidin which are greatly increased in some parts. Granular cells (*Mastzellen*) abound in great numbers. Around the bone at several points we see a notable accumulation of mononuclear and polynuclear round cells disposed around the vessels which are filled with blood.

THIRD STAGE (VERY ADVANCED STAGE OF THE DISEASE).

Inferior Turbinate.— The epithelium is very thin, consisting at the most of but one or two layers of flat cells with decidedly colored elongated, almost fusiform nuclei. These cells are not in intimate contact, but seem to be imbedded in a homogeneous stroma formed by the apparent fusion of these cells. Between the epithelial cells we make out quite a number of small round cells which in certain spots almost conceal the epithelial cells. The epithelium is limited upon its outer surface by a finely granular, amorphous layer of cells. In the sub-epithelial tissue we find very large cells with irregular, vesicular nuclei, refractory to stains with a very rarified chromatic reticulum; the layer just alluded to is not continuous here and is formed of but two or three rows of cells.

The stroma of the mucous membrane is fibrous containing many thick fibers in which we see but a few fixed cells; most of the cells are replaced by mobile corpuscles. Blood vessels are rare; only a few capillaries are to be made out (Fig. 4, b) superficially; in the interior we see numerous perivascular cells of Waldeyer, few of which are normal. Not a trace of a gland can be found. The only suspicion of the existence of a gland in this place is a clump of proliferated cells here and there which might at one time have been the site of a glandular structure. (Fig. 5. a). By Weigert's stain, we make out numerous cocci and some rare bacilli only on the surface of the epithelium. By Nicolle's method of staining we see the short forms of the bacillus mucosus.

Middle Turbinate.—The superficial epithelium is well-preserved, consisting of from eight to twelve rows of cells; the deep cells are cylindrical or elongated with well-

Fig. 5.—Section of atrophic mucosa upon left inferior turbinate body (very advanced stage of ozena).
   a. Dilated glands which are occluded.
   b. Capillaries with infiltrated walls.
   c. Destroyed epithelium and parvicellular infiltration into the stroma.

colored nuclei; the middle layer presents more of a circular form while the superficial layer consists of flat cells. Between the epithelial cells are found white blood corpuscles infiltrating the superficial layers. The stroma of the mucous membrane is limited by the epithelium and shows an enormous vascular dilatation especially of the venous capillaries, with considerable accumulation of red blood cells.

In the course of these histologic examinations it became apparent that in the first two stages the bone becomes thinned out and sensibly eroded by the process of resorption, while the mucous membrane is but slightly changed.

In the third stage it was impossible to cut a single shred of bone, the whole osseous structure having almost completely disappeared, and all that was left was a fragment of mucous membrane. The lacunae of Howslip and the osteoclasts predominated in the second stage but they were also numerous in the first stage and we could make out a trace of vascular absorption at this time. The parvicellular and perivascular infiltration was conspicuous in the first stage when the large *Mastzellen* noted by the German authors were present in great numbers; finally, the presence of hematoidin in the medullary spaces and in the periosteum were signs which demonstrated that the process is accompanied by a slight but progressive inflammation chiefly of the periosteum, but also of the medulla. This was especially noticeable in the first stage. In the advanced stage, there is a condition of chronic atrophic inflammation which obscured the traces of the preceding inflammation. While we could make out very easily when the process was recent by the condition of the bone and periosteum, the parvicellular infiltration of the mucous membrane occurred when the affection had reached an advanced period. The apparently normal cells of the first stage began to desquamate in the second stage; cylindrical cells changed their character to flat cells with their nuclear axis oblique to the superficial parts. The basement membrane, which was normal in the first stage, becomes infiltrated in the second, coincident with a penetration of the masses of leucocytes into the epithelial layers and especially around the sub-epithelial glandular tubules. The epithelium had disappeared by the third stage; one or two rows of cells only are to be seen; the stroma remains unchanged with the exception perhaps of a slight separation from the epithelial layer.

In the first stage the adenoid layer is more or less profoundly infiltrated but nevertheless, here and there are to be seen normal spaces where one recognizes the fine fibrillary tissue which is prolonged down to the beginning of the cavernous tissue. This infiltration becomes more pronounced in the second stage, principally around the capillaries. In a more advanced phase of this period, the mucous membrane is almost uniformly infiltrated. At first the glands which are lined with flat cells appear to be dilated, in the superficial portions of the mucous membrane, while in the deeper portions, we see the glands arranged in clusters, lined with cylindrical cells; the intertubular spaces are markedly infiltrated. In the second stage we see the partial or total occlusion of the glandular orifices by the cellular proliferation and there remains not a trace of the cluster-arrangement of these glands as noted above. In the more advanced stage, we see only a trace of isolated obliterated glands.

In the beginning of the disease the vessels are still numerous in the medullary spaces, but are not quite normal in regard to the venous channels commonly observed. The cavernous tissue is visibly atrophic while the sinuses which are normally dilated appear to be contracted and surrounded by bundles of elastic fibers; there is also a thickening of the walls of these sinuses. The adenoid layer contains many capillaries. No trace of the blood vessels is to be made out in the medullary spaces in the second stage; only capillaries are found in the periosteum. The cavernous tissue is poor in vessels and the amount of fibrous tissue has increased to such an extent that the parts appear "stroma-like." In the third stage, we see capillaries only in the mucous membrane and nothing remains of the cavernous tissue.

The periosteum is thickened in the first stage. In some parts we may see the original arrangement of the periosteum in fibrous layers, separated from the bone by proliferation of the osteoblasts, among which we distinguish fusiform cells. Many osteoclasts can be made out here and there in the bone.

The results of my researches have led me to the belief that the process of ozena has its point of beginning in the bone; this explains the tenacity of the affection as seen even when the patient is in the very first stages of the disease. This opinion is in accord with what I have for a long time maintained, i. e., that no case of ozena arises without a favorable individual predisposition, without a favorable substratum. It is a case of individual predisposition, without a favorable substratum. It is a case of individual predisposition, that is to say, a certain morbid state of the origninsm which has been designated by the term scrofula; this term is improper, perhaps, but it indicates the whole pathogenesis of the affection in a concise way. This affection, beginning as an atrophy of the medullary blood vessels and especially of the arterial capillaries of the periosteal zone, can easily excite a periosteitis or a rarefying osteitis.

We may here speak of the comparison made by Zuckerkandl, who likens the circulatory system in the nasal mucous membrane to a long tube with a ball midway between the two ends; the artery would represent the afferent vessel, the vein the efferent vessel and the ball would represent the muscular erectile tissue. By means of this comparison we may explain the atrophy of the whole mucous membrane consecutive to proliferation of the periosteum. A part of the arterial capillary system being occluded, the turgescence of the cavernous body would be diminished owing to the fact that the arterial vessels comprise more of the cavernous tissue than the venous vessels (Zuckerkandl).

By degrees the turgescence would diminish, the venous sinuses would atrophy, and a few capillaries alone would keep up the vitality of the nasal mucous membrane. According to Zuckerkandl, the glandular tubules possess a rich capillary blood supply which has for its function the occlusion of the glandular orifice when the gland is in a state of repose; when the supply of blood diminishes, the mechanism of the gland is interfered with, so that the walls relax and secretion goes on with less activity; moreover owing to the loss of nervous activity consequent on the deficiency of circulating blood to nourish the parts, the chemical composition of the secretion of these glands is altered, so that instead of having a neutral or slightly alkaline reaction, as it normally has, it now has an excessively alkaline reaction.

Dilatation of the nasal fossa increases the evaporation of moisture from the nasal mucous membrane; the current of air is decreased in velocity; there is a relaxation in the intensity of the respiratory variations of the intra-nasal aerial pressure, which normally gives an important accessory impulse to the blood circulating through the nasal mucosa (Sänger); thus the circulation of the blood in the nasal mucous membrane is retarded.

Because of the fertility of the parts in leucocytes,

because of the alkaline nature of the fluid secreted. because of the destruction of the ciliated epithelium, the secretion which is poured out, under the influence of increased evaporation, becomes stagnant in contact with the nasal mucosa and furnishes an ideal soil for the growth of bacteria. This soil is especially conducive to the growth of the bacillus mucosus, which, by its viscid and adhesive nature, gives the special characteristic to the ozenous crusts of detaching themselves almost spontaneously; at the same time, the growth of this bacillus or the conjoint growth of others, gives the nauseating odor to this process. It is but natural to see how the fermentation which takes place under these conditions is capable of producing by its irritation an infiltration of all the tissues of the mucosa; this explains the progressive infiltration of the sub-epithelial structures which occurs simultaneously with the development of the disease.

We can also attribute to this irritation the fact that we often find hypertrophic zones in the inferior turbinate instead of atrophic zones; thus we see even a polypoid hypertrophy of the middle turbinate of the same side as the ozena. While my observations confirm his in this respect, I cannot agree with his theory of the etiology of this affection; my theory is based on both histologic examinations and clinical findings. Ozena is neither a consequence of other affections, nor is it a rhinitis: in fact, we find in an advanced period of the disease, a practically normal mucous membrane or epithelium upon the inferior turbinate body—the site of predeliction of this affection, while the bone is greatly eroded, the periosteum proliferating, etc. The middle turbinate is rarely the site of atrophy and this clinical finding is in perfect accord with the embryology of the parts; the inferior turbinate has its origin from a process of the superior maxilla while the middle turbinate arises from the ethmoid bone.

From my clinical and histologic investigations I am confirmed in the belief that the veritable essence of ozena resides in the bone and not in the mucous membrane. The affection is rebellious to both antiseptic treatment and to curettage. The bacillus mucosus while not regarded by me as a specific micro-organism of the disease, is nevertheless met with in nearly all the cases where ozenous crusts are present. That ozena is not an essentially infectious disease is proved by the fact that in spite of the epithelial desquamation, no trace of a micro-organism can be found in the tissues either in the beginning or in an advanced stage of the affection. The histologic communication of Drs. Belfanti and della Vedova, that a pseudo-diphtheria bacillus has been separated from ozenous tissues has been contradicted by many writers, among whom we may name Lautman (Annales des mal. de l'oreille, etc., March, 1897). who says, that he has but little confidence in the methods of these two investigators. Lautman affirmed that they found the bacillus in the mucous membrane, and not in the ozenous turbinate bodies. His examinations of the ozenous turbinate bodies, like mine, were without results.

My conclusion is, that the bacillus mucosus is the etiologic factor in the production of two of the most disagreeable *symptoms* of ozena, viz., fetidity and crusts, but it is by no means to be classed as the specific etiologic agent of ozena, as I maintained in the discussion of the bacteriologic etiology of ozena at the International Congress of Otology in Bale, in 1884. The etiology of the bone atrophy and, in consequence, of the mucosa also, can be found in a nutritive alteration of the tissues of the turbinate bodies or of one turbinate body. This change begins in the bone and is often associated with a congenital, general, systemic disorder. I wish to emphasize the fact that the ozenous patient is born *ozenous;* that is to say, the child which afterwards suffers with the full manifestations of the ozenous affection comes into the world with a special predisposition for these nutritive changes, which determine an erosion of the bone and its ultimate destruction, and an atrophy of the mucosa of the turbinate bodies.

LARYNGEAL CHOREA (?) OF REFLEX NASAL ORIGIN. REPORT OF A CASE.*

J. A. STUCKY, M. D.,

LEXINGTON, KY.,

LARYNGOLOGIST TO THE GOOD SAMARITAN HOSPITAL.

The case I am about to report is one of that peculiar unique type which is puzzling and frequently a source of embarrassing annoyance, because of its obstinancy in yielding to treatment. It is presented for your consideration not only for this but also for the reason that we are liable to find a cause for some of the most obstinate diseases with which we have to deal, where we least expect or suspect it. Pathology has shed little light on the possible cause of reflex nasal troubles, and the clinical phenomena are so numerous and deceptive, furnishing only a number of hypotheses and theories.

I think this case may correctly be reported as one of laryngeal chorea, because there were, in addition to other marked hysteric and nervous disturbances, some choreic twitchings. The characteristic symptoms, we are taught, is the "spasmodic barking cough," which disappears entirely when the patient sleeps. In this case the cough was worse at night, and instead of sounding like the ordinary bark of a dog was more like the yelp of a hound pup, which made the patient a nuisance and source of embarrassment to herself and to every one within hearing distance.

Following is a brief history of the case:

Miss ——, aet. 23. Fairly well nourished, intelligent woman, of distinctly neurotic type. Menstruated when twelve years of age. This function remained normal for five years. At the end of this time her health became

---

*Read at the Eastern Section American Laryngological, Rhinological and Otological Society held in Washington, D.C., February, 1899.

impaired (called general debility) and menstruation did not return for three months. Had a convulsion while in class at school, which was followed by scant menstrual flow. For several years from this time she was an invalid, suffering for days with dysmenorrhea at irregular intervals.

A gynecologist dilated the cervix uteri, but gave only temporary relief—later on both ovaries were removed. This stopped the dysmenorrhea but left her a physical wreck. An operation was performed for the relief of vesical calculi, and six months later large hemorrhoids were removed. During this period of five and a half years, she had been under the care of twelve physicians and gynecologists, and from the number of operations performed, I concluded that everything within the abdominal and pelvic cavity, had been removed that could be, with any safety to life.

At time of consulting me (May, 1898) she says she has been troubled with a peculiar barking cough for four or five years, sometimes losing her voice entirely, for days. At first the cough only lasted a few days and was not of enough consequence to demand treatment. This has gradually increased, until for the past few months it has been most persistent, the convulsive seizures being more frequent and prolonged in duration. One peculiarity is very noticeable, in this differing from a true chorea— viz.: sleep does not prevent the cough unless produced by a full dose of morphine. Also the recumbent position aggravates the trouble. She has been subject to "head colds" and "sore nose," since childhood, but nothing was thought of it and no especial attention given to it.

She complains of little tickling and some soreness in the throat, and she is easily fatigued by talking. Examination revealed marked hypertrophy of both inferior turbinates extending from "tip to tip" (anteriorly and posteriorly), more marked anteriorly. There was little stenosis or obstruction to respiration on account of large roomy nasal passages. The pharynx and arytenoids presented a congested granular condition. evidently due to irritation produced by constant coughing, and disease in nasal cavities. The ventricular bands were thickened, chordæ vocales hyperemic, and in attempts at phonation were driven forcibly toward the median line by spasmodic action of the adductors.

The case being one of peculiar interest, and markedly neurasthenic. no local treatment was employed, and a modified "rest cure" was begun. assisted by tonics, massage and hypnotics. At the end of three weeks this treatment had accomplished nothing, save in improving the general condition. The cough not only had "gotten no better fast" but was worse, being aggravated by the continued recumbent position. Things were getting desperate, as the hoarse yelping cough was not only distracting the patient, but every body else within hearing distance. Just at this stage she was taken with a violent attack of acute rhinitis, in which the turbinals became so swollen as to block the passages completely. For relief of this I applied a ten per cent. solution of cocain and much to the surprise and delight of myself, my patient, and the long suffering nurses. the cocain not only relieved the nasal discomfort but stopped the cough as long as the effect lasted.

Encouraged by this, chromic acid crystals (fused on probe) were applied thoroughly to both inferior turbinals throughout the entire length. There was immediate cessation of the cough, and the neurotic symptoms began to disappear. In two weeks from date of first application, a second was made. The patient remains entirely relieved and her general health better than since childhood.

REPORT OF A CASE OF ACUTE, PURULENT ENDO-MASTOIDITIS DEVELOPED IN THE COURSE OF A CHRONIC OTORRHEA; FOLLOWED BY AN EXTRA-DURAL ABSCESS WITH SLOUGH-ING OF THE DURA MATER—METAS-TATIC ABSCESS OF LUNG WITH SPONTANEOUS EVACUA-TION OF THE CAVITY— RECOVERY.*

P. M. PAYNE, M. D.

MEXICO.

RESIDENT SURGEON EYE, EAR, NOSE AND THROAT HOSPITAL, NEW ORLEANS, LA., 1898. NOW OCULIST AND AURIST TO AMERI-CAN HOSPITAL, CITY OF MEXICO.

On December 9th, 1897, a Spanish cabin boy A. L., age sixteen years, was brought to the office of Dr. A. W. De Roaldes, with a temperature of one hundred and six degrees, and a history of chronic suppuration of the middle ear, (left side) recurrent acute otorrhea lasting over a period of eight years. Two years ago in one of these attacks patient was threatened with mastoid complications. External examination showed no swelling, no redness nor was there any appreciable pain except on pressure over the mastoid region. Otoscopic examination showed the canal to be free from all acute inflammation. Chill and dizziness were complained of during the examination.

Notwithstanding the almost complete absence of external mastoid signs, the diagnosis was made of acute purulent endo-mastoiditis with probable cholesteatomatous masses obstructing the additus ad antrum. Drum was extensively destroyed and middle ear partly disorganized, showing comparatively very little discharge, in fact only a moist condition rather than a purulent accumulation;

*Read at Nineteenth Annual Meeting of Louisiana Medical Association.

there was however considerable odor. Probe revealed necrosis of tympanic cavity, and the hearing was almost destroyed in that ear. Operation was declared urgent and patient sent at once to the New Orleans sanitarium.

The next day, the parts having been well prepared, a Schwartze's operation was performed by Dr. Augustus McShane in the presence and under direction of Dr. De Roaldes.

Communication with the middle ear was secured after removal of granulating tissue and caseous foul-smelling epithelial masses. Wound was now irrigated and packed with iodoform gauze and patient put to bed. Slight chill on following morning, (eleventh) maximum temperature of the day being 101 1-5°.

December 12th.—Temperature 102°. Reference to temperature chart shows irregular daily rise and fall of temperature, and on the 13th, three days after operation, it was 103 3-5° F.

December 13th.—Dressing was ordered removed, wound irrigated, peroxide freely used and nurse instructed to renew same at night as considerable pus was found in canal and antrum. Dressing ordered changed twice daily. He complained of pain in left side of chest.

December 15th.—Patient was seen in consultation with Dr. R. Matas, who after discussing the indication of another intervention, agreed in the main line of the treatment.

December 17th.—For the last four days he has continued to suffer from septic rigors followed by rise of temperature during which spells he experiences a sharp and well localized pain in chest with troublesome cough, these symptoms abating during the period of febrile remission. Meanwhile suppuration continued to be foul and so abundant that an extensive Stacke's operation was done, with an exploration in the direction of the lateral sinus, although there were no marked signs of a thrombosis. After curetting vigorously the large cavity resulting from a radical Stacke's, it was found that there existed granulation tissue between the two plates of bone in the rear of the antrum, and in the direction of the lateral sinus. In following this granulation tissue considerable necrosed bone was found and chiseled out, and finally the sinus was exposed, and a large quantity of very offensive pus was evacuated from an extensive extra-dural abscess, the pus having found its way from the cavity of the abscess between the tables of the skull through a fistulous opening in the inner table. The exposed lateral sinus was found pulsating and otherwise unaffected.

December 19th.—Notwithstanding this intervention septic symptoms continued to be manifested by frequent rigors and rises of temperature with an occasional subnormal registration. Patient complained to-day of pain and swelling in anterior temporal region.

December 22nd.—A large abscess starting from the last mentioned region was opened midway between the ear, and the eye, and found to correspond to a purulent cavity extending under and below the zygomatic arch, the bone of the temporal fossa being denuded and rough. Anterior and posterior wounds were made to communicate by drainage tube.

December 27th.—The condition of patient is very little changed and he has become very pale, very weak and very much emaciated. Notwithstanding all efforts at careful drainage and thorough disinfection the wound continued to show an unhealthy action and to discharge foul-smelling pus.

The means of the patient having been exhausted he was removed to the Eye, Ear, Nose and Throat Hospital where he continued to be attended to by the writer. On the night of this transfer, although carefully executed by two of the resident surgeons in a closed carriage, patient gradually grew very weak, considerably depressed and somnolent. At 8 p. m. thermometer registered 96 2-5, with a pulse of sixty-four. These distressing symptoms were properly met with by alcoholic stimulants, digiti in ,strychnin and warming of the extremities.

December 28th.—A thorough investigation was made: minute examination of chest and mediastinum gave no positive information as to the exact location of the abscess from which this bad smelling pus was being coughed up daily, especially during the rise of temperature, the only appreciable auscultatory symptom being a prolonged expiration in the left supraspinous fossa. The doses of 'creosote et morrhuol' which had been administered from the beginning were increased and patient highly nourished every few hours.

December 29th.—At 8 A. M., he had a very severe chill followed by a temperature of 106°, with a return to normal at 10 A. M. At 2:30 P. M. he had another chill with temperature not so high as morning, but he complains of severe pain just below left nipple. At night, wound was dressed, the gauze being saturated with usual foul-smelling pus. Deeply imbedded in the posterior wound a mass was observed to pulsate and apparently increase in volume. On removal with forceps and close examination it was found to be a slough of the dura mater presenting on one side some of its characteristic appearances. viz: a white, smooth, glistening densely fibrous surface excepting certain spots which were of a darker color and evidently necrosed. The other surface presenting a fungating mass of sloughing tissue partially covered by foul-smelling purulent discharge. Thorough irrigation resulted in the expulsion of much fetid pus from region of lateral sinus.

December 30th—With the exception of light chill patient seemed a little better, as he has failed to-day to cough up his usual fetid expectoration, and the appearance of the posterior wound was much improved and discharge very much lessened.

December 31st.—Patient was seen to-day by Dr. Ernest Laplace of Philadelphia, who suggested the withdrawal of drainage tube in front part of wound, change of antiseptic lotion with the very best and most highly nutritious diet, expressing in a general way a hopeful view of a final recovery.

January 1st. 1898. —All drainage tubes having been removed and strips of iodoform gauze substituted, the patient who had for thirty-six hours previous been free from all cough, expectoration and fever was seized with another rigor sending the thermometer to 104° F. This last paroxysm was accompanied by vomiting and severe abdominal pains presumably caused by overfeeding. These symptoms were quickly relieved by an evacuation of the bowels by enema. Temperature and pulse were normal at midnight. This was the last febrile manifestation. From this time all wounds assumed a healthy cicatrizing action and patient proceeded rapidly towards recovery.

January 27th.—Patient was discharged from indoor department, and for some time continued to attend the outdoor clinic. Before sailing his hearing was carefully tested, and was as follows:

March 4th, 1898.	R. E.	L. E.
Watch	40 in.,	5 in.,
R. T.	(plus) - 35	12
L. V.	20	18
Whisper	20	10

Weber St. Test. Lateralized in left ear.

I read letter from the patient written from Liverpool, March 30th, acknowledging himself in perfect state of general health with complete recovery of local affection.

THE TECHNIC OF LARYNGECTOMY.*

W. W. KEEN, M. D.,

PHILADELPHIA, PA.

The operation advised by the author, is as follows:

1. The general preparation of the patient is the same as for any other operation.

2. As in all cases about the mouth, nose, pharynx and larynx, he is particularly careful to make a systematic attempt for two or three days beforehand to secure at least partial disinfection. While partial disinfection is not as good as complete, yet the results in the treatment of fractures of the base of the skull, in the extirpation of rectal tumors, etc., shows its great value. The teeth are very carefully cleansed with the tooth brush. If there are any old stumps of teeth present, it is better that they be extracted, and the operation deferred a few days until these wounds heal. For two or three days before the operation, every two hours, when the patient is awake, the mouth and each nostril should be sprayed with a solution of boracic acid, listerine, or both.

3. Nearly all authors recommend a tracheotomy either as the first step of the operation, or more frequently 10 to 14 days before the operation. In the few cases in which dyspnea is great, he should be disposed to do a tracheotomy, say two weeks before the laryngectomy, not, however, with a view of preventing the entrance of blood and wound fluids into the lungs by the introduction of a tampon canula, but for the purpose of improving the general condition of the patient. In the case which is the basis of this article, he did a tracheotomy at the time of the operation, but removed the tracheal tube at the termination of the laryngectomy, immediately closed the wound in the trachea, and obtained absolutely primary union. In any future case, he is strongly of the opinion that it would be better to omit tracheotomy entirely. It was not in his

*Abstracted by W. Scheppegrell, M. D., New Orleans, La.

opinion needful, and by omitting it we would eliminate one cause of septic pneumonia.

4. The entire operation, after the trachea is invaded, is done with the patient in the Trendelenburg position. He is quite persuaded that the majority of surgeons do not appreciate to its full the advantages which this posture possesses in all operations about the upper air-passages. As he stated in a previous paper, he employs it in epithelioma of the lip, in extirpation or other operation on the upper and lower jaw, in removal of the tongue, in cleft palate, in operations on the tonsils, and pharynx, and all similar operations. Blood will not run uphill any more than water, hence, if we employ this posture in laryngectomy, we would avoid one of the chief reasons for tracheotomy and the employment of a tampon canula.

The disadvantages of tampon canulae are very great. Kocher, as also the author, has lately dispensed with them entirely. The three most commonly used are those of Trendelenburg. Hahn and Gerster. Of the three, Gerster's is distinctly, in his opinion, the best. It can be more accurately adapted to larynges of varying sizes, and is much less likely to injure the parts either by undue pressure or by difficulty of introduction. In one case, in his attempts to introduce a Hahn canula, the rings of the trachea were considerably torn. The objections to Trendelenburg's canula are: arrest of respiration, which sometimes follows its introduction, the production of pressure gangrene in the trachea, obstruction in the lower end of the tube by the rubber ballooning into the trachea beneath it, the bursting of the rubber or its being cut, and if none of these mishaps occur, the air often gradually escapes and thus renders it useless.

The Hahn canula cannot be made aseptic as easily as others, and, as Lennox Browne has pointed out, it sometimes requires 20 minutes for the expansion of the sponge after its introduction.

5. Anesthesia is at first done through the mouth, and is so continued until the larynx or trachea is invaded. A large tracheotomy canula (12 millimeters in diameter) is then introduced and held in place by disinfected tapes tied around the neck. The inner tube of this canula is removed and the metal tube of a Hahn canula, which precisely fits

it, introduced. A rubber tube connects this with the ordinary funnel for the administration of chloroform.

6. The operation proper, the author described in connection with the case reported, and then indicated the improvement in technic which he purposed adopting in the next case he might have.

7. The after-treatment: (1) Posture; the patient is kept in the Trendelenburg position by placing a chair under the foot of the bed. This position prevents any wound fluids from running down (or rather up) into the lung. This position is to be maintained for a day. On the second day the bed is lowered to the horizontal plane; on the third, he is allowed to sit up in bed on a bed-rest; on the fourth, to get out of bed and sit in a reclining chair, and on the fifth day he may walk about the room.

(2.) Food: For two days nutritive enemata only are to be used. After that a teaspoonful of liquid food is given, at first every half-hour always followed by a tablespoonful of sterile water to wash away any food that might possibly leak into the laryngeal wound. At the end of a week, full diet as to quantity may be given, but no solid food until the tenth day. A catheter or esophageal tube is not required. In his case, the patient could swallow from the first.

(3.) Dressing: The primary dressing was described in connection with the case. On the day after the operation the small gauze drain should be removed. Half of the stitches may be taken out on the fourth day, and the remainder on the sixth day. In the case reported, the temperature on the day after the operation rose to 101.8 F., and fluctuated between the normal and 101 for a week, when it fell to the normal.

8. The only objection to this method which occurred to the author, is that it absolutely precludes the use of any artificial larynx. The possession of voice is nothing when compared with a speedy recovery and a greatly diminished danger of a fatal result. Rutsch believes such a larynx is very unsatisfactory. The author realizes the fact that one case does not prove the value of any method, but its advantages were so striking in this case that, as laryngectomy is a relatively rare operation, he has not thought it best to wait until he could accumulate a much larger

experience before bringing it to the attention of the profession.

9. As to the final results, it is too early to draw any inference as to the possibility of recurrence in this particular case, nor is this his purpose. His intention is rather to demonstrate a method of laryngectomy which would diminish more especially the immediate mortality of the operation, and secure speedy recovery by primary union. The chances of recurrence are no greater, nor yet any less, after operation by this method than by any other. This particular patient has been able to go out in all weathers during the past extremely severe winter, though living as far north as Waterville, Me., where the thermometer has been many degrees below zero.

ABSTRACTS FROM CURRENT OTOLOGICAL, RHINOLOGICAL AND LARYNGOLOGICAL LITERATURE.

I.—EAR.

Cerebellar Abscess of Otitic Origin. Autopsy.

McConachie and Hartwig. *(Journ. Amer. Med. Assn.,* April 8, 1899.) The diagnosis of cerebellar abscess must be founded upon the complexity of the symptoms, viz., severe headache, nausea, vomiting, vertigo, a staggering gait, facial paralysis, choked discs with retinitis, slow pulse, temperature low, slowing of the respiration, Cheyne-Stokes' respiration, yawning, slowness of cerebration and general apathy, irritability, intolerence to light, delirium, rigidity of the neck, motor and sensory paralysis.

When complications exist, as sinus thrombosis and leptomeningitis, other symptoms supervene to make the diagnosis difficult. In sinus thrombosis, rigors and chills with high temperature and increased heart action are almost invariably present, with tenderness along the course of the jugular vein. In leptomeningitis, there are high temperature, rapid pulse, general irritability and marked acuteness of special senses.

A cerebellar abscess usually terminates in death when operative procedures are not used. The abscess contents escape and a new inflammatory action is set up. Abscesses have become encapsulated and remained quiescent for years without giving rise to serious trouble, but such cases are rare. We should never anticipate such a result. Our duty is to operate early if a successful result is to be hoped for. The time to operate is when we have made our differential diagnosis—a deep problem and sometimes a very speculative one. *Scheppegrell.*

Drumhead Perforations, Their Site and Significance.

Potts, B. H. and Randall, B. A. *(Jour. Amer. Med. Assn.,* March 4, 1899.) The authors give the result of a study of 1000 cases from private and clinical record books. The findings are conflicting with those of Moos and other writers, in that the large proportion of perforations was found in the lower posterior quadrant. The statistics of actual perforations from private records, with their larger proportion of acute cases, have as large a per cent. in the upper and posterior quadrant (38 per cent.), which is 16.4 per cent. of the whole number.

As regards the Shrapnell region above the short process, the presence here of a discharging opening in the flaccid membrane in 10 per cent. of the cases fully accords with what has been urged as to the dry pin hole "foramen of Rivinus" at this point, which has been found in 25 per cent. of adults, although never seen in early life. Bad as are the severe cases of attic and antrum disease with perforations at these points, there is no sense in including under their ill-repute the many cases which are of no exceptional severity and in no need of radical treatment by operation.

The anterior-inferior perforation, commonly stated to be the most usual, appears in but 25 per cent. of the cases. The authors take a very conservative view of the perforation of Shrapnell's membrane. When one sees such cases by the hundred and secures more satisfactory results without operation than reliable men gain from their ossicular incisions, it is easy to see that more room yet remains for conservative and thorough-going treatment. *Scheppegrell.*

Mastoid Complications of the Exanthemata of Children.

DENCH, E. B. (*Pediatrics*, Vol. VII, No. 12, 1899.) The two diseases in which aural complications are most frequent are measles and scarlet fever. In the former disease, the complication is often of a mild character, yielding readily to treatment, whereas in the latter it is not only extremely common, but progresses with great rapidity, and extension of the inflammatory process to the mastoid can be prevented only by the most radical measures in the early stage of middle ear inflammation.

Where post-auricular abscess develops, simple incisions through the soft part is not sufficient. Even in young children the mastoid antrum should be entered in every instance, as the cranial bones are very thin and infection of the intra cranial structures may occur through the external surfaces of the temporal bone, as well as through the tympanic roof or through the posterior wall of the mastoid antrum. *Scheppegrell.*

Fifty-One Mastoid Cases.

GREENE, D. M. *(Journ. Amer. Med. Assn.,* May 20, 1899.) An interesting synopsis of 50 operations, including four brain abscesses and one perforation of the sigmoid sinus. *Scheppegrell.*

Acute Inflammation of the Tympanic Cavity.

SMITH, S. M. *(Phil. Med. Journal,* May 6, 1899.) The most common cause is extension by continuity of a catarrhal inflammation from the throat or nasopharynx. Treatment of these parts should therefore not be neglected. The various degrees of inflammation are then outlined, and their respective treatment described. *Scheppegrell.*

A Modified Siegle's Pneumatic Aural Speculum.

BURNETT, CHAS. H. *(Journ. Amer. Med. Assn.,* June 3, 1899.) This instrument is practically a Gruber speculum made of metal, to which is fitted a glazed lid that transforms it into the Siegle pneumatic speculum. It is nickel-plated both within and without. There are two small openings on its inner wall, at the point connecting it with the air tube, which act as a sieve to prevent the drawing up of particles of cerumen or dirt into the mouth of the operator. *Scheppegrell.*

Syphilitic Perichondritis of the Auricle

PACKARD, F. R. *(Phil. Med. Journal,* June 24, 1899.) Perichondritis auriculae of syphilitic origin is quite rare, as literature on this subject shows but few cases. The case reported was that of a negro of 25 years, who presented all the typical symptoms and who was promptly cured under specific treatment. *Scheppegrell.*

A New and Successful Treatment for Certain Forms of Headache, Deafness and Tinnitus Aurium.

VANSANT, E. L. *(Journ. Amer. Med. Assn.,* June 24, 1899.) The clinical notes of 18 cases form the basis of a report which show the remarkably prompt and prominent result from the simple expedient of syringing the nasal accessory sinuses and the Eustachian tube with a stream of dry hot air under pressure. The relief in many cases was immediate, and headaches of a month's duration were relieved after a single treatment of a few minutes with hot air.

The effect of injections of dry hot air into the middle ear, for the relief of tinnitus and the improvement of the hearing was very marked. Better results were obtained in cases where catarrhal rather than sclerotic changes were present in the middle ear and Eustachian tube. The instrument used in these cases is like that employed by the dentist for a similar purpose. *Scheppegrell.*

Mastoiditis, Diagnosis and Treatment.

DAVIDSON, JNO. P. *(Journal Amer. Med. Assn., July 15, 1899.)* A synopsis of the diagnosis and treatment of mastoid inflammation. Hot applications, leeches and Wild's incision are condemned. *Scheppegrell.*

Catheter Inflation of the Tympanum; Its Value and Technique.

RANDALL, B. A. *(Journ. Amer. Med. Assn., May 27, 1899.)* A careful description of the technic and advantages of this valuable procedure. *Scheppegrell.*

Deaf-Mutes, with Clinic.

STAPLER, M. M. *(Jour. Amer. Med. Assn., May 27, 1899.)* The author believes that he has obtained good results by cauterizing the fibres of the tensor palati muscles which preside over the opening of the Eustachian tube. *Scheppegrell.*

Hydrochloric Acid Application to the Bony Walls of the Tympanic Cavity and Meatus.

ARD. *(Journ. Amer. Med. Assn., July 22, 1899.)* This remedy was first proposed by Dr. Bull, of Christina. He believes that the treatment is not indicated if the ossicles are diseased. When dead bone is visible, he applies cotton soaked in the acid, four per cent., and leaves it in contact. In a cavity, he introduces the cotton into it, removing it the next day, these applications being made a week apart. The acid gradually decalcifies the affected bone and acts like curettage. It is a strong antiseptic and cured from one-third to one-half of all cases treated.
Scheppegrell.

The Pathology of Catarrhal Deafness.

HINKEL. F. W. *(Phil. Med. Journal,* July 15. 1899.) The hypertrophic form of middle-ear disease is responsible for a large percentage of cases of chronic catarrhal deafness. Early treatment is generally successful. Proper attention to throat affections in childhood will usually prevent these complications. *Scheppegrell.*

Closing Perforated Ear Drums.

PELTESOHN. *(Journ. Amer. Med. Assn.,* July 22, 1899.) A description of the method of Okuneff in applying trichloracetic acid to the margin of the perforation.

Scheppegrell

Chronic Middle-Ear Suppuration, with Permanent Retro-Auricular Opening; Radical Operation.

GLEASON. E. B. *(Journ. Amer. Med. Assn..* June 10. 1899.) The case reported was complicated by permanent retro-auricular opening. A radical operation was performed and the antrum found packed with cholesteatomatous material. *Scheppegrell.*

Affections of the External Ear.

RANDALL, B. A. *(Journ. Amer. Med. Assn..* March 4. 1899.) External affections make up 25 per cent. of all ear diseases, one-half of this number consisting of impacted cerumen—about 14 per cent. Of the remainder, diffused inflammation of the auricle and canal, which may be called eczematous, makes up about five per cent., and furunculosis two per cent., leaving for remaining diseases about four per cent. *Scheppegrell.*

Suppurative Disease of the Ear; The Presence of Polypi and Granulations Therein not an Unfavorable Indication.

LAUTENBACH. L. J. *(Journ. Amer. Med. Assn..* March 4. 1899.) It is not intended to infer that in every case of polypoid complication the results are better than in those where neoplasms do not exist. Yet the average amount of benefit derived in the latter class of cases is much greater than in the former. Granulations form an evidence of the activity of the inflammatory action in a benign manner and exhibit an effort of nature to extend her healing processes to as great a measure as possible. They allow the retention of necessary vascularity, the nerve innervation remaining, until such time as through artificial or natural means everything is ready for a quick healing, and all we need is to take advantage of nature's open door. *Scheppegrell.*

Adenoid Growths, Their Relation to Deaf-Mutism.

GETCHELL, A. C. *(Journ. Amer. Med. Assn.,* March 4, 1899.) In three cases of deaf-mutes operated upon, there was improvement in the hearing of the first, but not in the other two. Adenoid growths and enlarged tonsils should be removed from deaf-mutes if there is any probability that they contribute to the deafness, if they make more difficult the acquirement of spoken language by the oral method, and to improve the general health of the child.
Scheppegrell.

Necessity of Early Recognition and Treatment of Non-Exudative Inflammation of the Middle Ear.

PILGRIM, M. F. *(Phil. Med. Journal,* March 11, 1899.) The author begins his article with an excellent criticism of the methods of patients who call on the aurist during the interim of a short visit to the city, and expect to be cured by some magic process of an affection of 20 years' standing before their return home.

In non-exudative inflammation of the middle ear, he has had good results from treatment with aural massage, and believes that it should be given a fair trial. He very properly objects to the criticisms of aurists who have never given the method a trial and who are therefore not competent to judge of its merits. *Scheppegrell.*

Chronic Catarrhal Otitis; Treatment by Oto-Massage.

DUFOUR, C. R. *(Jour. Amer. Med. Assn.,* April 15, 1899.) The author has had good results from the use of oto-massage, especially with the electric pneumo-masseur. Two or three applications should be made weekly.
Scheppegrell.

Sarcoma of the Middle Ear.

BROSE, Evansville, Ind. *(Archives of Otology,* Vol. XXVIII, Nos. 2 and 3.) A girl, aged 3 1-2, seven months before coming under the author's observation first complained of earache. Two months later she again had earache which this time was followed by a discharge. A few weeks later the mother detected in the right ear a small reddish growth which slowly became larger. A fetid discharge filled the auditory canal and the supposed polyp was removed with a snare. After the growth had returned and had been removed a number of times, there was observed an enlargement over the mastoid which rapidly increased in size. This swelling had a soft doughy consistence, there was no redness and it was not painful on pressure. Operation was advised and undertaken. The soft reddish-gray fleshy mass over the mastoid was removed by the sharp curette. The outer table of the mastoid was eroded and with the curette the antrum was freely exposed. The external ear polyp on microscopic examination proved to be a small round and spindle-cell sarcoma. Recurrence again speedily took place and the child died in convulsions about nine months from the date of the primary earache. *Campbell.*

A Contribution to the Statistics of the Dangerous Complication of Suppurative Ear Diseases and of Operation on the Mastoid Process.

TEICHMANN, Berlin. (*Archives of Otology*, Vol. XXVIII. Nos. 2 and 3.) Of 1750 cases more than half (56.6 per cent.) belong to the periods 6 to 30 years while 71.8 per cent. belong to the years 0 to 30. According to the returns three-quarters of the dangerous complications follow chronic purulent otitis and only one-quarter the acute. The suppuration in influenza-otitis almost always runs an acute course up to the inception of the dangerous complication, while in the otitis of scrofula, tuberculosis, diphtheria and measles, it runs a chronic course.

Campbell.

Influence of Sea Climate and of Surf Bathing on Aural Affections and Hyperplasia of the Pharyngeal Tonsil.

KÖRNER, Rostock. (*Archives of Otology*, Vol. XXVIII. Nos. 2 and 3.) The author's opportunities were not sufficiently great to form conclusions of any value.

13 children with very large hyperplasia of the pharyngeal tonsil gained 1778 grms.
15 " a " " " 2142 "
18 " a moderate " " 1741 "
98 " without any " " 2183 "

Campbell.

The Magnifier in Otoscopy.

BOENNINGHAUS, Breslau. *Archives of Otology*, Vol. XXVIII, Nos. 2 and 3.) Its advantages are, enlargement of the image of the tympanic membrane, stronger illumination of the same, generally and especially of its extreme edges (prismatic effect) and bringing out more clearly the dimention of depth.

The lens the author uses has a focal distance of $7\frac{1}{2}$ cm., that is of 13 dioptries, which, when using the reflector, produces a magnification of about $2\frac{1}{2}$ times. *Campbell.*

Intestinal Disturbances Produced by Otitis Media of Infants.

HARTMANN, Berlin. *Archives of Otology*, Vol. XXVIII, Nos. 2 and 3.) From recorded observations the author deduces the following:

1. Acute febrile otitis causes a diminution in weight or arrest of increase in weight.
2. Otitis accompanied by grave septic symptoms probably causes diarrhea.
3. An acute febrile otitis occurring during intestinal diseases may act upon the general constitution, and by reducing the vitality, aggravate the intestinal affection, or retard recovery.
4. Whether there is a direct relation between atrophy and an otitis, must be reserved for further observations.

Campbell.

Percussion of the Mastoid Process.

EULENSTEIN, Frankfurt. *Archives of Otology*, Vol. XXVIII, Nos. 2 and 3.) The author in 1894 published the results of the examination of ten cases of acute disease of the mastoid, in which he arrived at the following conclusions:

I. By means of percussion (compared with that of the other side) a positive diagnosis of a diseased condition of the mastoid can be made—provided dullness is elicited.

II. Dullness on percussion indicates the presence of a diseased area near the surface of the bone, the degree of dullness depending upon the extent of the area involved.

III. The absence of dullness is no proof that the bone is not diseased.

IV. Where other symptoms of mastoid disease are present and there is no dullness on percussion it indicates that the diseased area is either very small or deep seated.

In this paper, he gives the histories of ten more cases, which support his earlier conclusions, and he states, that by percussion, we are enabled to recognize mastoid disease earlier, and that it is a valuable adjunct to the indications for opening the mastoid. *Campbell.*

Two Cases of Otitic Sinus Thrombosis.

KNAPP. New York. (*Archives of Otology.* Vol. XXVIII, Nos. 2 and 3.) CASE I. A girl, aged ten, with a history of intermittent otorrhea since her first year. She complains of pain in the left ear, forehead and occiput, has vomited and had a shaking chill followed by fever. There is hyperesthesia of the skin and she is constipated.

Mastoid on both sides appears normal, free from redness, the left slightly larger than the right.

Fetid pus comes from the left ear. On this side the region below the mastoid and in front of sterno-mastoid muscle is very painful to the touch. Temperature 105 F., pulse 120, respiration 40 to 45. Choked disc readily seen in both eyes.

The mastoid antrum and middle ear were opened and the attic cleansed with a sharp spoon. The ossicles were absent. The wound was extended by chiselling the bone in a backward direction and at a depth of 3 to 4 mm. the lateral sinus was reached. It was deep-black and plainly pulsating. A hypodermic syringe withdrew dark blood without odor. The wound was packed with sterile gauze and the child put to bed. Four days later when dressings were changed a large portion of necrosed sinus and contents with gangrenous odor were removed. Fresh blood came from above, not from below. Cough, pain in back. Temperature 105° F., pulse 145, respiration rapid and laborious. On the sixth day she died.

On autopsy the sinuses were found filled with dark blood, the sigmoid sinus was destroyed. The jugular bulb was filled with a dirty whitish-yellow clot, which, toward the beginning of the vein, gradually contracted, adhering to the wall. Immediately below the clot, the jugular vein was empty, with smooth walls and normal calibre, but thence it rapidly contracted to a narrow tube with an even diameter of 2 to 3 mm. down the whole length of the neck. The walls of the tube had the thickness of an artery of small calibre, its inner surface was smooth but the lumen was interrupted by round greyish pellets at intervals of 2 to 2.5 cm., adherent to the walls of the vein. They had the appearance of coagulated fibrin.

CASE 2. A man, aged 18, who had occasional severe attacks of earache as far back as he could remember. On a number of occasions growths removed from the right ear. A mastoid operation had been performed, the antrum exposed and the posterior wall of the ear canal removed. This operation wound never fully healed, a fistulous canal led deep into the substance of the bone. At this time he came under the author's care, the Mt. was gone. Bare bone was felt at the bottom of the middle ear, the probe passed under the lateral wall of the attic and considerably backward. Temperature 100 F. The right ear was totally deaf. The diagnosis was: Extensive chronic caries of the mastoid, attic and tympanum extending into the labyrinth.

A radical operation was done, removing all carious bone. The patient did well for three weeks, when general symptoms of pyemia with articular metastases developed, but there was freedom of lungs and meninges.

Two further operations were done and in the last a puriform thrombus was removed from the transverse sinus. The patient made a slow but complete recovery.

The author makes mention of the well known fact that sinus thrombosis which commonly show articular metastases gives a better prognosis than thrombosis with pulmonary metastases. *Campbell.*

Fracture of Malleus and Annulus Tympanicus.

ALLPORT. (*Archives of Otology.* Vol. XXVIII, Nos. 2 and 3.) The author reports a case of rupture of Mt., fracture of the handle of the malleus and fracture of annulus tympanicus caused by being thrown from a carriage. *Campbell.*

The Operation for Otitic Brain-abscess with Special Reference to its Clinical Value.

ROPKE. (*Archives of Otology,* Vol. XXVIII, Nos. 2 and 3.) The author reviews the literature, describes the technic of the operation and gives the results of 141 cases operated on which he has been able to collect. Of these 141 cases 57, or 40.4 per cent., were permanently cured.

Of 26 cases following acute otitic 11, or 42.3 per cent., were cured. Of 109 cases following chronic otitic 47, or 43.1 per cent., were cured. Hence the prognosis of the operation is about the same in either acute or chronic cases.

The symptoms which give the indications for the operation do not help as much as regards prognosis. Cases with normal or subnormal temperature are, in general, more favorable than those which run a violent course as the virulence of the infection is likely to be less and there is less likelihood of complications.

The presence of focal symptoms before operation is of no value in estimating the future course.

Site and size of abscesses are very important for prognosis. Small abscesses are situated usually near the surface of the brain and naturally offer a better prognosis than those larger and deeper seated.

The contents and walls of the abscess play an important role. Color and odor of pus gives us no clue as to its virulency. Bacteriologic and microscopic examination will reveal whether or not it contains pathogenic germs.

Cases where deep depressions and hollows exist in the walls of the abscess with no living membrane are unfavorable.

Eighty-one of the cases were operated on through the squama. Of these 38.3 per cent. recovered.

Forty-three operated on through the tegmen 40.2 per cent. were cured, and 7 cases where the combined opening was made 70.1 per cent. recovered. *Campbell.*

Extensive Laceration of the Auricle and Complete Section of the External Auditory Canal, with Partial Detachment of the Sterno-cleido-mastoideus Tendon and Splintering of the Tip of the Mastoid by a Blow from a Brick. Operation for Restoration of the Auricle and Canal.

BURNETT. (*Archives of Otology.* Vol. XXVIII, Nos. 2 and 3.) The author comments on the relative infrequency of injuries to the auricle. In all injuries to the external auditory meatus, one must preserve the lumen of the canal. *Campbell.*

II.—NOSE AND NASO-PHARYNX.

On the Importance of Nosebleed as an Early Sign of Softening of the Brain, with Consideration of the Relations of Both Diseases to Arterio-Sclerosis.

KOMPE, CARL, DR., Friedrichroda, Germany. *(Fraenkel's Arch., IX, 2, 181.)* Softening of the brain (encephalomalacia) is, to-day, generally ascribed to local anemia of the brain tissue, caused by occlusion of arteries by thrombosis or embolism. The etiologic factor is either arteriosclerosis or specific endarteritis. Only the former is the considered in this article. Before the first indications of sclerosis of the cerebral vessels appear, sometimes premonitory symptoms are observed, which give warning of the early approach of softening due to arterio-sclerosis. To them belong all the early signs of arterio-sclerosis at the heart and periphery (cardiac hypertrophy with aortic changes, tense radial artery, tortuous temporal arteries, etc.)

Sclerosis of the vessels of the brain may be accompanied by the same condition in the vessels of the nose, as they are, partly, branches of the same main artery. As sclerosed blood vessels always rupture *in front* of the thickened and stenosed portion of the vessel, the reason why the hemorrhage must be copious is thus explained: a thrombus not forming, and occlusion not being possible on account of the loss of contractility of the walls. Furthermore, the vessel is not completely severed, but only opened on one side, which always leads to more dangerous hemorrhages. In the author's five cases, the severe nosebleed occurred when arterio-sclerosis had not yet been diagnosticated. It is immaterial if the hemorrhage was arterial or venous as the veins also may be sclerosed.

Spontaneous nosebleed in individuals above forty years of age, which cannot be traced to one of the well recognized local causes, is a suspicious sign of general arteriosclerosis, and calls for a consideration of all symptoms of sclerosis of the carotid artery or, rather, of the vessels of the brain. If the ophthalmoscope confirms this, a fairly positive diagnosis of partially advanced sclerosis of the brain vessels may be made; and from it, of incipient softening of the brain. Incipient arterio-sclerosis can be influenced, as far as the cellular infiltration is concerned.

but not in regard to the retrogressive process in the vessel walls. Auchard recommends daily doses of 1 to 3 grammes of sodium iodide and of potassium iodide alternately. Vierordt has had surprising results from iodides, especially in sclerosis of the coronary arteries. Hygienic treatment must be used in addition. *Morgenthau.*

On the Removal of Posterior Hypertrophies of the Lower Turbinal.

OSTMANN, PROFESSOR, Marburg, Germany. (*Fraenkel's Archiv.* IX, 2, 200.) In the removal of hypertrophies by means of the cold snare, it is not uncommon to meet with occasional not inconsiderable hemorrhages as well as the difficulty, in some cases, in looping the wire over the base of the hypertrophy; to facilitate the operation by introducing the finger into the naso-pharynx is certainly very unpleasant to the patient. If the hypertrophic tissue be unexpectedly tough, the cold wire may not be able to sever it. There is the same difficulty in grasping the hypertrophy with the galvano-caustic snare; bleeding is here avoided by drawing the wire home slowly and gradually, but the rather large eschar in the neighborhood of the pharyngeal tubal ostium may infect the middle ear, and lead to a, usually severe, middle-ear suppuration. Combining galvano-cauterization with the cold snare is to be preferred to these two methods, because it is done more easily, and does not endanger the ear nearly so much. Hypertrophy of the posterior ends of the lower turbinals large enough to demand removal on account of the respiration is, usually, accompanied by such thickening of the rest of the mucous membrane of the lower turbinal that it is best treated by galvano-cauterization. This method serves both ends, thus permitting treatment of the case in one sitting. By painting with 10 per cent. cocain solution, the anterior two-thirds of the lower and as narrow a strip as possible of the hypertrophy are anesthetized in order not to have too much shrinking of the tumor.

With a flat cautery point a curvilinear furrow is drawn, from behind forward, immediately above the lower margin of the turbinated bone, down to the bone. In order to avoid all bleeding, it is advantageous not to go down to the bony tissue at once, but to cauterize once or twice in the same furrow. If necessary, a second or third line may be drawn parallel to the horizontal portion of the first. The first cauterization separates the hypertrophy from its base just where the principal vessels enter; and it sinks by its own weight. If the cold snare, bent a little to the outside, be then introduced through the nose and, in drawing it back again, be turned with the curvature looking upward and outward or directly upward, the tumor can easily be engaged. As the wire is drawn home, it glides into the galvano-caustic furrow. After leaving the loop drawn taut for about a minute, even the toughest hypertrophies can generally be severed without loss of blood. *Morgenthau.*

Nasal Catarrh in Children Its Cause and Treatment.

RICE, C. C. *(Medical News,* April, 1899.) Nasal catarrh in children is most frequently due to abnormal conditions in the nasal chambers and naso-pharynx, especially to adenoids. The technique of adenotomy is given in detail and the usual treatment of these cases.
Scheppegrell

Polyuria and Incontinence of Urine; Symptoms of Adenoids.

HUBER, F. *(Archives of Pediatrics,* Apr., 1899.) A description of the cause and effect in these cases, and the obvious remedies. *Scheppegrell.*

Non-Diphtheritic Pseudomembraneous Rhinitis.

PRICE-BROWN, J. *(Journ. Amer. Med. Assn.,* May 6, 1898.) In simple membraneous rhinitis symptoms, with the exception of nasal obstruction, are largely absent. The disease is frequently unilateral. The author reports a case of a woman of 25 years, suffering from hay fever, in whom he cauterized both inferior turbinals at one sitting. Twenty-four hours later both nares were stenosed and the lining of the passages was covered with false membrane. Two years later the operation was repeated with similar result.

The author concludes that non-diphtheritic pseudomembraneous rhinitis does sometimes occur, and, though a very rare disease, it is probably as frequent as primary nasal diphtheria. On clinical grounds alone it is possible in the majority of cases to distinguish it from genuine diphtheritic diseases. Owing to a possible mistake in diagnosis, isolation in all cases should be imperative until a reliable bacteriologic examination can be made.

Scheppegrell.

Disfigurements of the Nose and Mouth, and their Surgical Treatment.

ROBERTS, J. B. *(Journ. Amer. Med. Assn.,* May 13, 1899.) The saddle-back nose may be improved by making an incision through the skin and slipping a plano-convex piece of celluloid under the skin so as to fill up the hollow and make the line of the dorsum of the nose straight or nearly so. In such cases it is preferable to separate a portion of the nasal process from the upper maxillary bone by means of a chisel introduced through the nostril. If this is done on both sides of the nose, the pieces of bone may be pressed upward and forward so as to elevate the bridge of the nose, and held in this position until union occurs by pins thrust transversely under them.

An angular nose is remedied by an incision over the bridge and chiseling away the redundant bone or cartilage. It is preferable to make the incision a little to one side of the median line of the nose.

Twisted and crooked noses can be corrected by remedying the deflected septum by intranasal operation, separating the soft tissues of the nose very freely from their bony support, bending or twisting the nose into a better shape, and retaining it in this position by means of pins or plugs. *Scheppegrell.*

Asthma of Nasal Origin and its Radical Cure.

PAYNE. R. W. *(Pac. Medical Journal,* May, 1899.) In a number of cases, the author has been enabled to relieve a paroxysm of asthma by applying a solution of cocain to the nostrils, which illustrates the influence of nasal irritation in developing these paroxysms. In cases of asthma, the nasal cavities should be carefully examined, and where pathologic conditions exist, they should be corrected. The majority of cases are thus benefited, if not cured.

Scheppegrell.

The Influence of Turbinal Hypertrophy upon the Pharynx.

SOMERS, L. S. *(University Med. Magazine,* May, 1899.) A review of the effects of mouth-breathing in developing pharyngeal disease. *Scheppegrell.*

Empyema of the Antrum.

OHLS. H. G. *(Journ. Amer. Med. Assn.,* July 15, 1899.) After a careful review of the etiology, pathology, symptomatology and diagnosis of antral sinusitis, the author reports in detail two cases, the first a personal experience of a mild acute catarrhal sinusitis, and the other a typical severe inflammation due to a root-abscess and followed by recurring frontal headaches. The article concludes with a careful bibliography of the subject. *Scheppegrell.*

Treatment of Chronic Empyema of the Accessory Sinuses.

STOUT, GEO. C. *(Journ. Amer. Med. Assn.,* June 24, 1899.) Caries of the teeth is more frequently the result than the cause of antral disease. A radical operation is indicated in the majority of cases, although many cases of both frontal and maxillary sinusitis are curable by parcentesis through the nasal chambers. Whatever operation is selected, thorough drainage is the chief object sought in the treatment. *Scheppegrell*

Some Points on the Symptomatology, Pathology and Treatment of Diseases of the Sinuses Adjacent and Secondary to the Orbit.

BULL, C. S. *(Medical Record,* July 15, 1899.) Empyema of the antrum is usually the result of dental or alveolar disease, although it may follow inflammation of any of the adjoining parts. Treatment in all cases of empyema should consist of free opening, curetting, irrigation and drainage. Of malignant growths, sarcoma is the most frequent. *Scheppegrell.*

Urticaria and Odors.

JOAL. *(Revue Hebd. de Laryngologie, etc.,* June 10, 1899.) Three cases are reported in which certain odors produced urticaria. In one, the odors from aromatic essences used in the manufacture of liquors; in another, of iodoform, and in the third, the odor from roses, lilacs and hyacinths, accompanied in the two last cases with symptoms of hay fever. *Scheppegrell.*

Nasal Bacteria, the Relation they Bear to Disease.

KYLE, D. BRADEN. *(Journ. Amer. Med. Assn.,* June 10, 1899.) The results obtained conform with the results published by Drs. Park and Wright. The author believes that there is no doubt of the fact that the accumulation of secretion in the nasal cavity forms a favorable nidus for bacterial proliferation. *Scheppegrell.*

Arm-to-Arm Vaccination; Nasopharyngeal Syphilis a Result Thereof.

DONNELLAN, P. S. *(Journ. Amer. Med. Assn.,* June 10, 1899.) At the age of five years the patient was vaccinated on the right arm by arm-to-arm vaccination. A few months later the submaxillary gland on the same side became enlarged, but did not suppurate. Four years later, examination showed loss of both nasal bones, with arrest of development of the cartilaginous septum, which resulted in the saddle-shape deformity of the nose. The mucous membrane of the nose and throat showed the effect of syphilitic disease. *Scheppegrell.*

Hypertrophic Rhinitis in Female Patients at Puberty, and the Abuse of the Cautery in the Treatment.

ROBINSON, W. K. *(Journ. Amer. Med. Assn.,* May 27, 1899.) It is well known that women with healthy nasal organs suffer from engorgement of the turbinated corpora cavernosa at each menstrual period. The author gives the history of three cases in which atrophy in the nasal chambers resulted from cauterization during the development of puberty. *Scheppegrell.*

Malignant Disease of the Nose and Throat.

WRIGHT, JONATHAN. *(Journ. Amer. Med. Assn.,* June 10, 1899.) There appears to be some exception to the usual fatality of malignant disease, since the author has observed several instances of recovery without operation in which the diagnosis of malignancy had been made by competent authorities. It is possible that in these cases there were errors in the microscopic examination. It is a fact, however, that such growths in the nose present a clinical picture entirely different from that met with in similar neoplasms treated in the larynx. Adenoma in the nose is rare, and when there usually becomes converted into adeno-carcinoma or adeno-sarcoma. Such tumors are usually of slow growth and would seem to present a favorable chance for operation. He has found that they are very apt to occur in the aged and feeble, and the operation is often a most difficult one. In many cases it is almost impossible to secure a fragment of tissue for microscopic examination. Clinically, there is very little difference between adeno-sarcoma and adeno-carcinoma. In some cases it is very difficult to distinguish between syphilitic granuloma and small round-cell sarcoma. the most malignant of tumors. The surgeon should, however, always bear in mind the possibility of the existence of syphilis, and in any doubtful case place the patient on iodid of potassium.

Sarcoma of the tonsil is by no means rare, and he had himself met with 44 cases of the small round-cell variety. When affecting the tonsil, it is sometimes very difficult to make the differential diagnosis between sarcoma and simple hypertrophy or syphilitic hypertrophy.

Scheppegrell.

Acute Rhinitis.

WARREN, W. *(Med. Age.* Jan., 1899.) Small doses of calomel are advised as being abortive in the early stage. In the later stage, when the patient cannot visit the physician's office, a powder consisting principally of cocain and zinc oleate is recommended, the preparation to be used every three or four hours.

[In view of the many dangers attending the use of such a drug as cocain in the hands of the patient, the method of prescribing it for the patient's own use should be severely condemned. S.] *Scheppegrell.*

The Histopathology of Hypertrophic Rhinitis in Children.

GOODALE J. L. *(Journ. Amer. Med. Assn.,* Mar. 11, 1899.) A histologic description of three cases of hypertrophic rhinitis in children. As the subject is one of much interest, the following case, which is essentially similar to the other two, is given with considerable detail:

The patient, a poorly nourished girl of 10 years, had suffered from nasal obstruction for one year. The adenoid tissues were not much enlarged, but the lower turbinals on both sides were enlarged and in contact with the septum. An examination with a probe showed that the hypertrophy consisted of the soft tissue covering the bone, the latter being unchanged in size. The color of the mucous membrane was much paler than usual; cocain caused but slight shrinkage; there was a moderate amount of mucus in the nasal passages.

A portion of the lower turbinate on the right was removed with the snare and hardened in Zenker's fluid. Under the microscope, sections showed a furrowing of the mucous membrane, which varied in thickness from three to four cell layers on its exposed surface to 15 or 20 cell layers along the side and bottom of the furrows. The cells were columnar in type and showed nothing abnormal in the cytoplasm or nucleus. The epithelium lining the furrows was ciliated, while that on the exposed free surface showed no cilia. In the intercellular spaces were a few scattered polymorphonuclear neutrophiles.

Below the mucous membrane was a loose connective tissue extending to the periosteum. The transverse diameter of this connective tissue was from three to four times greater than normal. It appeared as an irregular wide-meshed, exceedingly delicate fibrous net-work, in which were contained scattering clusters of glands, a few thin-walled, more or less tortuous blood vessels, here and there a large sinus, together with numbers of free lying cells. The relative proportion of these constituents varied, as it does normally, in different parts of the sections, the free lying cells being most numerous near the mucous membrane, while the glands are more deeply situated. Furthermore, at the extreme anterior end of the turbinate the reticulum is loosest and the blood vessels, sinuses, round cells and glands are fewest in number, while a short distance posteriorly the fibrous reticulum is more compact, and these structures are proportionately more numerous.

The free lying cells are most numerous just below the mucous membrane, but are nowhere abundant or gathered to form groups. Those nearest the endothelium of the blood vessels have the character of the small mono-nuclear leucocytes of the blood. In the immediate neighborhood, plasma cells are found in small numbers, together with cells intermediate in form and staining properties between small mono-nuclear leucocytes and plasma cells.

These two types of cells with the intermediate forms are

also found in fair numbers in the immediate neighborhood of the glands, and are as well sparingly scattered throughout the whole of the connective tissue. Small numbers of polymorphonuclear neutrophiles occur irregularly distributed, together with a few eosinophilic polynuclear leucocytes. A relatively large number of *mastzellen* are found, most numerous near the mucous membrane. These are comparatively small with a few small grabules.

The sinuses are sparingly found chiefly in the central portions of the connective tissue. They are elongated, oval in shape, are lined with endothelium, and contain a varying amount of amorphous granular material, together with, in some cases, red and white blood-corpuscles.

The blood vessels are few in number, traverse the connective tissue in all directions, and possess a wall of one or two layers of endothelial cells.

The glands found in this region present no noteworthy alterations. *Scheppegrell*.

Vertigo, Especially as Related to Nasal Disease.

STEIN. O. J. *(Phil. Med. Journal,* Jan. 7, 1899.) A man of 49 years suffered from repeated attacks of vertigo. Inspection showed enlargement of the turbinals, especially of the middle. Politzerization gave a slight hissing sound from the left ear, although no perforation could be detected. Turbinectomy of the right middle turbinal and subsequent cauterization of the inferior turbinal resulted in rapid cessation of the vertigo.

The author believes that the impulse from the nostril sent along the different branches of the trigeminal nerve is reflected along the vasomotor nerves, producing an alteration in blood pressure in that region of the brain presiding over equilibrium and co-ordination.

Scheppegrell.

Empyema of the Frontal Sinus and Intracranial Infection.

GIBSON, C. L. *(American Journal of Med. Sciences,* Mar., 1899.) A man of 32 years noticed a swelling at the anterior angle of his left orbit. Five days after its appearance it was lanced and pus evacuated. A sinus persisted, and a year later the left frontal sinus was trephined and drained externally, no attempt being made to establish drainage through the nose. The discharge continued and polypi were removed from the nostril at various times during the past two years.

The patient complained of persistent frontal headaches, more marked on the right side. He lost 15 pounds. Under ether anesthesia, a horizontal incision was made in the line of the left eyebrow, and the fistulous opening enlarged by means of a gouge. The cavity was freely scraped with a Volkmann spoon, which was found to easily penetrate into the right frontal sinus, permitting the escape of considerable pus. The probe was passed upward from the nose and a rubber drainage tube inserted, which was brought out through the external opening and through the nostril.

Two days after the operation, the patient developed stiffness of the neck and exhibited the symptoms of meningitis. Three days later there was remission of the symptoms, but a recurrence developed and the patient succumbed nine days after operation with all the evidences of septic meningitis. The autopsy showed intense congestion of the pia; moderate amount of pus in the pia at the base of the brain, especially over the pons and the cerebellum. The cerebral wall of the right frontal sinus was entirely wanting. The cribriform plate was carious, especially on the right side. An angular mass of bone (cerebral wall) projected about three-quarters of an inch above the site of the left frontal sinus, pressing laterally against the crista galli. The upper surface of this bony mass presented two openings, one small, the other half a centimeter posterior to the first and communicating with the frontal sinus.

A diagnosis of empyema of the frontal sinus, caries of the frontal bone, and purulent meningitis was made.

Scheppegrell.

Deformities of the Septum Narium, their Classification with a View to Treatment.

CASSELBERRY, W. E. *(Journ. Amer. Med. Assn.,* Mar. 4, 1899.) A convenient classification of the various forms of excrescences or spurs of the septum, deviations of the septum, combined excrescence and deviations of the septum, and the various forms of operation advisable for each condition. *Scheppegrell.*

The Climate of Colorado; its Influence upon the Nasal Mucous Membrane.

BLACK, MELVILLE. *(Journ. American Med. Assn.*, Marc 4, 1899.) While the climate of Colorado is of the greatest benefit to pulmonary diseases, it has not the same beneficial influence upon nasal affections. This is due to the fact that the dry atmosphere tends to develop a vascularity followed by hypertrophy and then by sclerosis, the latter process taking place more rapidly than in a moist atmosphere. Myxomatous degeneration, however, is rarely acquired. *Scheppegrell.*

Disease of the Frontal Sinus.

PFINGST, ADOLPH. *(Medical Age*, March 10, 1899.) After a résumé of the pathology of frontal sinusitis, a case is described of a man who received a blow on the left side of the vertex of the head. Soon afterwards he began to have dull pains in the frontal region of that side, and a month after the injury a spontaneous perforation took place just under the supraorbital ridge, a little to the nasal side of the orbit. A modification of the Ogston-Luc operation was performed, the opening being made below the supraorbital ridge instead of above it, on account of the presence of a fistulous opening at that point. It was followed by good results. *Scheppegrell.*

Eruptions on the Face Due to Nasal Pressure.

MURRAY, G. D. *(Medical Record*, March 25, 1899.) In skin diseases of the face where the origin is uncertain, a careful search should be made for some irritation in the nose. Independent of pathologic lesions due to specific germs, such as syphilis, erysipelas, etc., there are many conditions due to pressure only, as from a spur hypertrophied turbinal, etc., which may leave their impress on the skin of the face. Several cases are given in illustration. *Scheppegrell.*

Epistaxis. Simple and Efficient Means for its Control.

CORNICK, BOYD. *(Journal American Med. Assn.*, Mar. 2, 1899.) A dry plug of prepared sponge about the size and length of the little finger of a 12-year-old boy is trimmed with scissors, soaked in boiled water, squeezed dry from unnecessary fluid and inserted in its full length along the floor of the bleeding nostril. It should not be allowed to remain longer than 12 to 24 hours.

Scheppegrell.

Adenoid Vegetations, How and When Shall we Operate for Them.

MAYER, EMIL. *(Journ. Amer. Med. Assn.,* March 4, 1899.) The Schleich mixture No. I is preferred for the anesthesia, the head of the patient being in a dependent position. *Scheppegrell.*

The Treatment of Ozena, with Special Reference to Cupric Electrolysis.

MCBRIDE. *(Edinburgh Medical Journal,* Vol. V.*)* The author reviews the literature on ozena and gives the histories of eight patients on whom he had employed cupric electrolysis. The strength of the current used was 3 to 10 milliamperes. and each sitting lasted about ten minutes. After cleansing the nostrils, cocain was generally used. The copper needle was attached to the positive pole and inserted into the inferior or middle turbinated—sometimes into the tissues of the middle meatus, while the platinum (or steel) needle was passed into the septum.

Four of the eight patients were practically cured for period extending to eighteen months. By cure is meant, absence of fetor and crust formation. In one case there was marked improvement. In one case apparent cure for some months, then syringing had to be resumed.

In two cases the improvement only lasted a few weeks.

Campbell.

Sarcoma of the Frontal and Ethmoidal Sinuses.

BURNETT. *(Archives of Otology,* Vol. XXVIII., Nos. 2 and 3.) A man aged 57, had a swelling at the upper and inner angle of the right orbit, which gradually increased in size. The upper lid was edematous and covered the pupil most of the time. The eyeball was displaced downward and outward. Operation was undertaken, and a mass of soft reddish material removed by spoon and finger nail. The large cavity thus revealed included the right and left frontal sinuses, and invaded the ethmoid on the right side.

Microscopic examination of the tissue showed it to be a round-celled sarcoma. *Campbell.*

III.—MOUTH AND PHARYNX.

Etiology and Treatment of Alveolar Hemorrhage.

HARTMAN, J. H. *(Journ. Amer. Med. Assn.,* July 15, 1899.) Hemorrhage may be due to traumatism, laceration or fracture of the bone, or to a hemorrhagic diathesis. Simple and direct pressure on the part is the best treatment, but tannin is the best agent for local application, and may be used in conjunction with pressure.

Scheppegrell.

Throat Cough.

BACH. *(Journ. Amer. Med. Assn.,* July 22, 1899.) Hypertrophy of the faucial, pharyngeal and laryngeal tonsils are frequently the source of irritation that may produce paroxysms of cough. Cauterization is the most effective treatment. *Scheppegrell.*

Frasnotomy Speculum.

SUTHERLAND, J. L. *(Journ. Amer. Med. Assn.,* Apr. 29, 1899.) A speculum devised by the author, which has been found useful in cases of tongue-tie.

Scheppegrell.

Extragenital Chancre, with Report of Cases.

BOSHER, C. L. *(Medical Register,* May, 1899.) Among the cases reported are two of the lower lip and one of the tonsil. The source of infection is not stated.

Scheppegrell.

Excision of the Tonsils.

PETERT, W. H. *(Journ. American Med. Assn.,* Mar. 11, 1899.) A description and illustration of a tonsullar snare devised by the author, with an explanation of its application and advantages for this operation. *Scheppegrell.*

The Faucial Tonsil; its Sphere as an Agent in Systematic Infection.

SWEENY, G. B. *(Journ. American Med. Assn.,* Mar. 25, 1899.) The tonsil is an open gate-way for the reception of all kinds of disease germs, and many cases of systematic infection find their origin in this portion of the glandular organism. *Scheppegrell.*

Black Tongue; With Photograph of a Case and Brief Consideration of its Etiology.

GOTTHEIL, W. S. (*Archives of Pediatrics*, Apr., 1899.) The patient, a child of two years, had the usual black-tongue discoloration, but without the papillary hypertrophy which has given this disease the name of "hairy

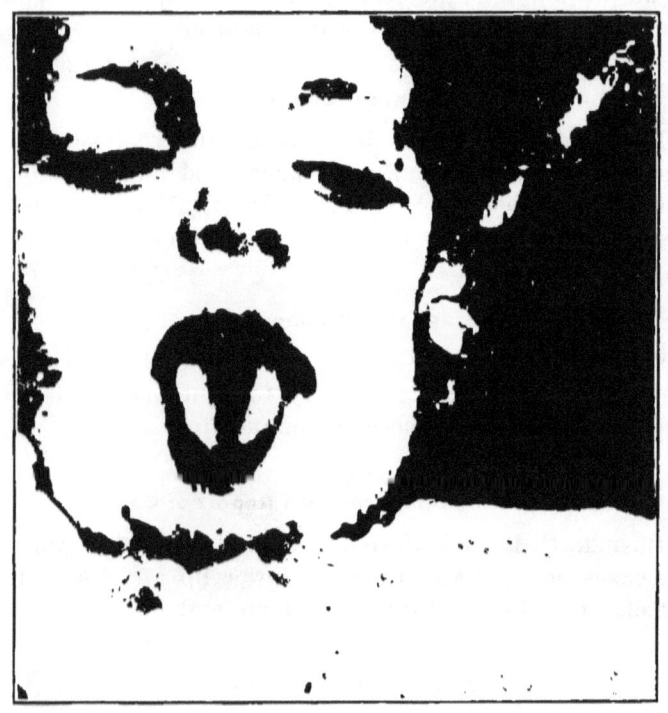

tongue." The affection is apparently harmless, save for the slight discomfort it occasions. By the use of a mouth-wash of a saturated solution of hyposulphite of soda the discoloration was removed in about a week, although recurrence is probable. *Scheppegrell*

The Submerged Tonsil.

PYNCHON, EDGAR. (*Chicago Med. Recorder*, Vol. XV.) A description of the symptoms and pathology, with a recommendation for tonsilotomy by electro-cautery dissection for its treatment. *Scheppegrell*

Gumma of the Tongue.

GOTTHEIL, W. S. *(Int. Med. Magazine, Dec. 1898.)* On examination, a woman of 24 years, was found to have a tumor occupying the middle area of the anterior part of the tongue, which measured one and a half inches in length

and three-fourths of an inch in breadth. The submaxillary glands were moderately swollen and quite hard. The anamnesis was negative as regards luetic infection, but specific treatment gave prompt relief. An excellent half-tone engraving illustrates the article.

IV.—LARYNX.

On Laryngeal Diseases in Syringomyelia.

BAUROWICZ, DR. ALEXANDER, Krakow, Austria. *(Fraenkel's Archiv., IX, 2, 292.)* The author observed a gradually developing paralysis of the left posticus muscle in a man of 31, followed by paralysis of the right half of the soft palate and by an, also gradually, developing paralysis of the right posticus. Antisyphilitic treatment, application of electricity, etc., were of no avail. The right half of the palate recovered partly, and deglutition became easier. The sensibility and reflex activity of the mucous membrane of the palate, pharynx and larynx always remained normal. The diagnosis of syringomyelia was confirmed by several authorities. This is the third case reported in which the nervous disease was accompanied by posticus paralysis. *Morgenthau.*

A New Intubator.

COWGILL, W. M. *(Jour. Amer. Med. Assn., April 8, 1899.)* The instrument is a modification of the "French" intubator. The modification consists of the angle put in the instrument. The angle at A in the shaft of the instrument brings the handle to the right and throws the hand of the operator out of his line of vision. The pistol handle B, which is turned downward and to the right, gives the operator an easy grasp of the instrument.

Scheppegrell.

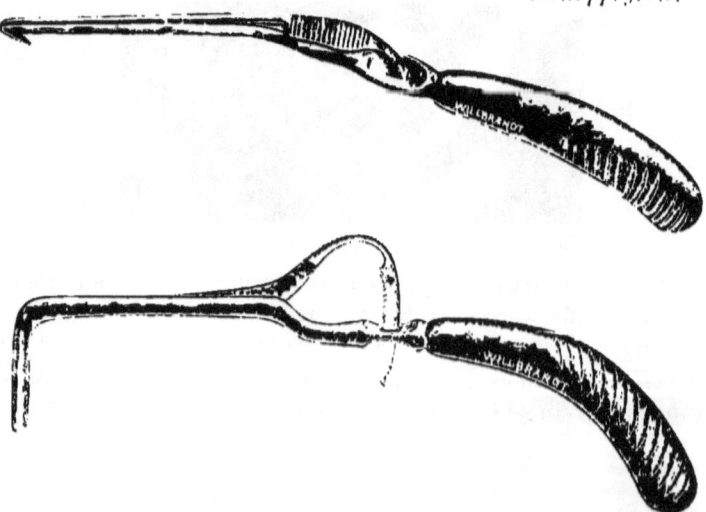

The "French" instrument with its straight hand and shaft throws the operator's hand into a very awkward position directly in front of him, and at the crucial moment, when the tube is about to enter the larynx, gives the arm and hand a rigidity not conducive to easy and accurate work. In the instrument described, these awkward positions are overcome, and make the operation of intubation easier. The instrument is made by Emil Willbrandt Surgical Mfg. Co., St. Louis. *Scheppegrell.*

Foreign Bodies in the Larynx in Children.

CLARK, J. PAYSON. *(Boston Medical and Surgical Journal,* June 1, 1899.) After stating that foreign bodies are more common in children than adults in the proportion of nearly two to one, he remarks that anything which causes a sudden strong inspiration when an article is in the mouth favors its entrance into the glottis. It may be arrested there or pass on into the trachea or a bronchus. Articles are caught in the larynx because on account of their peculiar shape they catch in one of the ventricles or stick into the mucous membrane somewhere. The most common objects are bones, beans, fruit stones, coins, needles, pins, seeds and grain. The statistics of Preobraschensky covering the period up to 1893 are quoted from at length. They show a mortality of operated cases of 27 per cent. while the unoperated cases had a mortality of 45 per cent. Since 1893, the author has collected the statistics of 34 cases of foreign bodies in the air passages of children of which only three died (less than ten per cent.). He reports in detail five successful cases from the Mass. Genl. Hospital, the last one his own. In all of these except the last tracheotomy was done and all recovered. In the author's case a five year old child had gotten a small tin tobacco tag into the throat three days previously. The tag was located in the larynx on laryngoscopic examination under ether. One-half of it was below the glottis and held in place by the lower end sticking into the mucous membrane. It was removed with forceps and recovery was uneventful. Dr. Clark urges that a laryngoscopic examination be always made in these cases, etherizing whenever necessary. He depreciates the use of emetics and says that inversion is of use only when the foreign body has weight enough to be affected by gravity and is not too firmly held by muscular contractions. Sulphate of atrophia is of value 1 200 gr. by mouth before etherizing as it lessens the amount of mucus secreted. If the dyspnea is urgent tracheotomy must be done at once but if not an attempt at laryngoscopic examination ought to be made and the foreign body removed through the mouth if possible. The laryngoscopic examination is less valuable after tracheotomy since it puts into complete repose the larynx and neighboring muscles relieving the spasmodic contraction which may hold the foreign body firmly fixed. *Richards.*

Statistical Contribution to the Question of the "Lateral Correspondence" of Laryngo-Pulmonary Tuberculosis.

MAGENAU, DR. CARL., Heidelberg, Germany. *Fraenkel's Arch.,* IX. 2, 304.) Krieg published an article (Ibid. viii. 3), on the Invasion of the Larynx by Tuberculosis, in which he traces the origin of the laryngeal tuberculosis, in most cases, to infection by the way of the lymphatic current from the lung to the larynx. Of 700 cases of laryngeal tuberculosis, 275 (39.3 per cent.) were unilateral; of these latter, 252 (91.6 per cent.) were synchronous ("corresponded") with a pulmonary affection on the same side; of the 700 cases, therefore, 36 per cent. were unilateral and corresponding. From these figures. Krieg concludes that the correspondence of location of pulmonary and laryngeal affection is not accidental, and rejects the theory that these unilateral laryngeal tubercular affections originate from the trachea (inhalation or sputum infection). He is of the opinion that in the large majority of cases of laryngeal phthisis, infection by way of the circulation is the rule; that is to say, by the lymph current from the lung to the larynx. The authors, as is known, differ as to the invasion of the tubercular virus into the larynx, as do the reports on the frequency of unilateral and corresponding laryngo-pulmonary phthisis. The author, who is assistant to Professor Jurasz, was instructed to examine all cases observed from 1891 to 1898. Only such were included in which the notes appeared to be accurate and trustworthy, altogether, 400. Of these, 274 were males; 126 females—a greater number, as generally accepted, belonging to the male sex; 280 patients (70 per cent.) were between 20 and 40 years of age.

In order to facilitate comparison, the author adopts Krieg's division of the cases:

1. Unilateral.
2. Bilateral.
    a. With predominance of one side.
    b. With demonstrable predominance of one side.
3. Median.

The last category embraces all cases with affection of the posterior wall or epiglottis alone, and those in which also other laryngeal parts were involved, i. e., all advanced cases, which have no bearing on the question at issue. These formed a considerable group, 182 (45.5 per cent.), of 400 cases. Purely unilateral disease was found in 65 (16.25 per cent.), not unilateral, 335 (83.75 per cent.). In these 65 unilateral cases, 26 (40 per cent.) were diseases on the same side as the lung; in 38 (60 per cent.), there was not correspondence (25 times both lungs were attacked); 11 times crossed lung and larynx tuberculosis was observed. Of the author's 400 cases only 26 (6.5 per cent.) were, therefore, on one side only, and on the same side (unilateral and corresponding). This result differs markedly from Krieg's 39 per cent., unilateral and corresponding, in 700 cases. Furthermore, Krieg refers to those cases of bilateral disease in which one side distinctly predominated. In 103 cases of bilateral tuberculosis, 81 (78.5 per cent.) showed greater tubercular changes on the same side in both the lung and the larynx. In the author's 400 cases, one side could be seen to be more affected in 61 instances; of these, 22 (36 per cent.) were corresponding. He does not, however, wish to draw any inferences from this disparity against Krieg's theory of the origin of laryngeal tuberculosis. The explanation of infection by the lymph current appears very plausible, but statistics do dot seem to be of value in confirming it.

The true proof of the possibility of the invasion of tubercle bacilli from the lung to the larynx by way of the lymph current has not as yet been brought; and its burden must be put upon physiology, anatomy, and pathologic anatomy. The questions which Krieg himself asks, at the end of his article, of these branches of science await answers.

	Unilateral.		Bilateral.		Median.
65	16.25 per cent.	153	38.25 per cent.	182	45.5 per ct.

Corresponding.	Non Corresponding.	Without predominance of one side	With predominance of one side 61 = 39 per ct.
26 = 40 per ct.	39 = 60 per ct.	92 = 60 per ct.	

Left, 12		Corresponding.	Non Corresponding.
Right, 14.		22 = 36.1 per ct.	39 = 63.9 per ct.

Morgenthau.

A New Intubation Instrument.

WHALEN, CHARLES J. *(Jour. Amer. Med. Assn.,* June 24, 1899.) The advantages claimed over the O'Dwyer

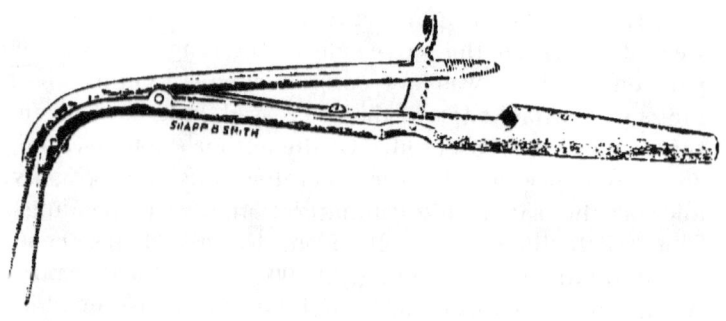

instrument is in the cheapness, the introducer and extractor being included in the same instrument.

Impaction of Foreign Body in the Larynx; Recovery After High Tracheotomy.

QUINLAN, F. J. *(Journ. Amer. Med. Assn.,* June 10, 1899.) A boy who was admitted to the hospital suffered intensely from dyspnea. Although the foreign body was lodged between the vocal cords, all efforts to remove it by intralaryngeal methods were unsuccessful. A tracheotomy was performed, by means of which the foreign body, a double-flanged screw, was removed without difficulty.

Scheppegrell.

Sudden Death from Foreign Body in the Larynx.

JENKINS, W. T. *(Jour. Amer. Med. Assn.,* June 10, 1899.) The first case, a boy, while playing in the park suddenly became cyanotic and died before medical aid could be summoned. The autopsy showed the opening between the vocal cords to be occluded by a bone collar button.

Several instances of instantaneous death have also followed the lodgement of a bolus of meat over the larynx. In another case reported, some of the teeth had been knocked down into the throat, an autopsy showing the teeth lodged at the bifurcation of one of the larger bronchi, and another death resulted from the rupture of a broken-down bronchial node into the trachea. *Scheppegrell.*

Intubation in Diphtheria.

MOORE, THOS. H. *(Phil. Med. Journ.,* June 24, 1899.) A record of 17 cases—12 of which were performed in the country—showing the advantage of intubation over tracheotomy. Antitoxin and intubation were instrumental in saving a number of cases that appeared critical.

Scheppegrell.

The Connection of the Female Generative Organs and Laryngeal Affections.

OPPENHEIMER, S. *(Phil. Med. Jour.,* Feb., 1899.) A clinical history of a number of cases illustrating the effect of diseases of the generative organs, menstruation and pregnancy on the larynx; also the effect of ovariotomy on the singing voice. *Scheppegrell.*

Phonation.

CUTTER, EPHRAIM. *Jour. Amer. Med. Assn.,* March 18, 1899.) An interesting exposition of this subject showing the relation of phonation to singing, which should be read in the original. *Scheppegrell.*

Formaldhyd: Its Use in the Treatment of Tubercular Laryngitis.

GALLAGHER, THOS. J. *(Jour. Amer. Med. Assn.,* March 4, 1899.) In using formaldhyd, the possibility of producing gangrene of the parts should be considered. It is safe to allow the patient to use a mild solution of 1 to 500, two or 3 times per day. The relief of the dysphagia is very marked, and in many cases formaldhyd is a good substitute for cocain. Its greatest effects are seen in the ulcerative and vegetative types. Stronger solutions, from 1 to 10 per cent., should be applied two or three times per week. *Scheppegrell.*

Paramonochlorophenol.

RICHARDS, GEO. L. (*Journ. Amer. Med. Assn.*, March 4, 1899.) The author has used paramonochlorophenol in a number of cases of a strength of 4 to 10 per cent. in equal parts of glycerine and water. Like lactic acid, it does not seem to injure any of the sound tissue with which it may come in contact. It acts more advantageously when used in connection with lactic acid. In the cases reported it seemed to have had good results, and even in severe cases improvement was noted.

Scheppegrell.

Falsetto Voice in the Male.

MAKUEN, G. H. (*Journ. Amer. Med. Assn.*, March 4, 1899.) The falsetto voice in the male is uncommon, and in an experience of more than 20 years as vocal teacher the author has not seen more than 10 or 12 cases, only five having reported for examination and treatment. These the author describes in detail. The direct cause of the defect in each case was found to be a faulty co-ordination of certain laryngeal muscles, this defect being mainly extrinsic to the larynx. By teaching the patient to use the voice in the chest register, all the cases were cured without difficulty. *Scheppegrell.*

Foreign Bodies in the Larynx.

KNIGHT, CHAS. H. (*Medical News*, April 15, 1899.) A boy, six weeks before application, had developed a violent fit of coughing and dyspnea, and it was surmised that he had swallowed a shoe hook. Many attempts were made to remove the hook, both with and without anesthesia, without success. A partial tracheotomy was performed and the hook was removed without difficulty.

Scheppegrell.

Catarrhal Laryngitis.

SLACK, H. R. (*Jour. Amer. Med. Assn.*, May 6, 1899.) The following formula first suggested by Dr. Jos. Holt, of New Orleans, is recommended:

    R. Chloralis .. gr. lxxv
        Potassii bromidi -------- gr. xlv
        Ammonii bromidi gr. xxx
        Aquae cinnamonii ad. ʒii

M. Sig. Teaspoonful, and repeat in 20 minutes if not relieved. *Scheppegrell.*

Recurring Laryngeal Stenosis After Intubation

FISCHER, F. (*Journ. Amer. Med. Assn.*, May 13, 1899.) The author indorses the opinion of the late Dr. O'Dwyer, that in the large majority of instances this condition is the result of traumatism. It may result from the use of improperly constructed intubation tubes, from allowing calcareous deposits to form on the tube, or from unskillful or rough manipulation in connection with the operation. Calcareous deposits do not form on rubber tubes.

Scheppegrell.

Some Experience with Intubation.

SAMPSON, F. E. (*Medical Herald*, June, 1899.) If the patient seems greatly exhausted, intubate in the prone position. If the case is more desperate than this, with slow, irregular and convulsive breathing, a tracheotomy should be done, using local anesthesia, or the operation should be performed without this. If the patient revives, intubate and close the tracheotomy wound. *Scheppegrell.*

Positions the Vocal Cord After Severing the Recurrent Nerve, and Posticus Paralysis.

BURGER, H., Amsterdam, Holland. (*Fraenkel's Archiv.*, IX, 2, 203.) In an article, which, because of its completeness, cannot be fairly abstracted, Burger contends that the facts known up to now warrant the following conclusions:

1. That a physiologic difference exists between the two antagonistic groups of laryngeal muscles, and that this difference embraces not only the muscles, but also the corresponding nerves.
2. That the dilators of the glottis obey other laws of stimulation than do the contractors; the latter, generally speaking, demanding a stronger stimulus.
3. That the dilators perish sooner, and that the contractors offer greater resistance.

With these physiologic facts corresponds the pathologic fact that the dilators are more liable to disease, i. e., Semon's law is confirmed. *Morgenthau.*

A Case of Hysterical Aphonia with Ventricular Band Speech.

HUNT. (*Journal Laryngology, Rhinology and Otology*, July, 1899.) The patient, a woman, age twenty-seven, eleven years ago became aphonic through fright. On laryngoscopic examination it was seen that the ventricular bands came firmly together on attempted phonation, so as almost entirely to hide the ligamentous glottis. However, it could be seen that the cartilaginous glottis was wide open and that the vocal processes remained apart.

All efforts to restore her voice proved futile, though at various times galvanism, hypnotism, and vocal drill were employed. A gradual improvement in the voice, however, took place in the last two or three years, until she could talk in a deep, rough but fairly powerful voice. Last September her voice became aphonic again, and she ascribes the loss of her voice to a fright.

Laryngoscopic examination now shows normal closure of the true cords on phonation, with considerable hypertrophy of the ventricular bands, so that on quiet breathing the cords are quite hidden. The points in this case which appear of interest are: (1) The severity of the original nerve shock; (2) the recovery after exposure to excitement when all the usual remedies had failed; (3) the development in the course of time of ventricular band speech to replace the lost natural voice. *Loeb.*

Remarks on Laryngeal Growths in Young Children.

MACKENZIE. *(British Medical Journal.* May 20, 1899.) The author presented a case in a girl, aged 17 months, who had suffered from increasing huskiness for about one year. Some respiratory stridor can be detected, especially when she struggles or cries. No hereditary tendency can be elicited and the child has suffered from no serious illness.

This condition of growths on the vocal cords at first affecting vocalization only, but ultimately producing respiratory symptoms also, arises at three different periods of child life.

First. It may be there at birth.

Second. It may be detected for the first time, at any period up to about the sixth month with no appreciable exciting cause.

Third. It occurs as a sequela of one of the exanthemata at any age up to about the fifth or sixth year.

After considering the various methods of treatment, the author urges tracheotomy because breathing is relieved and as time wears on the growths freed from irritation of coughing and phonation gradually lose their vitality and become detached from the vocal cords without any tendency to recur. As a rule six months is sufficiently long for the tube to be worn. *Campbell.*

MISCELLANEOUS.

Aqueous Suprerenal Extract; its Surgical and Therapeutic Uses.

MULLEN, JOS. *(Journ. Amer. Med. Assn.,* May 20, 1899.) The extract increases the anesthesia and the ischemia of cocain. It appears also to modify the postoperative swelling and facilitates the healing of the parts. The author has never seen any evidence of cocain toxemia when the extract had been taken. *Scheppegrell.*

Success in the Use of Diphtheria Antitoxin.

KOENIG, A. *(Phil. Med. Journal,* May 27, 1899.) During an outbreak of diphtheria at the Rosalia Foundling Asylum and Maternity Hospital, two cases already infected were injected with 800 antitoxin units. Cultures from all the children in the ward in which these cases appeared were examined and a large number showed the bacilli present. Of the throats of 74 children examined, 38 proved positive and of 22 adults, 14 were positive. All persons found infected were isolated and injected, only one case resulting fatally. The antitoxin used was the product of the Laboratory of the Pittsburg Bureau of Health. *Scheppegrell.*

The Diphtheria Bacillus; its Persistence in the Mouths of Convalescents

RUSSELL, H. L. *(Journ. Amer. Med. Assn.,* June 24, 1899.) In the first case described, antiseptic throat washes were used during treatment and convalescence, but, in spite of this, virulent diphtheritic organisms, with only occasional exceptions, were found four and a half months after the outbreak of the disease. In the second case they were found three and a half months after the development of the disease.

These cases are by no means exceptional. While it is impossible to maintain rigid quarantine for such length of time. a knowledge of these facts enables us to use greater precaution against infection. *Schappegrell.*

The Value of Thiol in Nose and Throat Practice.

WELLS. W. A. (*Phil. Med. Journal*, Apr. 15. 1899.) Thiol is an artificially produced ichthyol free from the objectionable odor, and having in addition the advantage that it does not cause irritation nor bleeding of the eroded surfaces. It has no toxic effects and its stain is easily removed. It is prepared in powder and liquid form. The following formulas are recommended:

R. Thiol liquid - - - 2 grams.
   Water - - - 100 "
M. Sig. To be used as a spray or gargle.

R. Thiol liquid - - - 10 grams.
   Glycerine - - - - 50 "
M. Sig. For swabbing the throat.

R. Thiol siccum - - - 5 grams.
   Vaselin - - - 10 "
   Lanolin - - - 20 "
M. Sig. To be used as ointment.

R. Thiol siccum - - - 5 grams.
   Amylum - - - 20 "
M. Sig. To be used as a dusting powder.

Schappegrell.

The Serum Treatment of Diphtheria in New York Foundling Hospital During 1898.

NORTHRUP. W. P. (*Medical News*, Apr. 29. 1899.) The fatality was 12 1 2 per cent., which the author considers an extremely good result in infants. As an immunizing agent, he believes that antitoxin has attained almost ideal results. *Schappegrell.*

Summer Care of the Tubercular.

PORTER, WM. (*Journ. Amer. Med. Assn.*, April 29. 1899.) The northern lake region, especially near Petoskey, Mich., has given the author good results in cases of tuberculosis and chronic broncho-pneumonia during the summer months. This refers especially to the Les Cheneaux country, some 10 miles north of Mackinac Island.

Schappegrell.

The Use of Ozonized Air in Diseases of the Lungs and Air Passages.

PEAVY, J. F. *(Journ. Amer. Med. Assn.,* Apr. 29, 1899.) Ozone is recommended therapeutically for tuberculosis and is administered by placing the patient on an insulated stool connected with the positive-charged rod of a static electric machine.

[Where ozone is indicated, more effective results are obtained from the Siemens ozonizer, by means of which the ozone may be economically and scientifically prepared. S.] *Scheppegrell.*

Sarcoma of the Thyroid Gland.

MORF, P. F. *(Journ. Amer. Med. Assn.,* April 29, 1899.) Sarcoma of the thyroid gland is a rare disease and is more common in those countries where goitre is endemic. It is more frequent in persons whose thyroid is the seat of ordinary goitre. It is seen oftener in the male than in the female, and more frequently in late than in early life.

Statistics show the relative frequency of round-cell sarcoma, which corresponds with the clinical observations of the rapid course of these tumors. The cervical lymphatic glands are involved in a large number of cases. Metastases are relatively frequent in the lungs and in the bones. *Scheppegrell.*

The Maxillary Sinus, Chronic Empyema Thereof; its Treatment

STUCKY, J. A. *(Journal American Med. Assn.,* March 4, 1899.) In the treatment of maxillary sinusitis, if the middle turbinate is so enlarged as to encroach upon the middle meatus or enough to obstruct the ostium maxillare, the anterior portion should be removed. In the majority of cases, the author found the operation of turbinectomy of the greatest value. If the ethmoidal cells are involved in the suppurative process or are otherwise diseased, radical treatment is indicated.

In the treatment of maxillary sinusitis, a large opening is made through the alveolar process into the antrum by means of a Buck mastoid drill, the second molar having first been extracted. The cavity should then be thoroughly inspected by means of the speculum introduced through the opening, and granulations, necrosis or other diseased processes thoroughly dealt with. The cavity is then packed with gauze. A counter opening through the inferior meatus to facilitate drainage has not been found necessary by the author, and he does not think it advisable. *Scheppegrell.*

The Etiology of Phthisis; a Summary of Scientific Points Involved in Churchill's Theory.

GARDNER, R. W. (*Journ. American Med. Assn.*, Mar. 4, 1899.) The administration of an oxidizable and assimilable compound of phosphorus under certain determinate, pathologic and therapeutic conditions has the constant effect of preventing any fresh deposit of tubercular matter. To obtain this effect, however, the patient should be carefully watched; the doses of the particular indicated hypophosphite should be carefully graded to meet the immediate condition of the patient. Anodynes, narcotics, stimulants, cod-liver oil and other remedies designed to ameliorate certain prominent symptoms, such as coughs, night-sweats, etc., should be abandoned as they operate against the object sought to be attained by the hypophosphite treatment. Hypophosphites of soda, lime, and occasionally quinine are alone to be employed.

Scheppegrell.

The Relation of the Trachea and Bronchi to the Thoracic Walls, as Determined by the Roentgen Rays.

BLAKE, J. A. (*Amer. Journal of Med. Sciences*, Mar., 1899.) The radiographs and diagrams demonstrate the relation of the bronchial tree to both the posterior and anterior thoracic walls. They show the greater size of the right bronchus, which not only makes it capable of receiving larger bodies, but also throws the dividing spur between the bronchi to the left of the axis of the trachea. This is aided by the greater respiratory current and the direction of the right bronchus, which conforms to that of the trachea. The deviation of the trachea is caused by the aorta, which crowds it to the right side.

In making the radiographs, which are quite interesting, the greater transparency of the bronchial tree to the X-rays was overcome by the injection of proper substances impervious to the rays. *Scheppegrell.*

The Other Side of the Antitoxin Question.

HERMAN, J. E. (*Medical Record*, March 11, 1899.) A

careful discussion of the *pros* and *cons* of the antitoxin treatment, the verdict being against its use as a therapeutic measure. The author believes that the whole theory rests on unestablished ground, and is more surprised that physicians continue to use it than to the contrary.

Scheppegrell.

The Results in Administering the Antitoxin of Diphtheria as an Immunizing Agent.

ADAMS, S. S. (*Archives of Pediatrics*, June, 1899.) From an analysis of the cases under observation, it was decided that it was impossible to draw any definite conclusion as to the value of the immunizing dose of antitoxin. It is believed, however, that the dosage in the cases referred to was too small, and that if more units had been used, better results would have been obtained. The average duration of immunity conforms with that obtained by other observers. It seems probable that the larger the immunizing dose, the longer the duration of immunity. The immunizing dose of antitoxin has no injurious effect upon the kidneys. Urticaria appeared in two cases, this being the only pathologic effect of the antitoxin observed.

Scheppegrell.

Operative Treatment of Exophthalmic Goitre.

DOEPFNER, CARL. (*Journ. Amer. Med. Assn.*, June 3, 1899.) The patient who does not improve under a prolonged treatment of the reliable physician should be subjected to operation. Partial excision of the thyroid gland should be tried in the first place, and always where there is dislocation and compression of the trachea. *Scheppegrell.*

The Use of Schleich's Mixture for General Anesthesia in 110 Operations on the Nose and Throat.

MCLAUGHLIN. (*Journ. Amer. Med. Assn.*, June 3, 1899.) This method of anesthesia is very rapid and the usual excitement is entirely absent. When properly administered, these mixtures are safe, especially in short operations.

Scheppegrell.

Tracheocele.

WEST, J. P. (*Archives of Pediatrics*, Apr., 1899.) A well-developed child of 20 months presented a swelling in the neck, which was first noticed five months before. The last two months it has grown rapidly, the swelling now being the size of a hen's egg and on the left anterior neck just above the clavicle and well separated from the larynx and trachea. There is no fluctuation or crepitation. Continued pressure when the child is at ease causes three-fourths of the swelling to disappear, but it immediately returns to its original size when the pressure is removed. On prolonged or restrained expiration, as in crying, the original swelling increases to double its size.

Scheppegrell.

Surgery of the Trachea; Two Unusual Cases.

JONES, W. S. and KEEN, W. W. (*Journ. Amer. Med. Assn.*, June 10, 1899.) The first case reported is one of attempted suicide by cutting the throat, which was followed by complete closure of the trachea by a diaphragm above the cannula persisting for 13 months. Breathing through the larynx and speech were entirely impossible. The diaphragm was dissected away and the trachea sutured, the operation being followed by the restoration of normal breathing. The right recurrent laryngeal nerve was probably divided or wounded in the suicidal attempt, which accounted for the paralysis of the right vocal cord which persisted and resulted in a mild degree of hoarseness.

The second is a case of stricture of the trachea. The trachea was divided longitudinally, the mucous membrane deflected, the stricture excised, the entire wound being immediately sutured. Union resulted by first intention.

Notes on Cocain.

JENNY, W. P. (*Medical Record*, March 25, 1899.) Contrary to the opinion of most investigators, the author believes that cocain is absorbed through the unbroken skin to a sufficient degree to be of therapeutic value.

[Solutions of cocain are absorbed through the skin, but in so slow a manner as to be of very little service in the majority of cases. By cataphoresis, however, the absorption takes place with considerable rapidity. S.]

Scheppegrell.

Formalin in the Treatment of Whooping Cough.

OLLIPHANT, H. S. (*Journ. Amer. Med. Assn.*, March 4, 1899.) After experimenting with a solution of formalin for a year the author believes it to be a specific. The duration of the severest cases was less than a week, and several were cured after three applications. Only weak solutions should be used. *Scheppegrell.*

Headache and its Relation to Diseases of the Eye, Ear, Throat and Nose.

JERVEY, J. W. *(Medical Record,* March 11, 1899.) In cases of headache these organs should be carefully examined. Where pathologic conditions exist, surgical or therapeutic measures frequently give prompt relief. Several cases are given in illustration. *Scheppegrell.*

The Sterilization of Instruments with Formaldehyd.

REIK, H. O. *(Phil. Med. Journal,* Feb. 4, 1899.) The author believes that formaldehyd is a convenient and economical means of disinfecting instruments.
Scheppegrell.

How Shall we Quarantine Against Diphtheria.

BRACKEN, H. M. *(Journ. Amer. Med. Assn.,* June 10, 1899.) The author believes our present methods of quarantine are inadequate. In view of the persistence of bacteria in the throat of patients recovering from diphtheria and the danger of sending them to school while this is the case, he believes that there should be special schools in order to avoid the loss of educational advantages resulting from such subjects remaining from school for so long a period. An ingenious but somewhat impracticable suggestion. *Scheppegrell.*

Medical Efficacy of Nosophen and Antinosin in Eye, Ear, Nose and Throat Affections.

LYDSTON, J. A. *(Journ. Amer. Med. Assn.,* June 24, 1899.) The author has found nothing to equal nosophen as a deodorizer in the purulent secretions characteristic of suppurative otitis media. Both nosophen and antinosin have been applied in all classes of cases with good results.
Scheppegrell.

Serum Therapy.

BUTLER, GEO. F. *(Journ. Amer. Med. Assn.,* May 27, 1899.) The present status of diphtheria antitoxin may be said to have established itself as a specific in the treatment of diphtheria. During the past year the percentage of larger doses has become greater and it seems certain that better results have been obtained. *Scheppegrell.*

The Failure of Antitoxin in the Treatment of Diphtheria.

Herman, J. E. *(Medical Record,* May 27, 1899.) The author has collected considerable literature, showing that the mortality from diphtheria has not been lowered since the introduction of antitoxin, and also that the summary of statistics show that cases treated without antitoxin have had a lower mortality than those in which it was used. He believes that many lives have been sacrificed which might have been saved by the usual old-time remedies.

Scheppegrell.

PROCEEDINGS OF AMERICAN LARYNGOLOGICAL, RHINOLOGICAL AND OTOLOGICAL SOCIETY, CINCINNATI, O., JUNE 2, 3, 1899.*

MASTOID OPERATIONS.

Dr. Max Thorner, Cincinnati, exhibited several interesting cases on whom he had done the mastoid operation. The first showed an excellent result, the large cavity in the mastoid being well covered with epidermis. There had been no discharge for almost a year. The patient was so well satisfied with the result obtained that he did not care to have a plastic operation done for the cosmetic effect. The second differed markedly from the first in that the first case presented an extradural abscess of eighteen years' standing, while in the second the symptoms developed very rapidly. The doctor is not in favor of the early closure of the opening with suture at the time of the operation or soon thereafter. The third case had complained of pain over the mastoid on several occasions, with intervals of almost complete freedom. Finally the pain increased, the temperature ran up to 101 degrees, and some swelling developed. On operation the mastoid was found full of granulation tissue and pus was demonstrated throughout the whole mastoid. Recovery was complete in some eight weeks after operation. The fourth case presented a somewhat similar condition of otitis media with discharge, which would stop and then open up again. Finally the pain became unbearable, although the temperature was only 100. On operation there was found great destruction of tissue, and pus was demonstrated throughout the mastoid. The entire cortex was removed and the operator went deep down into the mastoid. The third and fourth cases were remarkably similar.

THE FACIAL NERVE AS IT AFFECTS THE AURIST.†

Dr. George L. Richards, Fall River, Mass., read this

---

*From the Journal of the A. M. A.
†See Annals of Otology, Rhinology and Laryngology, May, 1899.

paper and presented a number of specimens which he had opened with the chisel to find the location of the facial nerve with especial reference to the mastoid operation. There are many variations in the course of the nerve, showing that its exact location cannot be predicted with certainty in any case. In the discussion that followed, a number of cases of facial paralysis following the mastoid operation were reported, some of which seemed to be permanent. The agents generally recommended for the treatment of facial paralysis were galvanism, faradism, massage and the use of strychnia and tonics.

ASTHMA AS A REFLEX MANIFESTATION FROM ABSCESS OF ANTRUM.

DR. CHARLES W. RICHARDSON. Washington, D. C., spoke of the reflex disturbances in the respiratory tract and other organs caused by pathologic changes within the nasal chambers and neighboring accessory cavities, the disturbances being at times so remote as to make it impossible to decide whether the relief of the accessory disturbance is an effect or an incident in the treatment. In the past three years he had had two cases of antral abscess attended by marked evidence of asthma, in both of which the asthma seemed to be excited in a reflex manner through the abscess of the antrum, and in which the asthma was relieved by the evacuation of the pus within the antral cavity.

TRAUMATIC HEMORRHAGE FROM NOSE AND PHARYNX WITH REPORT OF CASES.

DR. ROBERT C. MYLES, New York City, read this paper. He considered five important points: Whether the surgeon has any means of finding out before operation the probabilities of hemorrhage; if there is reason for suspecting excessive bleeding, whether there ought not to be more limitation in the area and depth in making sections and excisions. After each operation where there is a reasonable possibility of delayed or secondary hemorrhage, the patient should be kept within easy reach of the surgeon. Careful instruction should be given the attendant to watch the patient during the night. Several cases have been seen in which dangerous and alarming hemorrhages had occurred, and the conditions were not discovered until the patients were nearly exsanguinated. As soon as the surgeon has evidence that he is dealing with a bleeder, he should apply expressed tampons of tanno-gallic acid. In the discussion that followed a number of cases of hemophilia were reported, in which obstinate hemorrhage came on within a few hours after operations on the upper air-passages.

REPORT OF CASE WHERE DEFECTIVE SPEECH RESULTS IN SOME INTERESTING DERANGEMENTS OF CEREBRAL FUNCTIONS.

Dr. G. Hudson Makuen, Philadelphia, reported the case of a boy 15 years of age, whose speech was so defective as to be scarcely understood. While he was apt in arithmetic, he had special trouble with everything in which the use of letters and words was required, being often unable to write the simplest words or to remember the names of letters and words that he had learned. His writing and spelling were very defective, but he was proficient in the use of figures. He was by no means feeble-minded, but there was some functional impairment in the action of the cortical visual word-center, resulting in partial word-blindness. When for any reason the development of the faculty of speech is delayed, there is a corresponding lack of functional activity in both the visual and auditory word-centers, as well as in those related areas of the brain employed in the use of language and thought. So great is the dependence of the functional activity of these special centers and their related areas on the use of the faculty of speech that its lack of development, due entirely to mechanic obstructions in the peripheral organs, has in some cases led to a diagnosis of imbecility and even idiocy. The author advised systematic speech training to determine whether the defective cerebration was of functional or organic origin. The improvement of his patient under this treatment had been marvelous, and illustrated the importance of special training as a remedial agent in certain forms of feeble mentality characterized by defective speech and other motor disturbances.

APPROPRIATE TREATMENT OF CERTAIN VARIETIES OF NASAL DEFLECTION AND REDUNDANCY.

Dr. D. Braden Kyle, Philadelphia, in this paper gave

the following varieties of deflection of the septum: 1. The split cartilaginous septum with bulging into both nostrils. 2. Dislocation of the columnar cartilage. 3. Simple deflection in which the cartilage is very thin. 4. The letter S deflection. 5. Deflection of the cartilage with involvement of the bony septum. 6. Deflection due to the splitting of the cartilage with bulging on one side only. 7. Deflection in which there is redundancy of tissue overlapping the septum and extending close to the floor of the nose. Under deformities of the septum the paper dealt with the deviation or deflection from disease, traumatic deflection and congenital deflection. Concerning treatment, the author said that each deflection required some modification from a given method; no one operation would answer in all cases, as shown by the many methods proposed. The conditions presenting necessitate a combination of methods rather than the following of any one method.

USE OF RUBBER SPLINTS IN TREATMENT FOLLOWING INTRANASAL OPERATIONS.*

DR. J. PRICE BROWN, Toronto, Canada, read this paper. For tubage following intranasal operations the author thought that rubber splints made from thin rubber sheeting were superior to silver tubes, and much to be preferred to hard or perforated splints. They are smooth, compressible and elastic, can be readily cut to the required shape, and can be obtained of any thickness desired.

PYEMIC SINUS THROMBOSIS COMPLICATING PURULENT MASTOIDITIS WITH MULTIPLE EPIDURAL ABSCESS, CAUSED BY AN ACUTE OTITIS MEDIA. OPERATION. RECOVERY.

Dr. J. F. MCKERNON, of New York, read a paper on this subject. Some weeks previous, the patient had been operated on for polypi, after which he suffered an attack of facial paralysis, followed by intense pain in the left ear. The author considered the acute otitis the direct result of the erysipelatous inflammation traveling through the Eustachian tube and infecting the middle ear, as the discharge showed the streptococcus present. The complete disintegration of the mastoid he ascribed to the virulence

*See ANNALS OF OTOLOGY, RHINOLOGY AND LARYNGOLOGY, May, 1899.

of the inflammation. He considered the thrombosis due to his wounding the sinus at the mastoid operation. After the sinus was evacuated of pus, the patient did not convalesce satisfactorily until the metastatic deposit in the intestines had been evacuated, carrying away the poison in the system.

PHARNGOMYCOSIS.

Dr. LEWIS CLINE, Indianapolis, read this paper. This is a term applied to a fungus growth usually found attached to the posterior wall of the pharynx, in the orifices of the crypts of the tonsils, in the folds of the mucous membrane, behind the posterior pillars and on the base of the tongue. It consists of leptothrix, which rapidly develop into a vigorous growth. The symptoms are generally local, causing various degrees of irritation. Mycosis is a rare disease, although it is claimed that leptothrix is found in the majority of the mouths of patients examined. The disease must be differentiated from follicular tonsillitis, tonsillar concretions and diphtheria. The best treatment is the galvanocautery.

Dr. RICHARDSON referred to an article by a German writer in which it was claimed that the symptoms in cases such as those described by Dr. Cline were to be ascribed to the presence of a concretion rather than to a mycosis.

Dr. FRIEDENBERG called attention to the fact that the leptothrix causes a precipitation of earthy salts from solution.

Dr. STUCKY had secured good effect by the use of a 25 per cent. solution of pyoktanin rubbed into the masses two or three times a week. In a few weeks they would entirely disappear.

Dr. PRICE-BROWN had observed that the leptothrix was found not only on the surface, but also invading the mass, and reported several cases, one in a man who did his own grooming, a second in a man who worked two or three months polishing cows' horns, a third in a girl who worked in a brush factory, and lastly a case in a girl who was engaged for some time in making pillows out of the fluffy part of a weed. The cases seemed to indicate that the disease may originate from the inhalation of particles of the material with which the individuals worked.

Dr. Levy called attention to the possibility that not all cases so described may be true cases of pharyngomycosis.

Dr. Phillips observed that germicides seem to have little effect on the disease.

Dr. Myles secured the best results by excising the masses.

Dr. Loeb had seen the disease twice in the same family, in a brother and sister, and had never seen a case of pharyngomycosis in which the lingual tonsil was not involved.

The society will meet in Philadelphia next year.

DR. MAX THORNER.

Dr. Max Thorner of Cincinnati, Ohio, an esteemed colaborator of the ANNALS OF OTOLOGY, RHINOLOGY AND LARYNGOLOGY died August 26th. His many friends of this country and Europe share with us in our sorrow, which is the more acute since he was called away long before he had attained the climax of his possible usefulness.

Dr. Thorner had the rare gift of energy, untiring and uncomplaining, coupled with a mind that was never satisfied with the knowledge before him. It was on this account that he was, in so short a time, able to accomplish so much. Coming as a stranger to Cincinnati 14 years ago, he soon took front rank, and in these few years he had attained for himself a reputation that it is safe to say was unsurpassed by any one in the west.

He was born in Geestemunde, Germany, April 2nd, 1859, and was the son of Jacob and Bertha Thorner. He received his medical education in the Universities of Heidelberg, Leipsig and Munich, and pursued the study of his profession in the Hospitals of Vienna, Berlin and London. In the latter place he served as clinical assistant in the Hospitals for Diseases of the Throat and Chest, under Sir Morrell Mackenzie.

He was a member of the Academy of Medicine in Cincinnati, and served as President for one term. He has been, for some time, a member of the staff of the Cincinnati Hospital, the Jewish Hospital and the Ophthalmic Hospital. He had just been elected a member of the American Laryngological Association and was a member of the American Laryngological, Rhinological, Otological Society. He achieved his greatest distinction through the mastoid retractors, which he devised, likewise from his contribution upon the use of Salol in diseases of the nose and throat which it is said was the earliest publication upon this subject.

During the last few years, he was very much interested in direct laryngoscopy as suggested by Kirstein. He translated Kirstein's book into English, and developed the study of this particular plan of laryngoscopy in this country. His contributions to medical literature was as follows:

1. Foreign Bodies in the Larynx. (English and German.)
2. Salol in affections of Throat, Ear and Eye.
3. New Galvanocautery Handle.
4. Haematoma of the Septum Narium.
5. Tinnitus Aurium and Nasal Obstruction.
6. Chronic Rheumatic Throat Affections.
7. Imaginary Foreign Bodies.
8. Laryngectomy for Cancer of Larynx.
9. Object of Medical Societies. Address.
10. Benign Tumors of Larynx.
11. Thrush of Nose and Pharynx in Adults. (English and German.)
12. Atrophy of a laryngeal Fibroma in a Child. (French.)
13. Modified Tuberculin in Tubercul. Laryngitis.
14. Rheumatic Throat Affections. (Clinical Lecture.)
15. Intubation followed by fatal Edema.
16. The Management of Foreign Bodies in the Throat.
17. Acute Pharyngitis. (Burnett's System.)
18. Pathological Conditions following Piercing of Lobules.
19. Multiple Papillomata of Larynx.
20. New Mastoid Retractor.
21. The Autoscope.
22. Syphilis in Nose.
23. Serious Complications of Middle Ear Suppuration.
24. Removal of Foreign Body with Autoscope.
25. Practical experience with Autoscopy.
26. Autoscopy of the Air-passages. (Book of Kirstein.)
27. Uncommon Accident following Operations in Nose and Throat.
28. Primary Syphilis of Tonsil.
29. Intubation with improved Instruments.
30. Adeno Carcinoma of Nose.
31. Naso Pharyngual Tumor of unusual size.
32. The direct examination of the Larynx in Children.
33. The Asch operation.

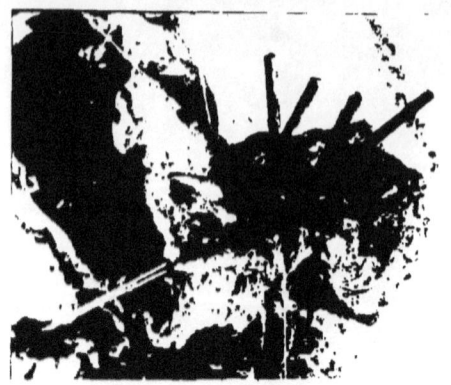

Plate I.

Plate II.

Tuberlocus of the Esophegus.

ANNALS
OF
OTOLOGY, RHINOLOGY
AND
LARYNGOLOGY.

DIABETIC ULCERATIONS OF THE THROAT.*

By W. FREUDENTHAL, M. D.,

NEW YORK.

Although in a disease like diabetes one might expect to find ulceration in almost any part of the body, it was a great surprise to me in looking for literature on this subject not to be able to find mention of one single case, not even in Seifert's excellent article on ulcerations of the larynx (Heymann's Handbuch der Laryngologie, etc). So it seems to me that these cases must be extremely rare.

The first case that came under my treatment was not correctly diagnosticated; first, because the patient did not tell me he was suffering from diabetes; and secondly, because I had never heard of diabetic ulcerations in the throat. As tuberculosis and syphilis could be excluded, I regarded the ulcerations as doubtful. I heard that he was suffering from diabetes only shortly before his death, without, however, attributing any importance to this fact.

When I saw the next case, the family physician told me that the patient was suffering from diabetes, and at once the question entered my mind whether these ulcerations might not be of diabetic origin. As I could not find any other cause for these ulcerations, I had to regard them as diabetic.

---

*Read before the New York Medical Union November 28, 1899.

Before entering into details I wish to state that it is my impression that there exist two different forms, viz: those of a malignant, and those of a benign character.

a. The *malignant ulcerations of the throat of a diabetic origin*, do not seem to yield to any treatment. Difficulty in swallowing increases constantly, and the patient dies in dreadful pain. It is natural that the exitus letalis is hastened by the lack of nourishment, due to dysphagia.

CASE I.—Mrs. N., 49 years of age; has suffered from diabetes for 9 years. Her throat commenced to trouble her about 6 months ago, when she thought she caught a cold. Soon she noticed pain in swallowing, which has constantly been growing worse. At present it is so bad that she can swallow only with great difficulty. Leaving out the other details of her case which present nothing unusual, I will say that on inspection the pharynx showed the following: There was a broad ulceration commencing on the left side of the uvula, involving this only slightly, but extending for about an inch down towards the pillars. Besides, one could plainly distinguish another smaller ulcer involving the anterior part of the left tonsil and the anterior pillar. Both ulcerations seemed to have originated from infiltrations, and showed somewhat elevated edges with ulcerative deepening at the centre. The whole picture impressed me as tuberculous. But there were absolutely no signs of this disease to be found elsewhere.

The lungs were normal, no bacilli in the sputum, nor were any to be found in the secretions taken *repeatedly* from the ulcerations. Nevertheless, these were curetted by me and treated with the much lauded lactic acid. The result was entirely negative. The patient had such pain after this "operation" that she could not sleep for the two following nights. She went from bad to worse, consulted many physicians and I only saw her twice more, a few days before her death; i. e.. about five weeks after her first visit. Another ulcer had developed on the right side, and the first ones had become larger and much deeper. She died of inanition.

CASE II.—Mrs. P. M., 59 years of age; has been married twice. Had a child with the first husband which died at ten months, of pneumonia, and two years later an abortion, which came on "in consequence of fright." Her physician, Dr. O., knew both husbands and is positive that they never had any specific disease, nor could any such symptoms ever be found on Mrs. M. Dr. O. discovered about two years ago that Mrs. M. was suffering from diabetes, and for three weeks previous to her coming to me, she had been suffering from dysphagia. On inspection one could readily see an extensive ulceration on the left side of the tongue extending over its edges, and corresponding to this, an ulcer on the mucosa of the cheek. The teeth of the patient were in good condition. No tubercle bacilli were in the sputum nor on the ulcers, no glandular swellings, nor any signs of syphilis or tuberculosis. Here, too, I applied lactic acid with no result whatever. A mild gargle and the application of cocain were about the only measures that gave her any relief. I saw her in consultation every few weeks and could notice that her condition was getting worse, the ulcerations growing deeper and larger, and finally new ones springing up on the epiglottis. She died about five months after I saw her first.

The next cases seem to justify me in classifying them among

b. The *benign ulcerations of diabetic origin*.

CASE III.—On May 7, 1899, I was called, as a last resort, to Mr. Wm. P., a merchant seventy years of age. Dr. Burgheim, the attending physician, told me that since he had known Mr. P., i. e., four years, he had been suffering from diabetes. For about two weeks he had complained of sore throat and dysphagia. He formerly had sore throat occasionally, but it improved rapidly as a rule. Now, however, it was getting worse constantly. When I saw him it was so bad that for three days he had been unable to swallow a morsel of food, not even milk. The patient was a tall, stout man, and so weak that he could not hold his tongue out. On examination, which was extremely difficult as Dr. B. had to hold the patient's tongue and the patient's head shaking constantly (tremor senilis), I found the whole larynx, as far as I could see, changed to a mass of ulcerations. There was a deep ugly looking ulcer on the laryngeal surface of the epiglottis, extending down as far as one could see, and the interarytenoid space was also ulcerated. Besides, there was a good deal of mucus in the larynx and trachea, which the patient was unable to remove. When I remarked that it was necessary to treat him locally, his children thought it cruel to bother a dying man with treatment, but, nevertheless, consented. I at once made my first application of orthoform, a procedure which took me about three-quarters of an hour. I wish to state, right here, that I attribute the most excellent and gratifying result in this case to this thorough treatment, which I continued for about four weeks daily, and to the use of orthoform. I cannot lay too much stress on two points; first, the ulceration must be clear, i. e., free from all débris and mucus so that when orthoform is applied, it is able to reach the nerve ends quickly, and thus be effective; second, the drug must remain on the ulcer untill it is absorbed. If a paroxysm of coughing sets in, naturally all the medicine is coughed up again and we have no effect from it. I, therefore, proceed in the following manner:

First, the throat is cleaned out with any indifferent spray. If the patient is as helpless as Mr. P. I generally swab the larynx out so as to get a clear view of the ulcer. Then the larynx is anesthetised thoroughly with cocain. After this is effected, I inject at intervals, as much of a syringeful of my orthoform emulsion as the patient can bear. Sometimes I commence with a quarter of a syringeful, equal to about ten grains of the solution or even less, and gradually give more. If we inject too much, or at too short an interval, the fluid runs down the trachea and a coughing spell results. I use menthol in the above emulsion from 2% to 10%: formerly I used only 10%, but found that in the beginning it is often too irritating.

My formula now is therefore:

R

Menthol	2.0	10.0
Ol. amygdal. dulc.	-	30.0
Vitelli ovi	-	about 25.0
Orthoform	-	12.5
Aqua. dest. q. s. ad.	-	100.0

Ft. Emulsio.

S. Shake *well* before using.

This preparation is somewhat weaker than that previously

in use, but it seems to me that nevertheless more orthoform is absorbed than from the other mixtures. I have had the best results with it, and would recommend it highly in this form. Occasionally, however, a very marked burning occurs after the injection of this mixture (I employ an ordinary laryngeal syringe), but this subsides after a few minutes and gives place to a feeling of euphoria.

These last remarks I made* some time ago and I can only repeat them here, in fact the effect of orthoform in this case was the most remarkable I have ever seen.

It was gratifying to me to see last summer that the above emulsion is being used in many clinics abroad, but since several prominent laryngologists abroad apply the emulsion on cotton, like lactic acid, which I do not consider correct at all, I have explained my procedure here more explicitly.

To return to our patient, Mr. P., he came to my office daily from May 7th to June 12th, when I left for Europe. By that time this extremely weak man had grown so strong that he could come to my office all alone, the ulcerations having healed entirely. I saw him again September 14th, when he complained about some " thickness " in the throat. He had felt well all summer until within a week, but even now he could swallow food easily. The laryngeal ulcerations were well cicatrized. In the pharynx there were superficial erosions on both sides of the uvula down to the tonsils. The whole soft palate looked somewhat edematous. All these symptoms disappeared within a few days and Mr. P. feels perfectly well now. He never had any internal medication.

Case IV.—This case was a physician, and I record it only from memory. The colleague, a strong, healthy man, 39 years of age, had gone through a follicular amygdalitis—not diphtheria—and three days later felt pain in the same place again. I found an outspoken ulceration of the tonsil. It was not examined for tubercle bacilli, as the doctor did not care to have it done, and it disappeared under mild treatment within six or ten days. Two weeks la-

---

*W. Freudenthal: " The Treatment of Dysphagia and Cough, especially in Tuberculosis." The Philadelphia Med. Jour., March 25, 1899.

ter the doctor was examined for life insurance and to our great surprise was rejected on account of diabetes. This diabetes was of a very mild form, disappeared within half a year not to return since, a period of nine years. He never had any signs of syphilis or tuberculosis, and is perfectly healthy and strong.

Case V.—Mrs. G. S., of Yonkers, 41 years old, had four children which are all healthy. For the last few years she has had "neuralgia" in the back. Her feet are swollen and she is very nervous. Three days ago her throat became sore and she thought she would suffocate. She feels now as if something were "lying on the left windpipe." I found an irregular ulceration on the left arytenoid cartilage, besides a slight laryngitis and bronchitis. As all these symptoms seemed to me suspicious, I told Mrs. S. to bring her urine next time, when I found outspoken diabetes. The examination of the sputum was negative regarding tubercle bacilli. I have not seen the patient again.*

In looking over these cases I think I am justified in drawing the distinction between benign and malignant ulcerations, although I have to admit that without orthoform Case III might also have run a malignant course, for he was surely very near his end when I saw him. Case IV, however was of such a mild nature that I am sure it would have recovered under any treatment.

In a disease like diabetes where we find trophic disturbances of the tissues all over the body, it is not surprising that we should also meet with ulcerations of the throat; yet I was unable to find any report of them whatsoever. They do not seem to have any special form or shape and look on first inspection at times like tuberculous, and again like syphilitic ulcers; nor is there any special place of predilection for them.

While we know that tuberculosis of the lungs is a frequent complication of diabetes, only *Loeri has seen two cases with tuberculous ulceration (infiltration) of the larynx—but no case of pure diabetic ulceration without tuberculosis has been reported.

*E. Loeri: "Veranderungen des Rachens und Kehlkopfes, etc." Stuttgart, 1885.

Dr. †Joal saw pharyngitis sicca frequently connected with diabetes. I see a great many cases of pharyngitis sicca during the year but cannot remember one single instance that a patient had diabetes at the same time. I do not doubt that it occurs among these patients, but surely not oftener than with other people.

---

†Joal: "De l'angine sèche et sa valeur sémélologique dans la glycosurie et l'albuminurie." Revue mensuelle de Laryngologie, etc. p. 161, 1882.

SURGICAL AND PATHOLOGIC FEATURES OF TUBERCULOSIS OF THE ESOPHAGUS WITH REPORTS OF TWO AUTOPSIES WITH TWO PLATES.

By Willard Bartlett, A. M. M. D.,

ST. LOUIS, MO.,

PROFESSOR OF PATHOLOGIC ANATOMY IN THE MARION H. SIMS COLLEGE OF MEDICINE.

The fact that Dr. Hans Schmaus (1) fails to treat tuberculosis of the esophagus in his work on pathologic anatomy may be taken as tacit evidence that he considers it uncommon. Green (2) dismisses the subject of tubercular inflammations of the esophagus with the statement that they "are very rare." In Kaufmann's (3) text-book is found a mere mention of tuberculous ulcers in this situation.

"Tuberculosis of the esophagus," says Stengel (4) "is extremely rare, and most frequently results from extension of tuberculous adenitis of the bronchial glands." He takes no notice of the various lesions which are possible in this connection. Orth (5) makes the statement in his "Diagnostik" that infectious granulomata in the esophagus are rare, though such of syphilitic and tubercular nature have been observed.

In an inaugural dissertation delivered at Munich in 1895 Hasselmann (6) could collect the reports of only 16 cases from the literature of the preceding 16 years. He himself had observed one case of tuberculous ulceration of the esophagus.

Glockner (7) described, in 1896, a case in which the only lesions were in the muscles, all other component structures of this tube being free from such affection. His observation teaches us nothing of a new disease nor of an unknown lesion; it is interesting simply on account of the location of the tubercle.

The above is sufficient to warrant at least two conclusions: (1) that tubercular affections of the esophagus are exceedingly uncommon, (2) that the authors who treat tuberculosis as an entity and dismiss the subject without even a mention of the esophageal lesions, fail to present a comprehensive view of the subject.

These lesions are: 1. The tubercle itself. 2, the ulcer, 3, the fistula, 4, the diffuse round-celled infiltration observed in any or all the structures in the neighborhood of one of the three first mentioned abnormalities.

One or more of the tunicae of the esophagus may become infected with tubercle bacilli in the following ways:

I. The sputum, as mentioned by Prof. Orth (5), is not uncommonly swallowed by those affected by phthisis pulmonalis.

II. Primary tuberculosis of the digestive apparatus is set up as a result of eating or drinking something contaminated with the specific bacterium; Ziegler (8) and other writers refer to milk as the most common vehicle in such cases.

III. As an example of this disease's spread to the esophagus "by continuity of tissue" it is only necessary to quote Orth (5) as authority for its occurence after a tuberculous ulceration in the pharynx.

IV. A similar affection of this tube might occur as part of a general miliary tuberculosis and be properly speaking, "hemetogenic."

V. I find nowhere a record of this disease in the esophagus being "lymphogenic" in character, still the possibility cannot be denied, for Gray (9) holds that the lymphatic channels from the esophagus communicate with those from the lungs, and the lungs, through the medium of these lymphatics, are commonly infected after tuberculous lymphadenitis in the nodes which Tillaux (10) describes as surrounding the bronchia like rings.

VI. Progress of the disease to the esophagus by "contiguity of tissue" is illustrated by the contents of softened peri-bronchial lymph nodes breaking into the lumen of this viscus.

A description of two specimens which affords multiple illustrations of the sixth etiologic possibility may not be devoid of interest in this connection.

I will for the sake of brevity include here only such notes of the two autopsies as are needful to a complete understanding of the cases.

I. The body was that of a large negro of 30 in a state of fairly good nutrition. Enlarged and softened lymphatic nodes in supra-clavicular regions. Slight edema on lower extremeties. Floating 9th and 10th ribs on both sides. Total splanchnoptosis. Large amount of sero-purulent fluid in left plural sack, plura thickly covered with a layer of fairly adherent fibrin. Complete atelectasis of lower left lobe. Obliteration of right pleural cavity through fibrous adhesions. Several ounces of sero-fibrinous fluid in pericardial sack, pericardium in great part covered with an eruption of granulamata mostly larger than the size termed miliary.

Lungs inflated but somewhat oedematous, a few healed scars at apex.

The spleen and a supernumerary spleen, of the size of a walnut, thickly studded with cheesy masses varying in size from a buckshot to a cherry. Slight evidences of amyloid degeneration in the cortices.

Liver, kidney and small intestines studded with gray nodules, miliary in size; enlarged cheezy lymph nodes in the region of the parta hepatis and root of mesentery.

Very few ulcerations in the lower part of the ileum, varying from a split pea to a dime in size.

The esophagus, the object of chief interest in the cases, presented no less than five fistulous openings in its anterior wall, the same possessing lumina varying in size from a goose-quill to a large lead pencil and leading into cavities within as many different peribronchial and posterior mediastinal lymphatic nodes.

In photograph No. 1 is seen the organ from within, the probes marking the location and general course of four of the fistulae just mentioned. Surounding the five openings referred to and completely covering the middle half of the mucous surface were countless ulcers varying in size from a small split pea to a dime. Their edges were elevated and hard, their bases clean.

Along the margins of the area thus described were a large number of small gray caseous nodules.

Judging from the very apparent age of the different lesions and from the fact that the patient had been rapidly declining for a few months only, I make a diagnosis of chronic lymphatic tuberculosis with tuberculous pleuritis and a secondary general miliary tuberculosis.

The esophagus seems to me to deserve in this case especial mention, (1) because the seat of a rare affection, (2) because of the great amount of esophageal tissue involved, and (3) because it presents all three manifestations, the tubercle, the ulcer and the fistula, this last lesion being responsible for the appearance of the other two, and itself present in such remarkable multiplicity.

II. The body is that of a tall negro of fair bony frame, greatly emaciated, cut muscles of a grayish red color and dry. Between anterior surface of liver and diaphragm are firm fibrous adhesions imbedded in which are caseous masses.

Liver extends one inch below free border of ribs. Serosa covering small intestines dotted here and there with flakes of fibrin and a large number of gray nodules varying in size from a pin head to a No. 4 shot.

On the right side above are five retrosternal lymph nodes, each of them corresponding in size to an almond and all containing caseous material. Lungs do not quite approximate each other, but are not retracted.

Both pleurae are the seats of fibrous and extensive adhesions, firmest over the apices. No abnormal contents on either side. Pericardium presents a few circumscribed fibrous patches; fluid contents slightly increased. Heart presents no abnormalities.

In the posterior mediastinum are many lymph nodes varying in size from a hazelnut to a walnut, the same showing on section caseous patches, those surrounding the bifurcation of the trachea being anthracocised.

The aorta shows a few limited areas in which the intima cells have undergone fatty metamorphosis.

Lungs on section present certain well defined patches of a deep red color and firm consistence, plainly hepatized;

also others of a yellow color in which caseation is evident. They are somewhat edematous.

Spleen is firmly adherent to diaphragm and at the hilus are a number of lymph nodes about the size of a hazelnut, the same undergoing retrograde metamorphosis. The capsule is thickened, the organ is somewhat large; its consistence soft, the pulp hyperplastic and scattered through it are a few well defined caseous patches, in size comparable to a split pea. The kidneys present nothing unusual.

Liver is seen from without to be thickly studded with gray nodules of various sizes. From without as on cut section the organ presents a finely granular surface, and in the porta hepatis lie a large number of nodes about the size of hickory nuts, many of them being caseous. Stomach is distended. Mucous membrane is clouded and marked by numerous small hemorrhages and covered by a thick viscid mucous. Small intestine and large intestine are free from microscopic lesions.

The mesentery lymphatic nodes are uniformly large, in size between a buckshot and a hickory nut, and most of them caseous. At the root of the mesentery lies a mass as large as the two fists of a man, consisting of numerous lymph nodes in all stages of caseation and softening, while upon it rests the head of the pancreas which is thus considerably displaced toward the front.

The esophagus which is here represented by photograph No. 2, presents on its anterior wall but one fistulous opening. From this a sinus leads upward and forward to a softened caseous lymph node situated just behind the bifurcation of the trachea. In the picture a probe is seen, one portion of it being within the esophagus, whose wall it pierces to become visible once more between the two halves of the above mentioned node, this latter lying just to the right of the esophageal border. This organ was absolutely free from further lesions, with the exception of four small superficial ulcers in the immediate vicinity of the fistulous opening.

The findings of the autopsy justify, to my mind, the anatomic diagnosis of chronic lymphatic tuberculosis, miliary tuberculosis of the liver, tuberculosis of the spleen, tuberculous pneumonia, tuberculous peritonitis and secondary ulceration into the esophagus, the result of a tuberculous destructive process in a peribronchial lymphatic node.

Years ago W. Ziemssen (11) made it known that secondary perforations (those from without) of the esophagus occurred most frequently as a result of tubercular destruction of lymphatic nodes. We have the same authority for the statement that such fistulæ are usually single, that two are uncommon, and that more than two are very rarely seen. In consideration of the above. Case No. I, with its five perforations, is even more interesting.

According to v. Ziemssen the tuberculous process does not as a rule extend far out over the mucous membrane, being limited usually to the immediate vicinity of the new opening. In this particular, my two cases illustrate both possibilities. In No. I. a greater portion of the inner aspect of the tube presents unmistakable lesions. while in No. 2. the mucous membrane was found, with the exception of a few small ulcers right at the opening, to be absolutely healthy.

Question might be raised as to whether case No. 1. was not in reality primary (perforation from within outward). No such assumption seems to me tenable, however, when one takes into consideration the very apparent extent and age of the lymphatic derangement, not to mention the fact that neither the pharynx nor stomach is affected, as might reasonably be expected if the primary focus were in the esophagus.

Both autopsies illustrate at least the tendency, mentioned by two of the authors quoted, to involvement of the esophagus in chronic lymphatic tuberculosis. The symptomatology does not however always give a clew to the presence of so grave a complication, as in neither of these cases, observed in a hospital, did clinical manifestations suggest to the attending physician the possibility of an esophageal lesion.

It is then I hope not going too far, in view of the dearth of diagnostic aids at our disposal, to suggest the use of the esophagoscope in every case of pronounced general lymphatic tuberculosis.

Most writers say with Tillmanns (12) that we are indebted to Miculicz for a really useful esophagoscope. The introduction of the electric light in this connection did much, however, to enhance its value, for which v. Hacker (13) must be given credit.

Aside from consideration of the abnormal, one can but regret that this instrument is not better known and more widely used, for v. Hacker (14) after his most extensive investigations warns us that pathologic manifestations cannot be understood until one has become well acquainted with the normal pictures presented in various parts of the esophagoscope.

This contrivance has been put to quite a variety of uses besides the mere examination of the mucous membrane. Kelling (15) mentions it in connection with extraction of foreign bodies, and cauterization both chemical and electric. It is clear, at least, that the removal of foreign bodies would be attended with less maiming, when conducted in full view of the operator. How often, as Kelling (16) suggests, might exploratory laparotomy for stomach diagnosis, be averted by the simple use of the esophagoscope. But the value of the instrument must not be measured by the foregoing alone. Those who have seen at an autopsy the evil effects of "sounding" some diverticula and strictures will agree with me that the old method of doing it in the dark, is open to improvement. In some individuals the use of this aid to diagnosis is attended with considerable difficulty. Kirstein (17) concluded that one-fourth of all individuals lack the anatomic requirements for such treatment.

The ideas of the different investigators vary somewhat as to the correct technique here involved. Kirstein makes use of the esophagoscope without an anesthetic and with the patient in a sitting posture. For v. Hacker (18), the one to be examined must lie on the side and submit first to the application of cocain, and then to the introduction of the Mikulicz instrument. Einhorn (19) prefers the last named instrument and a sitting posture, but uses no anesthetic. It is claimed by Kelling (20) that many of the difficulties experienced by others have been overcome by his jointed esophagoscope, which is to be extended after it has been introduced.

The writers have come to no agreement as regards an anesthetic. Ebstein (21), who has gone deeply into the matter, goes to neither extremity but takes the entirely logical position that no general medication should be indulged in, simply cocainization of the pharynx and entrance of the esophagus.

My conclusions as to the treatment of tuberculosis of the esophagus necessitate a hasty review of the surgery of that organ, embracing as it does many procedures of a nature little short of sensational.

Garre (22) resected in two patients portions of the esophagus together with the whole larynx. His operations were undertaken for the relief of carcinoma; the cut ends of the tube were joined by sutures, and both patients lived. Two cancer cases were treated by Narath (23) with resection of the upper portion of the esophagus. One patient lived one and one-half years and was able to swallow perfectly.

F. de Quervain (24), after describing a case of resection, expresses most clearly his deductions as to further treatment of the organ. The cut ends should be united if possible. When, however, this cannot be done the lower end of the upper segment must be sewn into the skin of the neck that a fistula may result, else the patient must cough up all the secretion that accumulates and at night be constantly exposed to the danger of aspiration pneumonia. The lower segment should at the same time be held up to the surface for the purpose of feeding, but if this be impossible it may be let go, as it soon becomes closed of its own accord, necessitating of course a gastrastomy.

The technique was from the purely theoretic standpoint somewhat enriched by Biondi in 1895. This investigator removed parts of two ribs, split the diaphragm and resected the cardia of dogs, a procedure which has, I may add, never been tried on the human subject.

A gastrostomy on a patient who gradually lost the ability to swallow is reported by Pepper and Edsall (26). The case is chiefly interesting on account of its pathologic aspect; for at the autopsy lung and larynx tuberculosis was found with two *cancerous* strictures of the esophagus and around the same, evidences of a tuberculous process.

From the above as well as from many similar reports it becomes apparent that the surgeons have confined their attention almost wholly to that portion of the tube which lies outside the thorax. We are then in a measure prepared for the statement of Levy (27) that, until very recently, 91.5% of all individuals afflicted with deep-seated growths of the esophagus were doomed to die without operative interference. The same author has a novel way of disposing of the lower end of the organ in dogs after resection; he draws it by a string into the stomach and out of the gastratomy wound, after which he cuts it off at the cardia. This inversion of the tube was found to be impossible on the human cadaver, the longitudinal muscle remaining in situ.

The first to recommend the opening of the undiseased posterior mediastinum for the purpose of operating on the esophagus was the Russian surgeon, Ivan I. Nassilov, but the first to carry Nassilov's daring idea into effect on the living was Rhen (28), who twice resected a portion of the organ for cancer. Stayanov (29), who gives us the most minute details of Nassilov's operation, states that up to March of this year the posterior mediastinum had been opened fifteen times with a mortality of 20% only, the same having been undertaken for the relief of the following conditions, tuberculosis of the mediastinum, acute phlegmon, lesions of the vertebra and cancer of the esophagus. Nassilov's idea was to approach the upper part of this tube from the left and the lower part from the right side, while Patarca (30), whose investigations on this subject have been profound, reasons that the organ on account of its relation to the aorta can properly be invaded from the right side only, no matter which portion of its intra-thoracic extent be involved.

No review of the surgery of the esophagus and posterior mediastinum would be complete without mention at least of the two French surgeons, Quénér and Hartmann (31), whose thesis, I regret to write, is not at my disposal.

From the foregoing it is seen that all portions of the esophagus are, under various pathologic conditions, amenable to surgical treatment. It then suggests itself naturally enough, that tuberculous perforation of this tube might, under the guidance of the esophagoscope be made the subject of similar therapeutics. Partial or total resection is made in typhoid and tuberculous perforation of the intestine in an endeavor to prevent peritonitis, so why may t a like procedure be undertaken in primary perforation of the esophagus accompanied by mediastinal abcess. In this case, simple drainage of the mediastinum without suture of the open esophagus, would be comparable to the treatment of chronic appendicitis by mere incision unaccompanied by the removal of the diseased appendix. However, the esophageal lesions in my two cases are of the secondary variety and the conclusions to be drawn therefrom are in consequence of another sort.

No idea of an operation could for a moment be entertained where the organs were so extensively changed as in case No. I. But in No. II the conditions presented are different; only one fistula existed and the remaining mucous membrane of the tube was practically normal. Here it seems to me that excision of the esophageal lesion would be productive of two directly beneficial results.

I. It would prevent a spread of the disease in the lining membrane of the organ. II. It would certainly obviate the possibility of food passing into the fistulous tract and there decomposing, the foundation of a mediastinal abscess.

I do not deny that caseous peribronchial nodes are usually multiple, that a second or a third might possibly break through into the esophagus, that unknown to us the disease may already at the time of operation have been transmitted to other parts of the body, but I should hope by such a procedure to place a barrier in the way of a progressively fatal malady and, at least, to prolong life.

3894 Washington Boulevard.

BIBLIOGRAPHY.

1. Hans. Schmaus.—Grundriss der Pathologischen Anatomie, 1898.
2. Green, F. Henry.—An Introduction to Pathology and Morbid Anatomy, 1898.

3. Kaufmann, Prof. Edward—Lehrbuch der Speciellen Pathologischen Anatomie, 1896.
4. Stengel, Alfred.—A Text-Book of Pathology, 1898.
5. Orth, Dr. Johannes.—Pathologische-Anatomische Diagnostik, 1894.
6. Hasselmann, H.—Ueber Tuberculose des Oesophagus. Centralblatt für Allegemeine Pathologie and Pathologische Anastomie, 1896. No. 18.
7. Glockner.—Ueber eine neue Form von Oesaphagustuberculosa. Prager Med. Wachenschrift. 1896. Nos. 11, 12, 13.
8. Ziegler, Ernst.—Lehrbuch der Allgemine, Pathologie, 1895.
9. Gray, Henry.—Anatomy Descriptive and Surgical, 1893.
10. Tillaux, P.—Traité d'Anatomie Topographique, 1898.
11. Ziemssen.—Encyclopedia of the Practice of Medicine, Vol. VIII, 1878.
12.—Tillmanns—Lehrbuch der Specillen Chirurgie. 1897.
13. v. Hacker.—Die Oesophagoskopie und ihre Klinische Bedentung. Beiträge zur Klin. Chirurgie, Bd. XX, Hft. 1.
14. v. Hacker.—Die Oesophagoskopie beim Krebs der Speiseröhre und des Magens. Beiträge zur Klin. Chirurgie, Bd. XX, Hft. 2.
15. Kelling.—Demonstration der Besichtigung der Speiseröhre mit Biegsamen Instrument. Allgemeine Med. Centralzeitung, 1898. No. 65.
16. Kelling.—Endoskopie für Speiseröhre und Magen.Münchener Med. Wochenschrift, 1898. No. 49 und 50.
17. Kirstein.—Uber Oesophagoskopie. Berliner. Klin. Wochenschrift. 1898. No. 27.
18. v. Hacker.—Uber der Technik der Oesophagoskopie. Wiener Klin. Wochenschrift. 1896. No. 6 and 7.
19. Einhorn.—Die Besichtigung der Speiseröhre und der Cardia. New Yorker Med. Monatschrift 1897. No. 12.
20. Kelling.—Endoskopie für Speiseröhre und Magen. Münchener Med. Wochenschrift, 1897, No. 34.
21. Ebstein, L.—Uber Oesophagoskopie und ihre therapeutische Verwendbarkeit. Wiener Klinische Wochenschrift, 1898. No. 6 und 7.
22. Garré.—Ueber Larynx und Oesophagusextirpation. Munchener Med. Wochenschrift, 1898. No. 18.
23. Narath.—Beiträge zur Chirurgie des Oesophagus und des Laryux. v. Langenbeck's Archiv, Bd. LV., Hft. 4.
24. De Quervain. F.—Zur Resection des Halsabschnitts der Speiseröhre wegen Carcinoma, v. Langenbeck's Archiv, Bd. LVIII. Hft. 4.
25. Biondi.—X. Congresso della Società Italiana di Chirurgia in Roma.
26. Pepper, W., and Edsall, D. L.—Tuberculous occlusion of the esophagus with partial cancerous infiltration. American Journal of the Med. Sciences. 1897. No. 7.

27. Levy, W.—Versuche über die Resection der Speiseröhre. v. Langenbeck's Archiv., Bd. LVI, Hft. 4.

28. Rehn.—Operationen an dem Brustabschnitt der Speiseröhre. v. Langenbeck's Archiv., Bd. LVII, Hft. 4.

29. Stayanov, P. I.—Les interventions Chirurgicales sur le Médiastin postérieur et les organs y contenus. Revue de Chirurgie, 1899. No. 3.

30. Patarca, G.—La Chirurgie intramediastinale postérieure, 1898.

31. Quénu et Hartmann.—Des voies de pénétration chururgical dans le médiastin postérieur. Bulletins et Mémoires de la Société de Chirurgie de Paris, 1891.

THE EFFECT OF ATMOSPHERIC CHANGES ON THE HEARING IN CHRONIC CATARRHAL OTITIS MEDIA.*

Seymour Oppenheimer, M. D.,

ATTENDING LARYNGOLOGIST TO BELLEVUE HOSPITAL, OUTDOOR DEPARTMENT; ATTENDING LARYNGOLOGIST TO UNIVERSITY MEDICAL COLLEGE DISPENSARY; SENIOR ASSISTANT TO THE CHAIR OF LARYNGOLOGY, UNIVERSITY MEDICAL COLLEGE.

It appears to be commonly accepted by the laity, and to a certain extent by physicians, that audition, especially when the auditory apparatus is the seat of a morbid process, is unfavorably influenced by atmospheric changes. With the view of ascertaining the detrimental action of barometric and thermal changes upon the already impaired hearing in catarrhal deafness, fifty consecutive cases of chronic sclerosis of the middle ear were studied over a considerable period of time, and the usual tests were used to determine the variations in the hearing under different atmospheric conditions.

The normal ear is practically uninfluenced by changes in the weather, excepting that, depending upon the amount of atmospheric moisture, the transmission of sonorous vibrations from a considerable distance is more or less clearly appreciated. This follows well known physical laws, and when clearly understood enables one better to appreciate the effect of the same atmospheric conditions upon the diseased ear. For the same reason the intimate relationship of the tympanic cavity to the nasopharynx and other portions of the upper respiratory tract must be taken into consideration. The auditory end organ, for it may be so considered, is but a specialized diverticulum of the respiratory areas, and partakes in all respects except

---

*Read before the New York County Medical Association, March 20, 1899.

in its special function of the histologic characteristics of the former. The mucous membrane of the tympanic cavity is but an extension of that of the nasopharynx and the physical conditions acting upon the mucosa of the latter affect in the same way that of the middle ear. This is not only anatomically demonstrated in health, but becomes especially marked in catarrhal conditions of the mucous membrane of the nose and pharynx, it being an almost constant rule to find diminution of auditory acuity following long-continued catarrh of the nose or throat.

The vast majority of cases of suppuration of the middle ear result from extension of inflammatory nasopharyngeal changes through the continuous mucous membrane of the Eustachian tube and subsequent infection through the same route. It is decidedly uncommon to observe a case of chronic sclerosis of the tympanum without evidences of nasopharyngeal catarrh antedating the aural affection. As a final evidence of the existence of this intimate connection, little can be done toward the improvement or restoration of the impaired hearing until the other portions of the upper respiratory tract are restored to an approximately normal condition. Therefore it becomes necessary for us to regard catarrhal states of the tympanum and catarrh of the nasopharynx as varying manifestations of the same disease, its phenomena varying only by the interference with the functions of the special organ involved, both being subjected to the same atmospheric influences.

As a rule all the conditions favorable to the production of morbid changes of the respiratory and tympanic mucous membrane are dependent upon a lowering of the general vitality, mainly due to the continuous action of unfavorable climatic or hygienic surroundings. This is of such common experience that it is not necessary to cite individual cases, but it was plainly brought out in the cases studied that in those individuals in whom the hearing was most adversely affected is unfavorable weather a lowering of the general health was present at such times. And when one is run down by disease or other causes, such as fatigue, the susceptibility to temperature and barometric changes is greatly enhanced. Unfavorable climatic conditions also, especially if their action on the individual is repeated or constant, produce a state of more or less systemic depression, which, in turn, reacting upon the mucous membrane, renders them less able to resist the morbid causes of catarrhal inflammation.

Primarily the changes in the atmospheric tension causes hyplealmia of the mucous membranes by chilling of the body surface, when the dermal covering is not sufficiently prepared for the action of cold or excessive humidity. This is true in a general sense of the entire mucous membrane system, and especially that of the respiratory tract, but locally the action of the lowered barometric pressure may be well observed in the nasal chambers, especially in an individual affected with catarrhal rhinitis. Under these conditions marked congestion and swelling, with increased secretion, result, the turbinal cavernous tissue becomes enlarged, and nasal respiration is decidedly diminished. In bright, clear weather, with a normal or rising barometer, capillary congestion does not take place, and there is no obstruction to the passage of air through the nares.

Besides the relation borne by the auditory apparatus to the upper respiratory tract, the tympanic cavity in its relation to the normal or the varied atmospheric pressure, is related indirectly to the various pneumatic spaces of the head. Changes in the normal atmospheric pressure act the same on the cavities containing air (the antrum of Highmore, etc.) as they do on the tympanum, the pressure exerted in these internal cavities being the same as the pressure exerted upon the external surface of the body. Practically, however, the pressure in the other head cavities does not vary, while the air in the tympanum changes to a certain extent with every act of swallowing, this portion of the pneumatic system being but temporarily in direct communication with the outer air. Upon this fact depends the restoration of hearing in certain cases of beginning catarrhal otitis. By the opening of the Eustachian tube the rarefied air in the tympanum is replaced with the normal pressure air. Diminution of the intratympanic pressure, even to a limited extent, seriously interferes with the transmission of sound waves.

For the reason that all the cases presented well-marked

evidences of sclerotic changes of the tympanic tissues, the term chronic sclerotic otitis media is here used as a characteristic designation in preference to the more common one of catarrhal deafness. In all these cases sclerosis and atrophy represent but later stages of a previous cell proliferation, and a complete understanding of pathologic changes occurring in the tympanum is necessary to comprehend the deleterious effect produced on auditory perception by atmospheric influences. The pathogenesis is constantly and slowly progressing from the initial catarrhal inflammation, with its cellular increase, exudation, and overgrowth, to sclerosis and contraction, and ultimately atrophy of the mucous lining. When this late stage is reached the blood supply is much limited, the tissues are firm and contracted, and the Eustachian tube and nasopharynx in the majority of cases partake in the more or less general condition. In the pure atrophic stage atmospheric changes in no way affect the greatly diminished hearing capacity; while in the sclerotic stage, where examination of the tympanic cavity shows patches of beginning or well-marked atrophy alongside of areas of cell proliferation or hyperemia, atmospheric changes invariably impair for the time being the already affected audition.

The degree of hearing impairment is influenced in proportion to the location of hyperplastic tissue situated at or in immediate proximity to the path of the transmission of sound waves. Should sclerotic or atrophic changes exist in practically all portions of the tympanum, except a small area of proliferating connective tissue situated at the point of articulation of the foot plate of the stapes with the oval window, any slight change in the barometric pressure would be immediately appreciated and impairment of hearing rapidly ensue from the additional succulency of this mass of tissue and the resultant interference with the transmission of sonorous vibrations from the ossicular chain to the important structures of the perceptive apparatus. Further, as a result of repeated acute attacks of catarrhal inflammation of the respiratory mucosa, resulting from atmospheric alterations, Eustachian salpingitis is of frequent occurrence, and from the consequent occlusion of the Eustachian tube stagnation of the air in the tympanum results: this air, originally of the same pressure as the surrounding atmosphere, is to a great extent absorbed by the blood-vessels and becomes rarefied. The usual pressure is then removed from the capillaries and they become engorged with blood, and exudation of serum takes place, greatly interfering with the hearing. The hearing is also influenced through atmospheric changes in other ways than this; such as the production of an acute syringitis superadded to the already existing chronic process, and occasionally acute inflammation of the tympanic mucosa may result and produce impairment of audition in the same manner.

In all, fifty cases were under observation, and of these there were thirty-one females and nineteen males.

As is well known, females are usually more susceptible to the ill effects of variable weather than males, especially as they are not, as a general rule, exposed as much to the elements as are the latter. Sex, however, bore no relation to the effect of barometric and thermal changes upon the hearing, the proportionate majority of those stating that the hearing was more impaired in damp weather being males. The youngest patient was nine years old, while the eldest was eighty-three. A careful analysis showed that although the hearing ability was less in those of advanced years than in the younger patients, yet all things being equal, the age of the individual was in no way a factor concerning the relation of the atmospheric variations to the impaired hearing.

The duration of time since the individual's attention was first attracted to the impaired hearing varied greatly, and from the insidious nature of this form of otitis many of the patients could not state the time when the impaired audition was first noticed. Of these, there were nineteen patients, and all stated that the aural affection had existed for at least two years, but other than this no definite data as regarded the duration could be obtained. In ten the disease had existed for a year; in eleven, between one and five years; in two, between five and ten years; in three, between ten and fifteen years; in three, between fifteen and twenty years; and in the remaining two, the affection had existed between twenty and thirty years. The duration of time intervening from the appearance of gradual impairment of hearing to the time when the patients were first seen varied considerably, as usually morbid changes in the tympanum are fairly well advanced before the impairment of hearing is noticed by the patient. A number of the cases bore evidence to this point; one, for instance, as a typical example, stated that the hearing became impaired but one week before he was first seen, but examination showed that the disease had existed for several years, and the impairment of the auditory function had undoubtedly existed for a considerable time, but attention was only directed to it by an attack of coryza caused by a sullen change in the weather.

The duration of the affection per se bore no relation to the increased impairment of hearing under unfavorable atmospheric changes, this point alone depending upon two factors: First, the amount and location of hyperplastic tissue in the tympanic cavity, and, secondly, the condition of the nasopharynx and Eustachian tubes. Of the first factor the location of the hypertrophied tissue has been referred to, and requires no further consideration here; the amount of new tissue or cellular proliferation is also an important factor in determining the influence of atmospheric changes upon audition in the specific case. Of the fifty cases, eleven presented every evidence of a general atrophic condition of the tympanum, and in all these the hearing was greatly impaired both for the aerial and bone conduction; four of these stated that sometime in the past the hearing was worse on rainy or damp days, but not to any marked extent, while in the remainder, atmospheric conditions had no influence at all. At the time the cases were examined, however, all gave evidence that barometric changes were entirely negative.

Twenty-one cases presented at the same time various phases of the pathologic changes incident to the catarrhal otitis, the examination disclosing that the atrophic condition was not as well marked as were the sclerotic and hyperplastic changes. All of these were influenced by thermic and barometric changes, varying in degree of the impairment of audition with the individual case; while in the remaining eighteen the hyperplastic condition predominated and the hearing was markedly diminished during long continued spells of damp, rainy weather, or when there were sudden atmospheric changes.

In this affection especially, the three morbid stages mentioned do not always follow each other in sequence; many are sclerotic or atrophic from the commencement, while in others areas of hyperplasia may remain for a number of years and shrinking and absorbing changes not take place till after long periods of time. It will be seen, therefore, that the susceptibility of the affected ear to adverse weather conditions depends to a great extent upon the special pathologic condition present and not upon the duration of time the affection has existed.

Attention was not especially paid to the influence exerted by the upper respiratory tract; but, independent of the aural pathogenesis, it was found that thirty-two of the cases complained of nasal obstruction, especially marked in rainy weather, and when this occurred the hearing was more diminished than when the turbinal swelling did not exist and when the nares were free. Whether one or both ears were affected bore no relation at all. In thirty-nine cases both were involved; in seven, the right only, while in the remaining four the affection was limited to the left side. The degree of hearing may fluctuate from day to day dependent upon the changes in the weather, and frequently, as a result of the catarrhal salpingitis usually present, there is an associated sense of fullness in the ears, referable to the Eustachian obstruction from the barometric changes.

Three factors in the climatic conditions exert to a greater or less extent their influence upon the aural tissues in morbid changes of the tympanum; these are the humidity, barometric pressure, and temperature. Of minor importance are the electric and other phenomena associated to a much less degree with the three prominent factors. The action of barometric pressure and humidity has been considered, while the effect of purely temperature changes is most difficult of explanation, but, so far as can be seen, practically exerts little influence. Generally, however, patients with catarrhal deafness will state that they are worse in the winter months, and during the heated term of summer the symptom complex, and especially the deafness is much less prominent. In all probability it is not the cold in winter that exerts a more deleterious action upon the ear than the higher temperature of summer, but the explanation seemingly lies in the excessive dampness of the winter months and the consequent tendency to the production of catarrhal changes of the respiratory and aural mucous membranes.

It may be here pointed out that catarrhal affections are most frequent in the transitional periods of autumn, and especially in the spring, during which times sudden alterations of cold, heat, dampness and moisture become unduly exaggerated. The deleterious effects of climatic conditions upon the auditory apparatus depend therefore not to such an extent upon the yearly range of temperature, humidity, and pressure, as upon the suddenness with which these conditions are apt to vary in a given period of time. For this reason catarrhal otitis is the most common in localities where sudden changes occur in varying rapidity.

CONCLUSIONS: 1. The hearing in at least seventy per cent. of cases with chronic catarrhal deafness becomes worse under adverse weather conditions.

2. The degree of impairment of audition, as influenced by atmospheric changes, is determined to a great extent by the location and character of the pathologic process in the tympanic cavity.

3. The morbid alterations most susceptible to barometric variations are those of hyperplasia.

4. In purely atrophic changes in the middle ear weather variations have little or no effect upon the auditory function.

5. Atmospheric influences also impair the hearing by unfavorably affecting catarrhal processes of the upper respiratory tract and Eustachian tube.

6. All things being equal, the impaired audition in chronic catarrhal otitis is diminished more (under unfavorable weather influences) in those whose general health is below par than in those otherwise healthy.

706 Madison Avenue.

DILATATION OF THE HEART COMPLICATING OBSTRUCTIVE LESIONS OF THE UPPER AIR PASSAGES.

JOHN A. THOMPSON, M. D.,

CINCINNATI, O.

It is well known to the profession that the lungs are chronically congested in valvular heart diseases, especially in mitral insufficiency. It is not so well known that the pharynx and turbinated bodies in the nose are also overfilled with venous blood in this disease. So long as compensation is perfect the condition of the heart does not affect any disease in the upper air tract. When compensation fails, the impaired circulation makes the cure of nasal and pharyngeal inflammations impossible. The effect of this venous congestion on the prognosis of diseases in the upper air tract has not been given sufficient attention in the literature of laryngology. There is not, to my knowledge, any suggestion by any medical writer that the contrary proposition may be true, that is, that obstructive disease in the nose or throat may produce a dilatation if not an actual valvular disease of the heart. Some cases I have treated suggest so strongly this teaching that I have concluded to put my observations on record. The great court of last resort in medicine, the general profession, may then decide whether or not the symptoms seen in these cases have been correctly interpreted. The following cases are chosen out of many I have seen because they are typical of different forms of this disorder.

W. J. T., aged 39, a large, robust appearing man, a farmer by occupation, consulted me in October, 1894, in regard to obstructed respiration. Slight hypertrophic rhinitis was found and a large spur in the left nostril growing from the septum nasi. No general examination was made at the time but the spur was removed under cocain anesthesia. There was temporary heart failure from the cocain at the time of the operation but the symptoms were not serious enough to be alarming. An examination of the chest made three days after the operation showed a slight passive congestion of both lungs and a markedly dilated left ventricle with the apex beat to the left of the nipple line. There was a loud mitral regurgitant murmur which was heard by Dr. Oliver as well as myself. The pulse rate when he was sitting quietly in the office was 120. The patient was given tincture of digitalis, ten minims four times a day, and a spray of menthol and camphor for the nose. He was placed in the care of his home physician until the pulse rate should be normal. He returned to his home one week after the operation. I examined him there one year later and found him with a pulse of 72, the heart contracted to its normal size, no murmur, no cough, no pulmonary congestion and no obstruction of the nose.

J. E. G., aged 23 years, a very tall, slender blonde, consulted me in August, 1898, in regard to an obstruction of the left nostril which interfered with free respiration. He had atrophic rhinitis. The septum nasi was deflected to the left and a large spur was in contact with the outer wall, touching the inferior turbinated body. This growth was removed by the ordinary method. As is usual in cases of atrophic rhinitis, healing was somewhat delayed. After all the obstruction had disappeared, the patient was still somewhat short of breath. Examination showed the apex beat of the heart in the nipple line. The left ventricle was considerably dilated. The pulse rate was 90 per minute. There was no valvular murmur. He was given tincture of digitalis, ten drops after each meal. October 12th the pulse rate was normal and the dyspnea was entirely relieved.

I was asked March 4th of the present year, by Dr. A. R. Walker, to see F. B., aged four years, for the purpose of removing hypertrophied tonsils and adenoids. Having great confidence in the Doctor's diagnostic skill, I went as requested, prepared to operate. The patient was an undersized, poorly developed child. With the exception of his throat trouble he had always been healthy. The family physician had not had occasion to examine the chest. Making the routine examination before deciding what anesthetic to give, the heart was found with the apex beat in the axillary line. The area of cardiac dullness was greatly increased in every direction. A loud systolic murmur was heard over the apex. The pulse rate was 118 per minute. The operation was postponed and the child given tincture of digitalis, eight drops four times daily. April 8th the pulse had been reduced to 100-108 per minute and was full and strong. The operation was then made under ether anesthesia. The operation and recovery were uneventful. The heart has contracted some but the murmur is still present.

January 7th of the present year I examined Miss J., aged seven years, for Dr. A. D. Birchard. Hypertrophied tonsils and adenoids were diagnosed and their removal advised. The child was nervous and frightened and to this the high pulse rate was attributed both at the first examination and on the day of operation. Two days after the operation it was noticed that the child frequently took long sighing respirations as though she was always short of breath. The mother then told me she had noticed this symptom for several months. An examination of the chest showed the lungs slightly congested and the heart considerably dilated. The apex beat was a half inch to the left of the nipple line. The pulse varied from 120 to 130 per minute. There was no murmur. Convalescence from the operation in this case was interrupted by an eruption of measles on the fifth day after the operation. When the fever from this intercurrent affection had disappeared she was given digitalis. The drug was not administered as long as was desirable. At a recent examination the heart was found to be smaller but has not yet returned to its normal size. There is no dyspnea.

Another phase of the clinical picture presented by patients with obstructed nostrils and dilated heart was shown in the case of Rosa R. She is a tall, thin, anemic girl of 23. She consulted me in November, 1898, complaining that all the symptoms of a hypertrophic rhinitis, for which I had treated her seven years before, had returned. In addition to the nasal symptoms there was marked dyspnea on exertion. The nose did not show on examination the color of hypertrophic rhinitis, but that of venous turgescence. The turbinated bodies did not contract quickly or completely when cocain was applied. Examination of the chest showed the lungs congested and the area of cardiac dullness increased. The apex beat was in the nipple line. The pulse rate was 120 per minute. There was no murmur. This was a case where the nasal symptoms were not due to a recurrence of a former disease but to impaired circulation. Local treatment had little effect on the nose until the pulse rate had been reduced by digitalis to 90 per minute. Then the nasal stenosis and accumulation of secretions were relieved.

The knowledge that obstructive lesions of the upper air passages are occasionally complicated by weakening and dilatation of the heart is important to every practitioner of medicine. The family physician can save himself embarrassment and damage to his reputation by a careful examination of the chest before referring such cases for operation. The operator must know the condition of the heart before inducing either general or local anesthesia. Local treatment of a mucous membrane filled with venous blood does little good. Success in the class of cases just described demands a combination of local and general treatment.

SUPPURATIVE ETHMOIDITIS AND ITS TREATMENT.

Frank S. Milbury, M. D.,

NEW YORK CITY,

BOROUGH OF BROOKLYN, SURGEON IN THE EAR, NOSE AND THROAT DEPARTMENT OF THE BEDFORD HOSPITAL, AND ASSISTANT AT THE NEW YORK EYE AND EAR HOSPITAL. (EAR DEPARTMENT.)

In diseases of the accessory sinuses of the nose, it is somewhat doubtful, in the opinion of rhinologists and otologists to-day, which is the most frequently involved, the maxillary sinus or the ethmoid: but I think the consensus of opinion will favor the latter, as recent critical research in pathologic conditions of the nose has shown trouble much oftener in this region than was ever suspected.

The condition may be either an acute or chronic catarrhal inflammation. The acute is not of serious import, generally subsiding quickly without much treatment, whereas the chronic form may end in suppurative ethmoiditis which usually occurs in the course of a catarrhal rhinitis in either its acute or chronic form. The radical view taken by Bosworth and Woakes that "the tendency in all cases, and the result in the large majority of instances of a suppurative inflammation of the ethmoid cells is necrosis of bone," and that "all ethmoiditis tends toward and usually develops, sooner or later, into necrosis" is an observation I think not experienced by rhinologists, generally, in their clinical work. When cases due to syphilis, tuberculosis, mercury, phosphorus or trauma are eliminated, it is not often that necrosis or caries is found.

Occasionally, suppuration has been observed in conjunction with facial paralysis, and it is an undecided question whether the suppuration is due to the paralysis or the paralysis due to the suppuration. Weichselbaum believes in the first, and quotes several cases to prove his theory, whereas Zuckerkandl is inclined to the latter.

Bosworth classifies ethmoiditis into three divisions: (1) Extracellular myxomatous degeneration, the disease being limited to the middle turbinated body: (2) intracellular myxomatous degeneration, in which not only the middle turbinated body, but also the ethmoid cells had undergone myxomatous degeneration; and (3) purulent ethmoiditis, which may or may not be associated with myxomatous degeneration of the ethmoid, but which usually is associated with nasal polypi.

The inflammation may extend to the cells from the nose, the orbit or any of the accessory cavities, or it may occur by a closure of the osteum ethmoidale by pressure of a turgescent or hypertrophied middle turbinated body, deflection of septum, nasal polypi, etc.

There is much difference of opinion as to whether ethmoiditis causes polypi or that polypi produces ethmoiditis. I believe that clinical experience teaches us that in a measure, both opinions are correct.

A severe inflammation is often followed by necrosis. Syphilis, no doubt, is one of the most frequent causes of necrosing ethmoiditis. Any or all parts of the ethmoid may be affected by an inflammatory process, which is bad enough at any stage, but when it reaches the suppurative type, becomes a most difficult affection to treat successfully. This disease may, and often does, extend to the frontal, maxillary and sphenoidal sinuses. Bosworth, in thirteen cases of ethmoidal disease, found antrum implicated in seven—over 50 per cent. Many cases of exophthalmia, narrowing of the field of vision and blindness have been reported, caused from the pressure of empyema of the ethmoidal and spenoidal sinuses.

I have found suppuration of the ethmoid quite frequent, but necrosis rather rare, excepting that due to syphilis, phosphorus, mercury, etc. If pain is experienced, it may be at the base of the nose, intraorbital at the back of the eye or in the eyeball, or a very general pain over the whole head, but oftener, there is no pain whatever. Mucopus will be seen to ooze from under the anterior extremity of the middle turbinated body, or the turbinated may be greatly distended and blanched in appearance with no flow of pus, as I have seen in several of my cases. Again, it may empty into the orbit and the pus be discharged from the inner canthus of the eye. It is generally attended with some fetor, but not so pronounced as is the case of empyema of the antrum. If only the ethmoid is involved, the fetor is not perceived by the sufferer, owing to the impairment of the sense of smell.

The health is usually much impaired; asthma, hay-fever and other neuroses often exist, also loss of appetite, nausea from the muco-pus secretion passing through the post-nasal space into the stomach. It is sometimes difficult to tell which cavity is the seat of the disease.

As before stated, a simple inflammation generally subsides spontaneously as soon as the acute nasal difficulty passes away, but a suppurative process may be most difficult to conquer, often lasting for months and years; occasionally never responding to the most skillful treatment.

It may happen that the whole intracellular structure is broken down and carried into the nose, thereby giving a good drainage and effecting a rapid and complete cure.

The treatment consists in thoroughly cleansing the nasal cavities with some antiseptic solution, removing all obstructions to free drainage. Among such are found deflected septum, hypertrophies, spurs, polypi, etc. By good care, a few will get well, but by far, the large majority will require surgical interference. This is accomplished by the removal of the anterior portion, or in some cases, the entire middle turbinated. This is quite readily done by the scissors, drill or various nasal cutting forceps, preferably those devised by myself which I have found to work better than any others. If they are used once by rhinologists, I believe will be continued in use until something better is invented. However, their merit or demerit will be demonstrated. After this is done, curetting or drilling with a bar to break up the intracellular structure is necessary in order to give free exit to pus, and access to the parts for proper cleansing. The wholesale removal of the turbinates, as practiced by Woakes, is absolutely uncalled for, as is proven by most of the leading rhinologists of the world. Recent cases respond rather quickly to treatment, but those of two or more years' standing are extremely obstinate, and I am doubtful if they ever get entirely well, no matter how cautiously handled. This seems to have been the experience of most men.

On referring to my history records, I find that during the last eight years, there have come under my treatment thirty-nine cases of suppurative ethmoiditis, a few of which I now report.

CASE 1. Mrs. J., 43 years of age, consulted me on April 3d, 1895, complaining of severe headaches particularly on left side, and great pressure at root of nose and in left eye. For eight years, has suffered intensely with asthma. An oculist had corrected a high degree of astigmatism, giving her some relief when reading or doing fine work. A surgeon had divided and exsected a portion of the left supraorbital nerve with negative results. She was advised to leave the sea coast and go inland in hope of relief which she did, spending sometime in Colorado, New Mexico, California and Honolulu with no benefit, excepting what resulted from diverting her mind in a measure, from her trouble. Eventually she returned to Brooklyn thoroughly discouraged. Was advised, for the first time, to consult a nose and throat specialist and referred to me by her family physician.

Examination shows great enlargement of the left middle turbinated body pressing on septum and outer wall of nose, which was broadened unilaterally. A small swelling appeared in inner canthus of eye. Some exophthalmia and contraction of the field of vision of recent date. The turbinated had a blanched appearance and was pultaceous to the touch, indicative of a cystic or suppurative condition, although no pus was visible. Otherwise, both nares were about normal. The general health was miserable.

With the cutting forceps, I removed the anterior extremity of the middle turbinated body, resulting in a free exit of fetid pus, curetted with a sharp instrument easily breaking down all possible intracellular structure which was carious; opened from the eye into the ethmoid cells;

cleansing thoroughly with 1 to 10,000 bichloride followed by an insufflation of aristol.

Treatment was continued with more or less regularity over one year, but pus formation has not entirely ceased. However, the asthma, headaches and all eye symptoms have absolutely subsided, and the lady is now in robust health.

The cause of this case is obscure, but it was probably a neglected intranasal affection.

CASE 2. Chas. B., 24, referred to me by Dr. Barker, October 9th, 1897, complained of constant occlusion of nose, being compelled to breathe through the mouth both day and night. Is a sufferer from hay fever and asthma, going to the White Mountains every summer to obtain relief. Has had on neck within six months thirty-two boils and one carbuncle of large dimensions. General cachexia, specific history at 18 years of age. Examination shows very large septal spurs right and left, and greatly hypertrophied inferior turbinates on both sides pressing on septum almost wholly filling vestibule. I could not see middle turbinates even under what contraction could be produced by cocain, which was but slight. Constant oozing of pus into nasopharynx. With saw, I removed spurs and later a portion of each inferior turbinated. In a few days, I was able to make a proper examination of the middle turbinates finding the right much enlarged and pus coming from beneath the anterior portion. Excepting the prophylactic measures necessary, I did nothing else for several weeks to give time for at least a partial recovery from the operations, at which time, pus was discovered flowing into the nose from the maxillary sinus. The question arose whether the ethmoiditis was caused by a backing up of the pus into the ethmoid from the antrum through the ducts, or the reverse.

However, I at once cut into the ethmoid and removed a large quantity of carious bone. The discharge from antrum soon ceased, proving that the antral trouble originated in the ethmoid. The boils promptly disappeared and no more have formed; also no hay fever or astma has manifested itself. Health excellent, and the patient has gained thirty pounds.

CASE 3. Mr. M., 54 years of age, consulted me on May 10th, 1893, complaining that for about seven years he had suffered severely with neuralgia; the pain starting from inner angle of right eye and extending first over the corresponding side of the head, and eventually over the whole head. If his nose became occluded, the pain was more severe. For many years, a flow of pus came from the right nostril, and the sufferer noticed that the pain was much greater when the pus ceased altogether or nearly so, than when there was free discharge. He had been treated by his family physician by both local applications, internal medication and nasal spray. A resection of the supraorbital nerve had been advised, but had not been carried out.

Rhinoscopic examination disclosed a much enlarged right middle turbinated body from around which a purulent secretion was coming. The ethmoid cells were opened by the forceps and curetted, and although filled with pus, no dead bone could be detected. The pain subsided immediately, and in two months' time the cure was apparently complete. In this case, only the anterior cells were involved.

In my thirty-nine cases, the maxillary sinus was implicated in sixteen. Eight were carious. Suppuration of the whole cellular structure in eleven and anterior involvement of twenty-eight.

215 Jefferson Avenue.

PLASMINE SOLUTION AS A RATIONAL CLEANSING AGENT.

By A. D. McConachie, M. D.,

BALTIMORE, MD.

ASSISTANT SURGEON TO THE PRESBYTERIAN EYE, EAR AND THROAT CHARITY HOSPITAL.

The necessity for cleansing solutions in the various diseased conditions to which mucous membranes are liable, has been productive of many and varied agents used in their treatment. Again in all wounds surgical or traumatic, septic or nonseptic, a necessity for some cleansing agent has existed, for the purpose of washing away debris or detritus and thus favoring resolution or healing, either by lessening the chances for infection or neutralizing infection already present. Sterile water alone, or water containing in solution various antiseptic agents has been our means hitherto. Yet such solutions are more or less incompatible with the normal vitality and activity of the tissues and cells of the parts affected. True in cases of virulent infection the need for the strong germicidal solutions is apparent, yet even here their use may be and doubtless is detrimental to the reparative powers of the tissues.

To environ the tissues with their normal pabulum should be our aim, hence our solution should be as compatible as possible with the tissues of the body and not laden with agents which while destructive to bacteria are also destructive to normal functioning of the part. With this end in view for the past year I have been using a sterile aqueous solution containing approximately the inorganic constituents of the blood in my treatment of all non-septic conditions of the eye, ear, throat and nose. Even in septic conditions of the same by mechanically washing away infective material I believe to be more conducive to rapid restoration than by using more toxic and hence more incompatible solutions.

In post operative conditions—e. g. after intra-nasal, pharyngeal, ocular or aural operations especially when done on non-infected parts, I know its use to be followed by more rapid and kindly repair. In the virulent infection of gonorrheal ophthalmia, when the stronger antiseptics are imperative, the urgent need of thorough cleanliness is seen, and here the solution answers well. That it has a more extended use in the post operative washing in general surgical and inflammatory affections I have no hesitation in predicting.

After using it in powder form put up by a local druggist in proper proportion for solution I had Parke, Davis & Co. prepare a tablet which when dissolved in two ounces of sterile water gives the normal inorganic salts in the proportion found in the blood and in solution of about the specific gravity of blood, but of course devoid of its albuminous constituents.

Each tablet (called Plasmine) contains
    Sodium Chloride.................5 1/2 grs.
    Sodium Carbonate...............1 1/2 grs.
    Sodium Phosphate............... 1/4 gr.
    Sodium Sulphate.................1 1/2 grs.
    Potassium Chloride.............. 1/4 gr.
    Potassium Phosphate........... 1/12 gr.
    Potassium Sulphate............. 1/4 gr.

I am satisfied that a trial of this solution will prove itself satisfactory both to the patient and the doctor in those irritative ocular and nasal conditions whether chemical, mechanic or operative requiring a lotion to wash away irritant material and thus favoring a rapid restoration to the normal.

That it is better tolerated than normal salt solution or sterilized water, borax drops, boric acid drops, etc., I feel convinced from the patient's statements. That it rapidly lessens infection I have proven by noting the lessened multiplication of various bacilli and cocci found in the various catarrhal conditions of the mucous membranes of the eye, nose and throat.

I trust others will give it a trial and report their experiences.

805 N. Charles St.

PERSONAL OBSERVATIONS IN THERAPY.—VALSALVA'S METHOD REVERSED.—MUCO-CUTANEOUS LESIONS.—HEMOSTATIC FOR MUCOUS SURFACE OPERATIONS.—A SPLINT FOR SEPTAL OPERATIONS.

By A. T. Mitchell, M. D.,
VICKSBURG, MISS.
OPHTHALMOLOGIST TO THE STATE CHARITY HOSPITAL.

This article is published in the hope that to some readers at least sufficient novelty may be found to justify publication, and the assurance is made that in the writer's modest facilities for practice the measures enumerated have at least done no apparent harm.

VALSALVA'S METHOD REVERSED.

A consideration of the pressure exerted on the membrana tympani by this effort at expiration with outlets closed, involves only the simple calculation of the individual's vital respiratory capacity and his capacity for forced expiration as indicated by some form of spirometer. With the first factor determined by individual measurement, and the second by spirometric registry, the normal atmospheric weight gives the third.

Nothing is quite so familiar to us as this way of forcing air to the tympanum by using the expiratory muscles' utmost effort to reduce the volume of the air in a previously expanded chest.

The result in forcing out the drum membrane can be as above shown accurately expressed in definite values.

Allowing for the difference in inspiratory and expiratory efforts in individual sets of muscles, I venture the claim that the same factors will figure in the estimation of the force exerted in the following manoeuvre for decreasing instead of increasing intra-tympanic atmospheric pressure.

If with the tragus pressed in the external auditory meatus as far as possible after forced expiratory effort, the anterior nares closed and the greatest possible inspiratory effort made, it stands to reason that there must be a decrease of pressure in the upper respiratory tract.

Granting that the space occupied by the air of an ordinary inspiration at the average in America, as being 24 cu. in., and that taken in by an extraordinary inspiration 100 cu. in., we add 124 cu. in. to the 100 cu. in. of air residing in the lungs after the forced expiration.

The atmospheric pressure being taken as 30 inches of mercury if we disregard the pressure of air in the external auditory canal, that is above normal when the tragus is forced in, we have the subjoined ratio.

224:100 X:30, and X--67.2 inches of pressure, an increase on the drum area of twice the normal at least.

When, after forced expiration, we close the mouth and nose, and inspire, *decrease* in the density and pressure of the gases concerned are FIRST felt in the nose and nasopharynx, because the volume of the confined air is *increased* FIRST in the lowest lung space by the descent of the diaphragm an the ascent of the elevators of the ribs. Such being the case, a comparative vacuum is made in the Eustachian tube and middle ear.

This condition would naturally draw everything downward, and in the event of fluid in those regions its movement towards aerial equilibrium is inevitable.

That this may be applied with benefit to exudates in the middle ear, is as I regard it, comparable to the question of deciding on the advisability of forcing air into the tympanum by catheter, bag or auto-inflation when desirable; or evacuating fluid by incision, or by auto-suction as I call this little procedure.

In my own experience, I have never seen the fluid line plainly marked on the membrane without making a free outlet.

In many, and the great majority of cases, when I was reasonably apprehensive of fluid mucus in that cavity, I have used this method tentatively with certainly no harm to the case, and yet with the feeling that in the absence of definite reasons for paracentesis, I had an anchor to windward.

MUCO-CUTANEOUS LESIONS.

In the treatment of all conditions of mucous surfaces one great condition for the repair of tissue cannot be fulfilled. This is asepsis, and in the futility of our meagre efforts in that direction, we every day see our wounds heal while bathed with nasal and oral fluids alive with pathogenic bacteria, and know that this rapidity of repair is because of the great blood supply. After using our best efforts to remove as many bacteria as possible from the spot in question, we have not the advantage of the general surgeon in being able to put an antiseptic compress on with a bandage over our operation wound.

That we may, in a measure, prevent excessive infection in our broken mucous surfaces, is. I believe, best done with the application of some *protective* covering. The principle that I wish to point out in the recommendation of the compound tincture of benzoin is the property of all gums to be precipitated from an alcoholic solution by the presence of water.

In splits of the nasal vestibule, abrasions, incisions for small abscesses of vibrissae follicles, for lesions of the labia, the stump of a shortened uvula, and for many other purposes where a tenacious, invisible coating is needed that is in itself antiseptic, it is effective.

I use it constantly as an application after every operation in the nose involving loss of blood as a hemostatic, and in the final stages of ulcers of the cornea, when the base of the ulcer is clear, I use it as a protective in lieu of the bandaged lids.

A small pledget of cotton on an applicator dipped in it will leave on any moist surface a precipitate of gum benzoin.

As a saturation for the ordinary gauzes, it is possibly more desirable than any, where the requirement is not so much drainage as the establishment of a permanent opening.

A HEMOSTATIC FOR MUCOUS SURFACE OPERATIONS.

The suprarenals of sheep desiccated as by Armour & Co. in a saturated extract in a saturated solution of boric acid will permit of an inferior turbinate being sawed off with no more loss of blood than two pledgets of cotton on tooth picks will remove.

A SPLINT FOR SEPTAL OPERATIONS.

Instead of fixed sizes and shapes of hard rubber splints, I have used in three late operations requiring the fracture of the septum, splints made of rolls of the common adhesive plaster doubled so that the adhesive sides are in contact with the mucous surfaces, and also holding the folds of the splint together. Small holes can be cut into it, and its cavity made to support the walls by stuffing cotton wet with an antiseptic solution into it. In dressing after the operation, it is only necessary to remove the cotton from its center, irrigate, and replace the antiseptic cotton freshly. It is clean, simple and so far successful that I shall continue to employ it.

ABSTRACTS FROM CURRENT OTOLOGICAL, RHINOLOGICAL AND LARYNGOLOGICAL LITERATURE.

I.—EAR.

Otitis Media and Earache in Lobar Pneumonia of Children.

MELTZER, J. S. (*Phila. Med. Jour.*, Aug. 5, 1899.) The author has observed a sufficient number of cases of lobar pneumonia in children at the beginning of which earache was a first and predominant symptom to make him think there is some casual relation between the two. He reports a number of cases as evidence of this, but in none of them did the earache terminate in suppuration. He thinks the pneumococcus may be the cause as it has been found in the discharges from suppurative ears or that possibly the earache is only a sympathetic pain of the chronically inflamed drum, since the initial earache nearly always occurred in ears which at some time or other had been the seat of an inflammation. *Richards.*

Conveyance of Infection Through the Medium of the Ear Syringe.

TODD, F. C. (*Journ. Amer. Med. Assn.*, Oct. 14, 1899.) Finding the ear syringes in common use to contain many bacteria after they had been in use for awhile and being unable to boil them on account of the leather packing, the author now uses a fountain syringe with a separate glass tip for each patient and also a bulb syringe with a valve at each end and terminating in a removable point. (If the point is all that is at fault, the common, large, metallic, piston syringe can still be used since its point is removable and can be boiled as readily as any instrument.—Reviewer.) *Richards.*

Treatment of Suppuration of the Middle Ear With Acetanilid.

LIBBEY, GEO. F. (*Med, News*, Oct., 1899.) A number of cases of suppuration of the middle ear are reported in which satisfactory and fairly rapid results were obtained by the use of finely powdered acetanilid as a topical application to the diseased middle ear. The ear is first cleansed with cotton on an applicator; this is followed by the use of peroxide and the ear and canal then wiped perfectly dry. Finely powdered acetanilid is now insufflated into the middle ear in as close contact to the suppurative membrane as it is possible to get it. The process is repeated daily. In the author's cases the time required to effect a cure varied from a few days to several weeks. Where it was possible to keep track of the cases the proportion of of recurrences was found to be small.   *Richards.*

Infective Sinus Thrombosis, Its Symptomatology and Diagnosis.

WHITING, FRED. (*Journ. Amer. Med. Assn.*, Oct. 28, 1899.) The principal question considered is that of symptomatology and diagnosis and their determining factors.

As regards the technique of operation there are two considerations of paramount importance, viz.. the control of hemorrhage and proportional reduction of shock and rapidity in the performance of the needful operative measures. No precaution can be too elaborate and no attention to detail too great in the matter of lessening the loss of blood. At the best it is frequently considerable and when this occurs no time should be lost in the performance of intravenous infusion of normal saline solution at a high temperature and in sufficient amount. Where toward the end of operation the heart action becomes labored and failure imminent, the intravenous infusion speedily allays the condition, the flagging energies of the heart are augmented and with the assistance of hypodermic stimulation thenceforth sustained. The operator should sacrifice every detail save cleanliness and thoroughness to the demands of time, a few moments more or less being important elements in a favorable or unfavorable termination.

Clinically the course of sinus thrombosis may be divided into three stages with local and systemic manifestations: the anatomic appearances of the sinus wall, the pathologic changes in the clot and the signs of circulatory obstruction may be denominated as local factors; while rapid and excessive fluctuations of temperature, frequently repeated rigors, peripheral or central metastases, embrace the essential systemic symptoms.

First Stage—Presence of a thrombosis, parietal or complete, not having undergone disintegration and accompanied by slight or moderate pyrexia, rigors being usually insignificant or absent.

Second Stage—Presence of a thrombus, parietal or complete, which has undergone disintegration with resulting systemic absorption, characterized by frequent rigors and pronounced septicopyemic fluctuations of temperature.

Third Stage—The thrombus has undergone disintegration with systemic absorption, rigors, rapid and great fluctuations of temperature, and central or peripheral embolic metastases, terminating usually in septic pneumonia, enteritis, or meningitis. The diagnosis in the first stage is seldom made preliminary to the operation for mastoiditis; its detection follows as a rule the recognition by the operator of extension of the carious disease through the inner table along the course of the sigmoid groove or at some point in the vicinity. Granulations from an eroded dura may be already protruding into the pneumatic spaces of the mastoid, the removal of which carries the operator to the parietal wall of the sigmoid sinus and to the uncovering of the same, and the finding of a thrombus. Recovery is here possible, though improbable, without operation on the sinus. If it take place it is by obliteration of the sinus lumen. Such a termination is not to be expected. The only safeguard against encountering the increased gravity of the second stage is to operate at once on recognition and the prognosis in skillful hands and in the absence of unfavorable complications is exceedingly favorable.

The interval between the first and second stage is brief and usually heralded by a sharp rigor. The symptoms of this stage (already given) are quite irreconcilable when associated with suppurative inflammation of the ear, with any ailment other than infective inflammation of the sinus. The general symptoms of this stage are those of septicopyemia and the manifestations are the results of the dissemination through the blood and lymph channels of pathogenic micro-organisms liberated for distribution by disintegration of the thrombus. In this stage the features of the patient assume a distressed and anxious look; the countenance is frequently suffused with copious colliquative perspiration; loss of appetite and constipation; respiration shallow and frequent. fluctuations in the temperature rapid and excessive, with repeated severe rigors. While the diagnosis is frequently difficult and the symptoms irregular there are certain pretty constant determining factors. These are rigors, which may be expected in four-fifths of the cases, though occasionally wanting. The cases in which the rigor was not repeated are few and the severity of the conditions in proportion. The persistent profuse sweating in the severe cases is one of the most distressing symptoms, weakening the individual and hastening prostration. These may keep up in fatal cases to a few hours before death. The fluctuations of temperature are significant. The marked pyrexia is subject to frequent remissions; at times the febrile period will be extremely brief, a space of two hours sufficing for a temperature range of 6° F. This high temperature is a valuable guide and warning to the septic complications to be expected. Of 95 cases of metastatic sinus thrombosis recorded by Hessler, but 12 exhibited temperatures of 106° F., and of 26 cases which were free from metatases, not one approached this degree. The pulse and respiration are at first moderately accelerated, until with advancing toxemia the pulse becomes feeble and reaches 180 or becomes too feeble to count; the breathing is similarly embarrassed reaching as high as 50 per minute. Vertigo and vomiting are present in some uncomplicated cases, but are more frequent when meningitis is present as a complication. Consciousness is a variable symptom, frequently remaining unimpaired till death; again there may be speedy loss of it associated with mild delirium. A mild form of delirium does not necessarily imply greatly increased gravity in the case, and may be due to a minute non-infective cerebral embolus, but if prolonged with occasional periods of violence the prognosis is distinctly bad, for coma supervenes, and speedy dissolution ensues.

Simultaneous with these manifestations appear what may be denominated local signs of circulatory embarrassment or obstruction. These symptoms are hemicrania varying in severity, radiating from the ear over the corresponding side of the head. Tenderness in the upper portion of the posterior cervical triangle, dependent upon phlebitis of the deep veins of the neck has been a pretty constant symp'om. The so called Griesinger's symptom is an edema of the occipital region extending downward and implicating the nape of the neck, due to phlebitis and obstruction of the mastoid and occipital veins. Gerhardt's symptom is elicited by laying the finger, with sufficient force, across the course of the external jugular of the affected side, to cause obstructive pressure, when it will be noted that the vessel exhibits but slight turgescence or none at all, while on the healthy side the external jugular becomes pronouncedly engorged on the application of pressure. Moderate edema of the eyelids has been observed, but is not constant. Neuroretinitis is present in about fifty per cent. of cases. Tenderness along the course of the internal jugular in the neck manifests itself in the late second or beginning third stage. In the third stage all symptoms are augmented with the additional ones due to the direct result of the dissemination of septic emboli. Here the hopes of successfully combating the disease rapidly diminish though so remarkable have been some of the cures that all hope should not be abandoned until the patient is in extremis. The disintegrated embolic masses are swept along with the blood stream until they meet an obstruction and there begin their infective inflammation. Should this stage be protracted a series of abscesses situated over the whole body may appear. These masses find their way into the general circulation through the medium of the jugular vein, hence the necessity for ligating this vein. Even when this fails to realize its full purpose it in no wise discredits the operative procedure. Metastatic abscesses of the abdominal viscera occur with less frequency than in other parts, as the bronchi and lungs, these being affected one and one-half times oftener than the combined other structures of the body. When possible evacuate the metastatic abscess and drain, allowing healing by granulation. Death in this stage occurs from pulmonary and pleural involvement, meningitis, abscess of the brain and general sepsis. Published statistics show that numerically by far the greater proportion of cases of sinus thrombosis are operated upon late in the second stage of the affection, and it may be expected that except in rare cases the first stage will escape detection. The earlier the diagnosis can be made the lower will be the mortality, hence the necessity for a more thorough study of the symptomatology. *Richards.*

Ossification of the Auricle and the Roentgen Rays.

WASSMUND. (*Deut. Medicin. Wochensch*, July 6, 1899.) Since Bochdalek's report of "A Physiologic Ossification of the Auricle" in 1866, only few similar cases have been mentioned in medical literature. These cases are those of Voltenini in 1868, Zudder in 1870, Schwabach in 1885, Linsmayea in 1889 and Knapp in 1892.

The histologic researches of Bochdalek and Knapp showed theirs to be genuine cases of transformation of cartilaginous into bony tissue. The rest of the above mentioned cases were diagnosed simply by the physical appearance of the auricle. The author has by the means of the X ray convinced himself of the genuineness of his case of ossification of the auricle, in a man 49 years of age.

Oppenheimer.

A Foreign Body in the Middle Ear for Two Years.

HAIKE, H. (*Deut. Medicin. Wochen. Berlin*, July 6, 1869.) A girl, ten years of age, who had two attacks of diphtheria in 1896 without any aural complication after a third attack of diphtheria in 1897, oozing of pus from the left ear was noticed which continued with some remissions until the present time. This condition was presumed to be a sequela of the several attacks of diphtheria. A careful examination of the ear, however, revealed a dark foreign body wedged in the middle ear, which could not be removed in toto. Small pieces were detached and submitted to examination. They proved to be pieces of cinnamon. The author concludes that the original foreign body was, no doubt, a small piece of cinnamon which entered the ear unnoticed, without causing any injury or even inconvenience to the patient. By a co-existing chronic otitis media the pus distended the small piece of cinnamon during the course of two years to such dimensions that it filled up the whole of the middle ear and caused dizziness and headache to such a degree that the parents brought the child to the Charity Hospital under the observation of the author.

Oppenheimer.

II.—NOSE AND NASO-PHARYNX.

Primary Sarcoma of the Nose.

HARRIS, THOS. J. *(Phila. Med. Journal,* June, 1899.) The author reports five cases of his own in detail and briefly fifty-seven cases, reports of which have appeared in current literature since Bosworth's report in his Diseases of the Nose and Throat of 1889. Dr. Harris' conclusions are:

"1. The cause of sarcoma of the nose is in no wise determined.

2. Degeneration of nasal polypi is strongly probable.

3. Sarcoma may occur at any age, but is most liable to occur between forty and fifty.

4. All forms of sarcoma are found in the nose, the round cell and spindle cell appearing with about equal frequency.

5. Sarcoma can spring from any portion of the nose, but the cartilaginous septum is the most common site.

6. Sarcoma develops insidiously, but obstruction to breathing and epistaxis are the chief symptoms.

7. Sarcoma is seen most frequently as a pinkish red tumor, rather soft, provided with a pedicle.

8. The prognosis is bad; over one-half die.

9. The round cell variety is the most fatal form.

10. Operation is indicated at the earliest moment."

Richards.

Anatomic Variations of the Nasal Chamber and Associated Parts.

CRYER, M. H *(Journ. Amer. Med. Assn.,* October 14, 1899.) It is absolutely impossible to do this paper, with its thirty-one illustrations any justice within the limits of an abstract. Touching as it does a field of great importance to every worker in the nose the reviewer would urge every reader of this journal to read the whole of Dr. Cryer's paper. *Richards.*

Nasal Catarrh; Its Surgical Treatment.

SUMNER, ARTHUR F. (*Jour. of Medicine and Science*, Sept. 1899.) A review of the physiology of nasal breathing and the symptoms which follow nasal obstruction accompanied by diagrams illustrating some of the more common causes of obstruction in the middle and anterior nares.

Richards.

Chronic Empyema of the Accessory Nasal Cavities With Report of 7 Cases.

STOUT, GEO. C. (*Phil. Med. Journal*, Aug. 26, 1899.) These cases were largely frontal combined with antral empyemas and illustrate the frequent coincidence of the two. A review of the various theories as to causation of these empyemas is given and the opinion expressed that most cases of empyema of the accessory cavities are caused by hypertrophic conditions of the nasal mucous membrane which close up the infundibulum, osteum maxillaire, and other openings. The author regards chronic antral inflammation as frequently coincident with trouble in the frontal sinus and says this should be carefully looked for before operating on the antrum. This fact is attested, on the post mortem table, by the frequent absence of any etiologic factor in the antrum, and by the action of gravity which would naturally cause pus from the frontal sinus to make its way into the antrum. The researches of Cryer and others have shown that these are frequently freely connected. A brief review of the various methods of operation are given. Paracentesis for frontal sinusitis is discussed somewhat fully and recommended in appropriate cases.

Richards.

An Anatomic Point in the Etiology of Naso-Pharyngeal Disease.

BROWNE, LENNOX. (*Phil. Med. Journal*, Aug. 26, 1899.) A reassertion of a statement first made in the author's text book (second ed. 1887) that "As a matter of experience I have long ago come to the conclusion that, while ease and completeness of postrhinal examination depend almost entirely on the amount of space at command between the uvula and the posterior pharyngeal wall, so also does this condition favor disease in the region under consideration— that is to say the wider the distance between the soft palate and the pharynx, the more surely one may expect on examination to find postnasal trouble." This statement which had been challenged by a reviewer in the *British Medical Journal* is confirmed by the report from Gerber's Poliklinik in Königsberg for the five years ending 1896 who says in referring to the etiology of ozena, "The nasal cavity is broader and shallower, the septum is shorter from before backward, and the depth of the naso-pharynx is increased. Measuring in 100 cases showed that in ozena the septum was shorter from before backwards by three milimeters, while the diameter of the naso-pharynx in the same direction was correspondingly increased." Lennox Browne explains this as due to a want of "correlation between the growth of the child and the ethmoid structures and began as an incident of embryonic life." He concludes by saying that atrophic rhinitis is associated with undue patency of the nasal orifice, nasal vestibule, nasal fossæ and of the naso-pharyngeal space. Opinions on this subject are invited by the author. *Richards.*

III.—MOUTH AND PHARYNX.

Tuberculosis of the Pharynx.

THEISEN, C. F. *(Journ. Amer. Med. Assn.*, August 12, 1899.) Two cases of pharyngeal tuberculosis are reported, one of which with pharyngeal uvular and tonsillar ulceration and with extensive lung involvement ran a rapid course and terminated in death six weeks after coming under observation; the other had a slight superficial pharyngeal ulceration directly back of the uvula with a small ulcer in the interarytenoid space. The pulmonary involvement was moderate. Under treatment with orthoform and lactic acid the pharyngeal ulceration healed. The author then discusses the differential diagnosis between this disease and carcinoma and syphilis and remarks that cases which clinically present all the appearances of tuberculosis frequently get well when iodide of potassium is administered (an experience which many others have had.) The question as to the part played by the tonsils and adenoid tissue in the etiology of tuberculosis is taken up and a preliminary report of some bacteriologic investigations given. Tubercle bacilli were found in two tonsils but none in adenoid tissue. The author, however, believes that if all extirpated tonsils and adenoids were subjected to careful histologic and bacterial examination, tuberculous conditions would be more frequently found. An exhaustive bibliography follows. *Richards.*

Accessory Thyroid Tumors at the Base of the Tongue.

SCHADLE, JACOB E. *(Journ. Amer. Med. Assn.,* Aug. 11, 1899.) The author records a case of accessory thyroid occurring in a woman of twenty-five who was anemic and complained of insomnia and gastric derangement. Although 133 lbs. in weight she was poorly nourished and evidences of nervous exhaustion were marked. At the base of the tongue was a tumor the size of an English walnut, covered with mucous membrane, intensely vascular, and at times almost entirely filling the fauces. During the menstrual period it was larger and more vascular. It was hard and immovable to the touch; there was no pain; speech was thick and non-resonant; the larynx and epiglottis were normal. Electrolysis reduced the size of the tumor somewhat but was followed frequently by considerable hemorrhage. The patient was referred to Dr. McBurney, who operated by a median incision from the symphysis menti to the hyoid bone, pushed aside the muscles of the base of the tongue and enucleated the tumor. Microscopically it was found to be a ductless gland of the thyroid type—an accessory thyroid. Dr. Schadle quotes some cases from the literature of the subject.

Richards.

Observations on Adenoids and Enlarged Tonsils and their Removal.

WISHART, D. J. GIBB. *(Phil. Medical Journal,* October, 1899.) A review based on 103 operations; males, 47; females, 56; faucial tonsils alone, 24; males, 6; females, 18; third tonsil alone, 31; males, 17; females, 14; faucial and pharyngeal tonsils, 48; males, 24; females, 24. Under five years of age 24%, between five and nine years 52% and over nine 24%. Nineteen cases were re-examined and twenty per cent. of these showed a persistence or recurrence of some portion of the growth. The author thinks the most satisfactory method of examination is by inspection through the nose and deprecates the examination by means of the index finger in the pharynx of the child, since it is disagreeable for both patient and physician, endangering the loss of the child's confidence. He objects to any palliative procedures, regarding them as a waste of time, if there are positive symptoms dependent on these enlargements. As a rule and always when adenoids are present he operates under profound anesthesia and usually under chloroform and says that thorough work cannot be done by a hasty scraping of the naso-pharynx in a struggling patient. The recurrence of the growth if it takes place is to be expected in those cases where no anesthetic is used. He operates in the recumbent position, and after the tonsils have been removed draws the head down over the edge of the table for the removal of the adenoids.

Two cases resulted fatally, one from an attack of scarlet fever and the other from the anesthetic. (This latter may be added to the growing list of deaths from chloroform in connection with operations for the removal of adenoid growths—Reviewer.) *Richards.*

Surgical Diseases of the Faucial Tonsils

STRONG, T. M. *(N. E. Med. Gazette,* July, 1899.) A review of the subject in which the following points are emphasized: the anatomic construction of the supratonsillar fossa and its liability to retain secretion in the lacunæ. As the lacunæ are largely out of sight they may be easily overlooked and an insufficient amount of trouble be found to account for the symptoms complained of. Small tonsils with obstructed lacunae may be more troublesome than much larger ones with wider open crypts. [The reviewer has several times had occasion to clinically verify these observations.] Cartilaginous or bony deposits in the tonsil are probably of congenital origin. There is a form of paroxysmal cough which has its origin in a pathologic condition of the tonsil, operating reflexly through the pneumogastric and a case history is given which evidences this. A young woman had a persistent hacking cough, evidently reflex. Careful examination failed to find its cause until the probe touched a point in the right tonsil, irritation of which produced a paroxysm of coughing. Repeated tests confirmed this although the tonsil was small and apparently not troublesome. Removal of the tonsil cured the cough. A brief resumé of the usual operative procedures is given and the claim made that pure hydrogen dioxide is the best styptic for hemorrhage in the throat or nose. *Richards.*

Peritonsillar Abscess.

COBB, FREDERICK C. *(Boston Medical and Surgical Journal*, July 27, 1899.) As a result of his anatomic studies the author has demonstrated that the pharyngomaxillary space can easily contain three or four drams of fluid and claims that many cases of peritonsillar abscess owe their clinical appearance to the accumulation of pus in this space, thereby forcing both the tonsil and peritonsillar tissues forward and upward. The pus is prevented from forcing its way backward among the great vessels of the neck by the septum formed of the styloglossus and stylopharyngeus muscles. Occasionally this burrowing has occurred. The cases seen were too far advanced for abortive treatment. The usual point selected for incision was in the median line above the tonsil, between the root of the uvula and the gingivopalatal fold. Some of the punctured cases closed too soon; this he regards as due to the varying direction of the muscle fibres of the anterior pillar and the superior constrictor, the openings sliding over each other so as to close the vent made by the knife. "By far the most usual point of passage of pus was found to be between the pillars of the fauces above this tonsil, suggesting that to puncture in this locality would be following nature's indication." He states that the date of relief was exactly the same in the unpunctured as in the punctured cases, viz., a fraction under seven days. Forty-four cases were examined and in none of them was there found to be any relation between rheumatism and peritonsillar abscess. The supratonsillar fossa and the infratonsillar space offer the surface for puncture most free from anatomic obstruction and puncture anterior to a plane passing through the posterior pillars cannot injure the great vessels if the knife be kept at all times anterior to such a plane. *Richards.*

IV.—LARYNX

Diphtheria and Membranous Croup.

POTTER, THEODORE. *(Phil. Med. Journal,* July, 1899.) The two diseases are considered to be one and the same and the difference in constitutional severity is explained to be due to the different anatomic construction of the parts involved. "Admitting the undeniable fact that general toxemia in the modern bacteriologic sense is comparatively slight in membranous croup, though the germs of diphtheria are also undeniably present, we may explain the fact without violence to reason or established pathologic principles. The larynx is a cartilaginous box, lined over its larger part with a thin and tightly placed mucous membrane. This box is much less freely supplied with lymphatics and lymphoid structures, in a word, with absorbing structures. The nose, naso-pharynx, pharynx and fauces including the various aggregations of adenoid tissue called tonsils, form a very hot-bed not only for the growth of germs and the accumulation of morbid secretions, but furnish favorable conditions for the absorption of such morbid products. An acute laryngitis with hoarseness and mucopurulent expectoration is usually accompanied by but slight disturbances of a septic nature; and this, though the expectoration is swarming with bacteria. The same is true of an ordinary acute mucopurulent bronchitis. Only when the bronchitis extends into the deeper parts of the lungs, where the anatomic condition is quite different from that of the larynx and bronchial tubes, where the absorption is free and rapid, only then do high fever and the other evidences of sepsis appear. And so above the larynx, an acute mucopurulent tonsillitis, the so-called follicular tonsillitis, is usually complicated by high fever, by general septic intoxication.

Is not this the key to the situation in the typical cases of membranous laryngitis? The conditions and agents for absorption are present in but slight degree. It is strange, therefore, that there should be but slight cervical adenitis, but slight fever, and an entire lack of the general disturbances and degenerations of muscle, nerve center, kidney, and other glandular organs, which in naso-pharyngeal diphtheria result from toxins absorbed from a larger, more active, deeper, and more freely absorbing diseased area?" The author thinks the isolated case of diphtheria occurs as often as the isolated case of membranous croup and regards both as primarily a local disease with secondary constitutional manifestations. *Richards.*

The Treatment of Incipient Laryngeal Cancer.

SCHEPPEGRELL, W. *(Med. News,* Aug. 5, 1899.) The only hope in any operative procedure is based on a very early diagnosis and complete immediate removal of all diseased tissue. If this cannot be done reliance must be had on palliative measures since "when the malignant disease has advanced to such a degree as to require the complete extirpation of the larynx for its removal the prognosis is not only very unfavorable but many operators have even doubted its justifiability on account of its attending danger, high mortality and the mutilated condition of the patient after the operation."

Successful endolaryngeal operations have been performed by several operators. It has the disadvantage that it is difficult to fix the limits of the disease and to totally eradicate. Thoyrotomy or some modification of it preceded by a tracheotomy seems to offer the best results with the least risk. If done early enough there is a fair chance of removing all the growth. Up to the present time the mortality in complete extirpation has been 84%; certainly a high mortality and sufficient to deter any but the boldest operator. *Richards.*

Dysphonia—Relief With the Galvanic Current.

CHRISTY, T. C. *(Phil. Medical Journal* Sept. 9, 1899, Oct. 21, 1899.) After citing the various causes of dysphonia the author reports a number of cases showing the remarkable improvement which takes place under the use of the constant current. Its advantages are that it is easy of application; soothing and agreeable to the patient; relieves the congestion, pain and irritation; does not excite pain or spasm of the glottis or trachea; relieves the swollen lymphatic glands; cures more promptly than any other agent, while the patients recognize its value and return regularly for its application. An electrode is used which adjusts itself readily to the episternal notch. The active electrode is the positive and is placed first over the trachea at the episternal notch while the negative is placed over the nape of the neck close to the hair line or over the thick muscular layers on either side of the neck. The posterior electrode is changed only to relieve the burning sensation which after long application sometimes becomes very acute, while the anterior is placed over the various portions of the larynx which are available. The strength of the current is adapted to the individual; it should not be painful. The operator should test the current before applying. As improvement takes place the strength of the current is to be weakened. The duration of the treatment depends on the strength of the current employed and on the degree of reddening of the skin which results. (5 to 10 minutes at a sitting.) It should be discontinued before producing pain. The electrodes are to be kept moist with warm salt solution and are not to be changed in position without first removing them from the patient. When the current is sent directly through the larynx, reduce the original current one half and tilt the electrodes on their sides. The aim of treatment "is to secure a current which is perfectly agreeable and soothing during the application and which relieves the condition existing if judiciously applied." (The reviewer can testify to the value of Dr. Christy's methods both as a treatment by itself and as an adjunct to local laryngeal applications. In sub-acute and chronic hoarseness with relaxation of the cords the application of the constant current as suggested by the author has produced better and quicker results than other methods, with the decided advantage that the treatment is not disagreeable to the patient. On the contrary they appreciate its value and return for its application.)

Richards.

Prognosis of Laryngeal Tuberculosis.

LEVY, ROBERT. *(Journ. Amer. Med. Assn.*, Sept. 16, 1899.) Dr. Levy regards laryngeal tuberculosis as more curable than it has been frequently considered. He admits that proof of cure is hard to get. He would regard a case as cured in which all active indications of disease fail to recur in two and in some cases after one year from their cessation. In the diagnosis, irregular spots of redness, characteristic anemia, typical infiltration and soft papillomatous excrescences are sufficient guides for the experienced. The cases of the infiltrative and papillomatous variety did better than the ulcerative. The pulmonary conditions may continue to grow worse at the same time that the laryngeal conditions are improving. No line of treatment is given; cases should be individualized and the treatment adapted to the needs of the particular patient.

Richards.

Remarks Concerning the Operative Treatment of Carcinoma of Larynx.

SENDZIAK, JOHANN. *(Monatschrift für Ohrenheilkunde,* Sept. 1899.) The author gives statistics concerning the operative treatment of 640 cases of malignant disease of larynx.

The year 1888, that of the death of Emperor Frederick III, he takes as a sort of culminating point and arranges his statistics accordingly as, "up to 1888," and "from 1888-1898." His classification of the various operations and results can be gathered from the accompanying schematic arrangement of his statistics as put together by the reviewer.

By *complete cure,* the author understands these cases with altogether favorable results for three years.

By *relative cure,* favorable conditions persisting for one year.

The operation of *partial excision* covers the cases where portions of the cartilages are removed, while *laryngo fissure* is limited entirely to operation on the soft parts.

His conclusions are as follows:

1. In the present condition of our knowledge of malignant disease of larynx, operative treatment is the only treatment justifiable.

2. Very favorable results are obtained when the operation is done during the early stages of the disease.

3. *Laryngo-fissure* and *partial excision* are the most successful methods. The first gives better results as to healing and is the safer, while the latter is more favorable as regards recurrence.

4. *Total extirpation* should not be considered too unsatisfactory although the statistics as to cure are not very favorable.

Allen.

RHINO-LARYNGOLOGICAL LITERATURE.

OPERATION.	COMPLETE CURE.			RELATIVE CURE.			RECURRENCES.			DEATH AFTER OPERATION.			INSUFFICIENTLY OBSERVED.
	9 cases, (25 pr ct.)			5 cases, (14 pr ct.)			14 cases, (39 pr ct.)			None.			11 cases.
Endolaryngeal. 36 cases in all. Up to 1888, 17 cases. After 1888, 19 cases.	1st Period 5 cases (14 pr ct.)	2d Period 4 cases 11 pr ct.+		1st Period 1 case (3 pr ct.)	2d Period 4 cases (11 pr ct.)		1st Period 7 cases (19.5 pr ct.)	2d period 7 cases 19.5 pr ct.					
	17 cases, 12.5 pr ct.			17 cases, (12.5 pr ct.)			78 cases, 57.3 pr ct.			12 cases, 8.8 pr ct.			16 cases.
LARYNGO-FISSURE. (Thyrotomy.) 136 cases in all. Up to 1888, 58 cases. After 1888, 78 cases.	1st Period 2 cases (1½ pr ct.)	2d Period 15 cases (11 p r ct)		1st Period 2 cases (1½ pr ct.)	2d Period 15 cases (11 pr ct.)		1st Period 56 cases (40.4 pr ct.)	2d Period 23 cases (16.9 pr ct.)		1st Period 3 cases (2.2 pr ct.)	2d Period 9 cases (6.6 pr ct.)		
	26 cases, 12.9 pr ct.			20 cases, (10 pr ct.)			63 cases, (31.3 pr ct.)			44 cases, (21.8 pr ct.)			27 cases.
PARTIAL EXCISION 201 cases in all. Up to 1888, 55 cases. After 1888, 146 cases.	1 Period 7 cases (3.5 pr ct.)	2d Period 19 cases (9.4 pr ct.)		1st Period 4 cases (2 pr ct.)	2d Period 16 cases (8 pr ct.)		1st Period 19 cases (9.5 pr ct.)	2d Period 44 cases (21.8 pr ct.)		1st Period 18 cases 9 pr ct.	2d period 26 cases 12.9 pr ct.		Some show a favorable result after 6, 9 and 11 months
	12 cases, (4.4 pr ct.)			24 cases, (9 pr ct.)			81 cases, (30.3 pr ct.)			94 cases, (35.2 pr ct.)			32 cases.
TOTAL EXTIRPATION. 267 cases in all. Up to 1888, 143 cases. After 1888, 124 cases.	1st Period 8 cases (3 pr ct.)	2d Period 4 cases (1.5 pr ct.)		1st Period 9 cases (4.4 pr ct.)	2d Period 15 cases (5.6 pr ct.)		1st Period 51 cases (19.1 pr ct.)	2d Period 30 cases (11.2 pr ct.)		1st Period 56 cases (21 pr ct.)	2d Period 38 cases (14.2 pr ct.)		In some the condition good after 9 months.

Gummata of the Larynx.

CORDES, H. (*Deutsche Medicinische Wochenschrift*, June, 1899.) The writer refers to the numerous clinical manifestations of syphilis and the difficulty often present in making a diagnosis without the aid of the microscope. By way of illustration he mentions the case of a woman, who consulted him for complete aphonia. Laryngeal examination showed infiltration of the left side of the larynx, involving the ary-epiglottic folds, arytenoid cartilage and ventricular band. Springing from the ventricular band a tumor was visible. No history of specific infection could be elicited and malignant disease was thought of. A small section of the growth was subjected to microscopic examination and iodid of potassium administered internally in large doses. Under this medication the growth disappeared, the infiltration became less and the patient made a complete recovery. *Oppenheimer.*

A New Instrument for the Application of Nitrate of Silver in Substance to the Larynx

CUBE. (*Deuts. Med. Wochensch.*, Aug., 1899.) The instrument consists of a piston syringe and a suitably bent canula, which is introduced into the larynx. The canula terminates in a solid button, with a shallow depression in the center. Melted stick silver is poured into the depression which is in the shape of a mould.

About the mould are a number of small openings through which a salt solution is injected after the silver nitrate has been applied. This protects the surrounding tissues from the action of the caustic. The author has been very successful with the use of the above instrument in the treatment of syphilitic ulcerations and small laryngeal polypi.

He cites the case of a singer whose voice became rough and hoarse in the middle and lower tones. On the left vocal cord was seen a condition of venous engorgement and dilatation. This well cauterized with a complete restoration of the singing voice. *Oppenheimer.*

MISCELLANEOUS.

Phenol-natro-sulfo-ricinicum in Rhino-Laryngology.

BAUMGARTEN, EGMONT. (*Wiener Klinische Wochenschrift*, No. 35, Aug., 1899.) The author gives his results with this new remedy so highly recommended by Prof. Heryng. The substance is a mixture of synthetic phenol and the sulpho-ricinate of sodium. It is a yellowish brown, syrupy fluid, miscible with water and with a strong carbolic acid odor. It is a powerful antiseptic and deodorizes and causes no pain when rubbed upon the mucous membrane of nose or throat. It is to be employed as a local application in 20-30-50% solutions in water. The paper treats exhaustively of its use in the various affections of nose and throat. The author obtained no particularly beneficial results in tubercular laryngitis, although especially recommended by Heryng for this purpose. It cleans up tubercular and syphilitic ulcerations and prepares the way for the application of lactic acid to the former. Acute and chronic rhinitis and laryngitis are not benefited. It is in atrophic rhinitis, with or without ozena, that the author finds the remedy of great value. He has had the generally unfavorable results obtained by all, with the other remedies, including massage and copper electrolysis, but finds the local application of a 30% sol. of this remedy to be most useful. Odor is relieved, and cleansing made much easier, the good results lasting for months. The atrophy of course remains the same. The very conservative tone of the article recommends it highly, and it is more than probable that the remedy will prove of value. *Allen.*

Beta-Eucain as an Anesthetic in Nose and Throat Work.

POOLE, WM. H. *(Detroit Med. News.)* Beta-eucain has all of the advantages of cocain except that of shrinking the tissues, and this is sometimes objectionable, without certain positive disadvantages of the latter drug. The eucain solution can be made sterile by boiling without destroying its activity. In operations in the naso-pharynx, pharynx, larynx or nose a four to ten per cent solution is used The effect of the drug is noticeable in two or three minutes, anesthesia is obtained in from five to ten and lasts from eighteen to twenty minutes. The author has never seen any heart depression or symptoms suggesting systematic poisoning. Following the anesthesia the disturbances of sensation have been much less defined and unpleasant than where cocain was used. An instance is cited where attempt to do an intranasal operation (Removing of polypi) under cocain produced alarming symptoms so that the operation had to be given up. The following day and several times after a four per cent. solution of eucain produced perfect anesthesia without the slightest unpleasant symptom. The reviewer has found the drug when applied in the nose to produce some smarting on its first introduction and its anesthesia to be not quite so complete as that of cocain. It does not seem to be followed by congestion nor does the patient complain of a "cold in the head" which congestion is so very disagreeable to some patients after the use of cocain in the nose. *Richards.*

Our Tuberculous Patients—Where to send Them.

McConnell, J. F. *(Journ. Amer. Med. Assn.,* Sept. 16, 1899.) Send very early cases to southern New Mexico and keep all late cases at their own homes. At an altitude of 3,800 feet pure air and an equable dry climate with 348 days of sunshine per annum is provided. This climate is an equable one. Even in July and August when the days are hot the nights are cool. Cases of the fibroid type which do badly at high altitudes will do much better in New Mexico. The author states that the accommodations are good. (One of the troubles with many places which are climatically good is that the accommodations for the invalid are so poor as to practically preclude the availability of the locality. The consumptive patient sent a long ways from home must be well provided for in the way of food and other creature comforts—Reviewer.)

Richards.

Some Fallacies in the Modern Treatment of Nose and Throat Diseases.

Roy, Dunbar. *(Medical News,* Aug. 19, 1899.) A plea for more thorough study of the individual case. By fallacies the author means the false recognition of pathologic conditions and the wrong remedy for the condition when it is recognized. He mentions the routine use of oily sprays; the abuse and improper use of the electric cautery and the needless removal of cartilaginous and bony spurs. All specialism should be preceded by several years of general practice. *Richards.*

A Case of Stammering With Demonstration of the Methods Employed in Treatment.

Makuen, G. Hudson *(N. Y. Med. Journ.,* Sept. 23.)

reports a patient, 29 years of age, who had stammered with varying severity since childhood. The chief characteristic was a spasmodic contraction of the muscles of the soft palate and tongue resulting in sudden closures during attempts at vocalization and articulation of the posterior palato-lingual cheek. The defect was more pronounced upon reading than upon speaking.

The primary neurosis responsible for the condition was located in the nerves supplying the muscles the respiratory organs, not in those of the pharynx as might be supposed from the location of the spasm.

The patient was given proper breathing exercises and was taught to compress the abdominal viscera by means of a voluntary action of the diaphragm and abdominal muscles and to make this compression greater or less according to the strength or intensity of the tone required. Exercises were also given for the control of both the levator and depressor thoracic muscles independently of voice and breath. When a combination of this newly acting mechanism was effected with the voice mechanism in the production of elementary sounds, he was instructed to carry the same principle into the enunciation of syllables. He has now not only overcome his difficulty but has at the same time acquired a more effective manner of speech.

Loeb.

TRANSACTIONS OF THE SIXTH INTERNATIONAL OTGLOGICAL CONGRESS AT LONDON.*

PRESIDENTIAL ADDRESS.

THE GROWTH OF OTOLOGIC SCIENCE.

BY

URBAN PRITCHARD, M. D. EDIN, F. R. C. S., ENG.,

LONDON.

PROFESSOR OF AURAL SURGERY IN KING'S COLLEGE, LONDON.

In the name of the British Organization Committee, and in the name, indeed, of all British Otologists, I wish to offer a very hearty welcome to our foreign colleagues and to their ladies.

We thank you most sincerely for coming here, in many cases hundreds—nay, even, I may say, thousands— of miles, in order to assist at this, the Sixth International Otological Congress, and I trust that your visit to London will be a very pleasant one; at any rate, I may certainly promise that we will do all in our power to make it so.

There is, however, one serious difficulty which, with all the good will in the world, cannot be removed. I refer to the fact that, owing to the immense size of this London of ours, so much loss of time is entailed in getting from place to place. When I remember how conveniently we were located during the pleasant gatherings of Congress at Basle, at Brussels, and at Florence, and the ease with which we were enabled to find our way about, I cannot help regretting that our vast metropolis cannot be, for the moment, brought within more manageable compass; but, as that is impossible, we must content ourselves with doing the best we can under the circumstances.

In bidding you welcome I have used the word "foreign" to our guests; but I do not like that designation in connection with our Congress. For Science acknowledges no

---

*From the Journal of Laryngology.

difference of nationality; she is herself all in all, and faithfulness to her the sole condition of citizenship in her kingdom.

Therefore let us regard ourselves, not as under our national flags, but as assembled in common brotherhood, marching together under the banner of Otology, and forming one part of that army, commanded by Science. which is engaged in overthrowing the foes of humanity, those foes which have Ignorance, Vice, and Prejudice for their leaders.

Personally, I feel a thrill of pleasure in seeing so many valued friends assembled again for conference, and of these may I be permitted to mention the names of Professor Politzer, Professor Guye, Professor Lucae, Dr. Arthur Hartmann, Professor Knapp, Dr. Ménière, and our last President, Professor Grazzi.

But it is a real grief to miss some old familiar faces. The genial President at Basle, Burkhardt Merian, dear old Sapolini of Milan, Moos of Heidelberg, and Delstanche (père) of Brussels—these are honored names which will long be remembered in the annals of Otology, though they themselves have passed "behind the veil."

Again, since our meeting in Florence, our branch of medical science has lost another faithful servant; I allude to Dr. Meyer of Copenhagen, whose name in connection with the discovery of post-nasal adenoids is so justly renowned. Lastly, among other names that must occur to each one of us, I will only refer to those of Professor Colladon of Geneva and Hewetson of Leeds, who were both to have taken an active part in our proceedings this week.

We deeply regret also to note the absence, from unavoidable circumstances, of several friends whom we should so gladly have welcomed among us to-day, and I am especially grieved that ill-health has prevented Dr. Charles Delstanche, our hospitable President at Brussels, from being at his accustomed place on this occasion. I believe that it is the first time that our Otological Congress has not had the support of his energetic and cheery presence.

Now, friends, it seems to me that at the opening of our Congress it is well that we should recall briefly the story of the birth and growth of Otological Science, and with your permission I will say a few words on this subject now, dwelling more particularly on the advances made in it during the last thirty years.

Although Toynbee is generally acknowledged to be the father of *modern* Otology, for the date of its birth we must go back some 3,400 years to the then flourishing country of Egypt. For Professor Roosa, in his excellent treatise, refers to a certain ancient papyrus (called, after its discoverer, the Papyrus Ebers), on which is written a monograph on "Medicines for ears hard of hearing," and "for ears from which there is a putrid discharge." And here, in our Museum, may be seen a confirmation of the fact that ear troubles not only existed in those days, but that they could be cured; for we have the good fortune to possess a curious old Egyptian relic, consisting of a wooden tablet on which is portrayed in bas-relief two effigies of the Sacred Bull and two auricles; this was undoubtedly a votive offering to the god Hathor from some "grateful patient."

In spite of its early birth, however, Otology, except perhaps with regard to its anatomy and physiology, did not make itself of great importance until the second half of the present century. It is true that here and there a surgeon might have been found who had turned his attention to some extent to this subject; and, indeed, our own Royal Ear Hospital in Dean Street, Soho, which is acknowledged to have been the first successful aural clinique in Europe —and, I believe, in the world—was established in 1816. But, speaking generally, we may safely assert that aural surgery continued to be more or less in the stage of infancy until between 1840 and 1860, when the study was vigorously taken up by Sir William Wilde and Toynbee, who thus gave a fresh impetus to the study of the pathology and treatment of diseases of the ear. Even then its importance was by no means generally recognized; indeed, only thirty years ago it was a favorite saying of more than one celebrated surgeon that "Ear d'seases may be divided into two classes: those which can be cured by any general practitioner, and those which, being incurable, may be relegated to the tender mercies of the ear specialist."

Is it any wonder, therefore, that in those days aural surgery was not only considered to be, but actually was, very much mixed up with the name of quackery; for, as scientific men refused to have anything to do with it, the door was left open for any charlatan to enter, and many strange stories gained credence as to methods of treatment which the patient was required to undergo. Indeed, one of my earliest boyish recollections of aural surgery was hearing the story of how a child, a deaf-mute, had been cured by a skewer having been passed through his head from one ear to the other. Although a somewhat better knowledge of anatomy has since made me doubt the accuracy of this statement, still it is certain that strange things were both said and done in the olden times, which did not redound greatly to the honor of the specialist.

In my own student days I well remember the sarcastic manner of Professor Partridge—Dickey, as we used to call him at King's College—when he said, "Ah, gentlemen, a little wax is a godsend to an aurist," meaning, of course, that its removal was an easy method of earning a reputation. And no doubt there is a certain truth in these words, though not exactly in the sense implied by the good old Professor; for which of us has not found that, by removing a plug of cerumen which has either not been diagnosed or which has resisted all the efforts of the general practitioner to dislodge, we have gained *kudos* and an appreciation which many of our more delicate operations have failed to secure.

Yes, Otology had indeed a hard battle to fight before it could be said to have won honorable recognition among men of standing in the medical profession, and I shall never forget the letter which one of these wrote to me in 1872 when he first learnt that I intended to devote myself to this branch of study. After lamenting my decision, however, he did conclude by saying, "*Now* I suppose that I must not regard *all* aural surgeons as quacks." And may I add, as a kind of commentary on this letter, that within a few years afterwards the writer of it came to me as a patient.

Things have indeed changed since then, for, instead of a few aural surgeons scattered here and there in Great Britain, we have now at least a couple of hundred, while the number of cliniques in London alone has been increased from two or three to near upon twenty. And in many other countries this branch of medical science is even more strongly represented.

As a natural result of the increased interest in the work, let me call attention to the unique Museum connected with this Congress, wherein is to be found the largest and most valuable collection of otologic specimens—a collection which could only have been brought together by the union of our international forces. The Museum is so complete that if you had come to visit that alone your trouble would have been repaid.

But in one respect there is still room for improvement. I refer to the need for the better recognition of otology by our universities and colleges. I am glad, however, to be able to report that one step has lately been made in this direction, for the University of Edinburgh has now made it one of the qualifying subjects for her medical degrees, and I look forward, with hope, to the time when her example will have been generally followed.

This "new departure" will, I trust, lead to a fuller recognition of the position of teachers of aural surgery. In this respect we, in the British Isles, are sadly behind other countries, where chairs of otology are numerous; whereas here, among all our universities and colleges, where so many able lecturers are to be found, in King's College, London, alone is the dignity of a professorship conferred upon its teacher of aural surgery.

Let me now pass in brief review the progress of the last thirty years.

So far as the *anatomy* and *physiology* of the auditory apparatus are concerned comparatively little has been added to the store of knowledge already gained, although a more intimate study of its parts has made that knowledge more complete and precise.

In *pathology*, as may be expected, there has been considerable advance.

In diseases of the meatus, although aspergillus was discovered before this period by Meyer, Schwartze, and Wreden, yet it was not elaborated with any fulness until later. Also, the nature and classification of exostoses have been worked out within this period.

Our knowledge of the changes in chronic middle-ear catarrh, and in sclerosis, has considerably advanced, although much here yet remains to be done.

The effect of pathologic conditions of the nose and

naso-pharynx upon the auditory apparatus, adenoid vegetations more especially, has practically been discovered. The world has yet to learn what it owes to Meyer.

In chronic suppurative catarrh, disease of the ossicles, the implication of the attic, the antrum and the mastoid cells have been worked out; also the intercranial complications which sometimes follow. The nature of the granulations and polypi are now better understood; and although Toynbee had already called our attention to cholesteatoma, its pathologic importance in connection with mastoid disease was not fully realized until quite lately.

In the pathology of labyrinthine disease there has not, perhaps, been so much advance; but Ménière's disease is now better understood; and Politzer has made known to us a disease of the bony capsule. Finally, the pathology of congenital syphilis affecting the internal ear has been partially worked out.

Our *means of diagnosis* have been considerably improved.

The diagnosis between affections of the conducting apparatus and the auditory nerve, which formerly was often confused, is now much more easily made out; this is chiefly due to the study of the tuning fork.

Methods of illumination have very greatly improved, to the immense advantage of the surgeon.

Bacteriology, again, has done much, and in all probability will do even more in the future, to help us in our diagnosis. Unfortunately, the essential apparatus is enclosed in such dense bone that the Roentgen rays have been of but little assistance.

In *treatment* there have been immense strides.

Even in chronic middle-ear catarrh and in sclerosis, those diseases which hitherto have baffled our most strenuous efforts, a distinct advance has been made indirectly, especially in prophylaxis, by treatment of the nose and naso-pharynx.

In suppurative disease there has been very great improvement in treatment. By means of boric acid, alcohol, and other suitable antiseptics, simple otorrhea has become much more manageable; and a far larger proportion of such cases are now healed, even without operation.

In the case of its complications—caries, granulations, and polypi—the advance made is most striking, and, in consequence, the large protruding polypus is now rarely seen; and no aural surgeon at the present time would be able to show so large a collection of these as Dr. Warden, of Birmingham, was in the habit of displaying some twenty-five to thirty years ago.

Curetting of carious spots, and the removal of ossicles, so important in the treatment of many cases, has only recently been introduced.

This brings us to the wonderful stride made in the treatment of antrum and mastoid disease, for which we have chiefly to thank Professors Schwartze and Stacke, although many others have contributed to the advancement. How much agony has been relieved, how many lives have been saved, by these operations!

And, gentlemen, this advance of surgery has carried us still further; for, by the joining hands of general surgery and otology, intercranial suppuration has been robbed of many of its victims.

But how, and why, is this? How is it that, formerly, our surgeons were unable to cope with those intercranial conditions? How is it that, now, we are able to operate on the tympanum, attic and mastoid, practically with impunity?

Gentlemen, this is due to the adoption of antiseptic surgery. May I beg your indulgence for proudly claiming to be pupil, colleague, and brother professor of him whom I regard as the greatest man living to-day—Lord Lister. Were it not that you would exclaim at my inconsistency, I should be tempted to add "compatriot" also. But yes, gentlemen, I will add the word. Not, however, in the sense in which I was just about to use it, that of English nationality; but with reference to that ideal country to which I alluded at the beginning of my speech, and of which we otologists are all naturalized subjects. Here, on the common ground of our chosen land, the land of science, we may all proudly claim Lord Lister as our compatriot, all rejoice to serve under such a leader in the battle against disease and death. The world does not as yet understand the full benefits which he has conferred upon mankind, but we, naturally, being his compatriots, have a better opportunity for doing so; and I can only add my earnest conviction that it is by faithfully following the counsels of our superior officer that our advancing column can best secure future victories.

Such, ladies, and gentlemen, is the brief, and therefore necessarily inadequate, record of the progress of otology which I desired to lay before you.

We have seen that this nineteenth century, which has brought to the world so many wonderful blessings in other directions, has not been unmindful of our branch of medical science. For, whereas at the commencement of the century the ear was regarded almost as a *terra incognita* scarcely worth consideration except as the seat of one affection only—that which was generally known as 'a deafness' now, at its close, this organ is fully-explored ground, and has been proved well worth the exploration. Otology has been raised from the rank of pseudo-quackery to an honorable position in scientific surgery, and its importance and bearing upon the body as a whole is now fully recognized.

But while we rejoice in the progress made in the past, we must remember that much still remains to be done. For instance, we have yet to clear away that opprobrium of aural surgery, namely, the chronic non-suppurative disease of the middle ear. Shall we, in the near future, be enabled to cope successfully with this hitherto invincible foe? Judging from the advance made in other directions I am bold enough, and sanguine enough, to think that we shall; and assuredly when that help comes we shall all unite in blessing its victor.

Now, it is the province of our Otological Congresses to take this and similar problems into consideration. But the real value of these gatherings is not to be measured merely by papers and discussions. This is one of their uses, it is true, for interchange of ideas is always good; still, the chief value of thus meeting together with others who are all interested in one common subject is the kindling of enthusiasm which is thus engendered, an enthusiasm which should serve to stimulate older and younger members alike to renewed efforts in the paths both of research and of practical treatment; and, therefore, in conclusion, I desire most heartily to wish that this, our sixth Congress, may be successful in all these directions.

A New Method of measuring the Quantitative Hearing-power by Means of Tuning-forks. By Dr. SCHMIEGELOW (Copenhagen).

Many experiments, he said, had been made in later years to find a reliable method. There were the methods of Hartmann, Gradenigo, and Zwaardemaker, which, however, could not be called satisfactory, as they did not give exact results. In order to use the time and vibration of certain tuning-forks in measuring the hearing power, it was necessary to know the vibration curve. If it were possible to measure the amplitude of each tuning-fork from the moment it was set in vibration to the moment when the tone died away, the difficulty in using forks as reliable tests of quantitative hearing would be solved. In the light of our present knowledge the amplitudes of the deeper forks only were measurable. Bezold and Edelmann had, by means of a verp cleverly invented instrument, constructed vibration curves of the deeper forks (from D^1 to F), and from these they constructed a standard curve. They furthermore presumed that this curve, being almost the same in all the deeper forks, must be the same for every fork, even the highest ones. It seemed, however, said Dr. Schmiegelow, that Bezold and Edelmann had started from wrong conclusions, and that the result of their experiments did not agree with theory. According to theory, the amplitudes decreased at an approximately geometric progression; that was to say, the logarithms of the amplitudes diminished directly with the time. This theory was no doubt correct, but only as far as the small amplitudes were concerned (Jacobson), or, in other words, the logarithmic decrement was greater and irregular at the beginning, but towards the end it became nearly constant. By a very carefully drawn mathematical diagram, Dr. Schmiegelow showed that in an examination of the curve found by Bezold and Edelmann it would be seen that the differences between the logarithms of the amplitudes corresponding to the time of 0-10-20, etc ,-100 seconds to begin with, decreased as they ought to do, but afterwards increased what they ought not to do. According to theory they should expect that the difference, after decreasing as it did to 0·151, ought to remain pretty nearly constant.

The difference, however, increased again, which meant that for some reason or other the vibrations were impeded at an increasing rate, and the curve therefore not correct. Everything tended to prove that the curve of the higher fork was different from that of the deeper ones, and that such fork had its own special curve. In order to find the curve of vibration for each tuning fork G. Forchhammer and I proposed the following method: A tuning-fork is struck, and the time during which it is heard at different distances from the ear is determined. The abscissas of the curve represent the distances, the ordinates the time of perception. The correctness of this method, said Dr. Schmiegelow, was founded on the fact that the amplitude was proportional to the distance at which the tone disappeared, the intensity of the tone being constant when the "Hörschwelle" was reached of the moment at which the tone ceased to be heard. The method was also practicable, in so far that instead of the microscopic amplitudes the macroscopic distances were measured, an advantage which was all the greater because the amplitude of the higher tuning-forks could not be measured microscopically. The forks examined were made by Edelmann in Munich, and were C G, c g, c^1 g^1, c^2 g^2, c^3 g^3, all of them unloaded.

The experiments were made under as good conditions as could possibly be procured in the open air at some distance from town. If, for instance, they were going to find the curve of the c^1 fork (261 vibrations), they would proceed in the following way: By six series of experiments they found that c^1 properly struck would be normally heard for 7 seconds at a distance of 160 cm. from the ear, 14 seconds at a distance of 80 cm., 23 seconds at 40 cm., 37 seconds at 20 cm., 62 seconds at 10 cm., 88 seconds at 5 cm., and 117 seconds when held as close to the ear as possible without touching it. According to the theory, the differences between the time at a distance of 5-10 cm. and the distances 10-20 cm. should be the same, because close to the ear, where they had to do with small amplitudes, the time increased at an arithmetic ratio (with constant differences) if the distance diminished at a geometric ratio. This theory was actually proved by the experiments. At the beginning of the curve (from 160-20 cm. distance) they found that the differences in time were smaller at the greater distances from the ear, that they increased up to about 20 cm. distance, and then became constant as far as the final part of the curve was concerned. The fact was that a tuning-fork did not emit the tone from the external surface of the prongs, but the vibrations were presumed to spread out from two points which were situated between the external surfaces of the prongs. By a series of experiments they had found that the distance between the tone center and external surface of the tuning-fork was about 1 cm. in the forks C G, c g, c^1 g^1, and c^2, whilst the distance was about 1·5 cm. in the forks g^2 c^3 g^3 c^4 g^4 c^5. As the distances were reckoned from that surface of the prong which faced the ear, they must therefore add to the distance 5·10 and 20 cm., the distance of the tone center from the external surface of the tuning-fork. With regard to the fork c^1 the addition would be 1 cm.

They were now able by means of calculated value of x and other experimentally found data to construct the curve for c^1.

If the patient heard the fork c^1, for instance, 7 seconds, the fork being struck powerfully and held close to the ear, it meant that the patient's minimum hearing amplitude, or his "Hörschwelle," was $\frac{160}{1·3} = 123$ times the normal for the distance. His hearing-power $\frac{1}{(123)^2} = \frac{1}{15129}$ of $\frac{1}{123}$ times. If the normal hearing power is equal to 1, the reduced hearing-power would be equal to 0·00007. Supposing, on the contrary, the patient heard the fork 62 seconds, his minimum hearing amplitude would be $\frac{11}{1·3} = 8·5$ times the normal for the distance. His hearing-power $\frac{1}{8·5^2} = \frac{1}{72·25}$ times 1 normal $\frac{1}{8·5}$ times the normal, and $= 0·0138$ if the normal hearing-power was equal to 1.

In this way they were able to construct the curve of every tuning-fork, and thereby to find how much the hearing power was diminished, if they only knew the time for which the fork was heard at a certain distance from the ear.

By comparing the curves of the different forks, they now

saw how greatly they differed. Some of them—the deeper forks—were steep and short; others—the higher forks—were flattened and long. In other words, the assumption of Bezold and Edelmann, that the curves were always the same was not correct, and one employing their method could not get at reliable results. This could easily be illustrated by some examples. For instance, the forks c-g^1-c^2-g^3-c^4. They were, according to his experiments, normally heard close to the ear during 328, 202, 162, 55, and 43 seconds respectively. Suppose they had a patient who heard these forks only for half the time, the normal hearing-power would, according to Bezold and Edelmann, for all hearing forks be equal $0.049 = \frac{1}{20}$. If, on the contrary, they used the special curve of each fork, the result would be quite different, because they found that the decrease of the hearing power for c would be equal to 0.026 $\frac{1}{30}$ of the normal hearing; g^1, $0012 = \frac{1}{144}$ of the normal hearing; g^3, $0.00006 = \frac{1}{17364}$ of the normal hearing; and c^4, $0.000025 = \frac{1}{40000}$ of the normal hearing.

The enormous difference between the results given by this and by Bezold-Edelmann's method was obvious. He therefore believed that if one wished to use the time in which a fork was heard to measure the quantitative hearing-power, it would first of all be necessary to know the curve of the forks employed. In order to find these curves, he hoped the method he had given would be useful.

Dr. SCHMIEGELOW, replying to questions by Professor POLITZER and Dr. DUNDAS GRANT, said the experiments he had carried out were in connection with the mathematic aspect of the hearing-power. In the clinical world they had used the very good and practical methods of Dr. Hartmann, but he thought they were far from reliable. If they wanted to compare the result of the hearing-power by the different tuning-forks, and to know the influence on the voice, they could not get any certain basis to work upon. He was only as yet on the fringe of the question.

A Scheme for the Uniform Notation of the Results of Investigation of Hearing-power. By Professor Dr. GIUSEPPE GRADENIGO (Turin).

The methods which he proposes have been already for

some time used with good practical effects in the Clinic and in the Polyclinic at Turin. The language employed is Latin. The various experiments are indicated by the initial letters of the authors' names who have described them. Here is the scheme:

$$\begin{array}{c} \text{AD} \\ \text{S (18\char`\") W } \quad \text{R (+16\char`\"), H. Hm. Ht. P, v, V,} \\ \text{AS} \\ \text{AD} \\ \text{C c } c^1 \, c^2 \, c^3 \, c^4 \, c^5. \\ \text{AS} \end{array}$$

Explanation.

AD, AS = Auris dextera, auris sinistra.

S = *Schwabach's* experiment (c=128 vibr.). Duration of normal perception with own tuning-fork c=18″.

W = *Weber's* experiment (c). An arrow designates the side towards which the lateralization takes place.

R = *Rinne's* experiment (C). Normal perception with own tuning-fork C = +16.

H, Hm, Ht = *Horologium*, watch per äer, ad mastoidem, ad tempora.

P = *Politzer's* acoumeter.

v = vox aphona, whispering voice; V = vox communis, conversational voice.

The results of the measuring of the hearing-power for the various tuning-forks are expressed in hundredths of the normal duration of perception.

The following example will better demonstrate the method:

$$\begin{array}{l} \quad\quad\quad /\text{AD}-I \quad \text{prope} \;\; + \quad\; + \quad\; \succ 5 \;\; 0{\cdot}30-0{\cdot}15 \;\; \succ 5 \\ \text{S (18)+6W} \quad \text{R (+16)} \;\; \text{H,} \;\; \text{Hm, Ht,} \;\; \text{P} \quad\quad \text{v} \quad\quad \text{V} \\ \quad\quad\quad \text{AS}-15\char`\" \quad\; 0{\cdot}05 \;\; + \quad\; + \quad\; \succ 5 \;\; 2{\cdot}00-1{\cdot}00 \;\; \prec 5 \end{array}$$

AD	12	42	72	95	100	95	100
	C	c	c^1	c^2	c^3	c^4	c^5
AS	50	80	87	95	100	100	100

Experimental Investigations on Acoustic Phenomena in Fluid Media. By Dr. R. KAYSER (Breslau).

The final sound-vibrations, Dr. Kayser said, which determine hearing take place in the cochlea, and therefore in a fluid medium. It has hitherto been impossible to investigate the conduct of vibrating bodies in fluids, because here has been no means of recognizing with any ease the vibrations of a body in water. He said he had, however, found a method of overcoming the difficulty. It consists in the use of a telephone, which is so modified that the plate of metal is surrounded on all sides by liquid. (Dr. Kayser then gave a description of this water-telephone.) By means of this method it has been easy to prove that spoken sounds, or the sound of a tuning-fork in front of the plate, throw the metallic plate into feebler vibrations than when there is no water present. Low tuning-forks from C^1 downwards, and high ones from C^4 upwards, are not heard at all. If we imitate the conditions in the ear, with two openings closed by means of membranes (fenestra ovalis and fenestra rotunda), and put one of these openings in communication (by means of a columlella) with a membrane corresponding to the membrana tympani, the following takes place: If the second opening is closed by means of any unyielding mass so that a distension of the fluid outwards is prevented, then the production of a sound in the telephone is no weaker than when similar distension of the fluid is present. It thus appears to result from this experimental proof that the molecular vibrations of the auditory ossicles have a greater significance than they were credited with according to the theory of Helmholtz, at present held. Further, it is proved by means of the water-telephone that the diminution of the intensity of vibrations is increased in proportion to the bulk of liquid which lies upon the metal plate, and the degree of its viscosity. In glycerine or milk the diminution of the intensity of the vibrations is markedly greater than in water.

Professor LUCÆ said: It is not surprising that the sound should get weaker whenever you put a sounding tuning-fork into the water; it is new, however, that certain sounds, the higher and lower ones, should get lost. If you put a sounding tuning-fork into the water, the sound gets lower up to the extent of an octave. Whether the human voice gets so much lower too is the next question. Because the sound gets so much lower by the pressure of the water, it does not necessarily imply this.

Dr. KAYSER, closing the discussion, said: It is a well-known fact that the tuning-fork loses in height under water, but I do not know whether it is as much as an octave. However, with the telephone under water this could be easily proved. The tuning-fork is made to sound under the water, and the receiver on the other end will give undeniable evidence. The lowering may be a fifth, but hardly an octave. The human voice is not influenced. It may be difficult to prove that actually, but so far I could not find any evidence of it.

A New Optic Method of Acoumetry. By Professor GIUSEPPE GRADENIGO (Turin).

If we paint at the end of one of the branches of a tuning-fork which vibrates with sufficient amplitude a distinct figure (say a tall triangle), this figure will appear more or less doubled. The duplicate images will overlap, the overlapping part being very distinct in outline and color *(field of double image)*, while the separate portions will be much paler and less distinct in outline *(field of single image)*. As the vibrations diminish in amplitude the "field of double image" becomes greater—the two images gradually merging into one. The growth of the field of double image corresponds to the diminution of the amplitude of the vibrations of which it thus becomes a measure.

When we choose a figure in form of an inverted V (Λ), black upon white ground, and if we mark it transversely with lines or steps forming various segments, we can in this manner obtain an exact index of the amplitudes of vibrations at any instant of the tuning-fork's decrement. Since the amplitude of vibration is directly proportionate to intensity of the sound, we have thus an excellent clinical method of acoumetry. Professor Gradenigo expressed his thanks to Dr. G. Ostino, Professor C. Raymond, Dr. C. Gaudenzi, and Dr. O. Pes for their valuable help in these researches.

The best results are obtained with forks whose branches make wide excursions (up to 60 vibrations a second); but the method can also be used with forks up to 250 vibrations.

As the examination with low notes is of great value in the study of the affections of the sound-conducting apparatus, the method is very useful in spite of this limitation.

Of the facts which he had been able to elicit, he wished only here to refer to the two following ones:

1. In the vibration's period measured with the said method, the decrement of amplitude goes according to the geymetrical progression in proportion to the time.

2. The individual mistakes in the appreciation of the duration of the sound-perception in persons not accustomed to this kind of researches—that is, in most of our patients—are much greater than one would believe without such a direct objective control.

Dr. W. MILLIGAN (Manchester). *Some Observations upon the Diagnosis and Treatment of Tuberculous Disease of the Middle Ear and Adjoining Mastoid Cells.*

Mr. President and Gentlemen,—The widespread interest which has of late been manifested in this and other countries in the endeavor to check the ravages of tuberculous disease in its numerous forms has an interest to the otologist, not only on account of the general merits of the case, but more especially on account of the frequency with which tuberculous lesions are met with in and around the middle ear.

The factors which come into play in producing tuberculous lesions of the middle ear and its adnexa are but imperfectly understood, and their investigation opens up a wide field for research and experiment.

Does the bacillus gain entrance to the middle ear by way of the Eustachian tube, or is it conveyed along vascular or lymphatic channels? What also is the relation between tuberculous naso-pharyngeal adenoid vegetations and tuberculous middle-ear diseases?

Questions such as these are not easily answered, and yet their solution must appeal to all as being of much importance.

For some years past I have been particularly interested in this subject, and as opportunity has presented itself, have endeavored to investigate these questions both in their practical and in their scientific aspects.

That a large proportion of the cases of suppurative middle-ear disease with accompanying bone lesions met with in practice are of a tuberculous nature will, I think, be admitted by all. and that the prognosis in such cases is not very favorable will, I believe, be conceded by those who have had large clinical experience.

The characteristic features of tuberculous middle-ear disease may be somewhat masked on account of an accompanying pathogenic infection, and an accurate diagnosis may be impossible if one relies upon finding the bacillus of tubercle in the secretion from the middle ear.

Time after time it has been my experience to examine cover-glass preparations of pus from the middle ear for bacilli, and with negative results, although the tuberculous nature of the lesion has been proved beyond all doubt by means of inoculation experiments and by the subsequent clinical history of the case.

In my experience primary tuberculous lesions of the middle ear and adjoining mastoid cells are comparatively common, especially among the children of the poorer classes, and I believe also that secondary tuberculous infection from such a primary focus is by no means of infrequent occurrence.

Amongst causes which may be considered predisposing are the following: (1) hereditary tendency; (2) unhealthy environment; (3) unsuitable feeding; (4) exposure to infection from tuberculous relatives; (5) the presence of tuberculous naso-pharyngeal adenoids.

The relation of nasal obstruction to tuberculous middle-ear disease deserves special consideration. In many of my cases post-nasal adenoids have been present, and in a small proportion have themselves been tuberculous. The almost constant degree of Eustachian catarrh which their presence implies produces a soil which is favorable to the growth of the tubercle bacillus, and once it has found a footing in the middle ear the conditions favorable to its development are present, viz., a suitable soil, absence of light, a more or less uniform temperature, etc.

In the early stages these tuberculous foci appear as slightly elevated yellowish points in the mucosa, after a time coalescing and breaking down to form superficial ulcers.

Should the deposit occur upon the inner aspect of the membrane, perforation ensues. Such perforations may be multiple, and the destruction of tissue is usually quite painless. The edges of such perforations have a pale, indolent-looking appearance, and the accompanying discharge from the ear is usually thin, ichorous, and frequently fetid.

Within the mastoid cells such deposits are also frequent,

and I am inclined to think that in some cases, at least, the disease begins first of all within the mastoid, and subsequently spreads to the middle ear. At a very early stage the bone becomes affected, and undergoes an amount of destruction which is almost inconceivable, considering the comparatively slight external indications present. In some cases which have come under my observation practically the entire cancellous tissue of the mastoid—occasionally of both mastoids—has been eaten away, leaving merely a bony shell upon which the middle fossa is poised. Owing to this early and extensive destruction of bone, the facial nerve in part of its course is exposed, with resulting facial paralysis. In fact, early facial paralysis in a case in which sthenic symptoms have been absent should, I hold, always be looked upon with suspicion and as a probable manifestation of an underlying tuberculous lesion. Early implication and enlargement of the glandular structures around the ear is also a most important symptom, and when masses of enlarged glands occur around the ear any discharge from the tympanic cavity should be microscopically examined for bacilli.

To definitely establish the fact that the aural lesion is of a tuberculous nature the characteristic bacillus must be found. This may be an exceedingly difficult task, but in all cases it is worth while staining and examining the secretion from the middle ear.

Should no evidence of its presence be found in this way, small pieces of granulation tissue may be removed by forceps pressed between two cover-glasses and stained in a suitable manner. Occasionally bacilli will be found in such preparations. The method which I believe gives the most reliable results, however, is the inoculation of guinea-pigs with small fragments of tissue removed from the middle ear or adjoining mastoid cells, and I believe that it is advisable to inoculate with fragments of bone and mucous membrane removed from an area where the disease is seen to be advancing. In many such cases when the mastoid has been opened for the purposes of treatment, a pultaceous-looking mass will be found filling up the cavity, but this material is practically valueless for experimental purposes, consisting as it does of broken-down tissue, inspissated purulent débris, and epithelial cells.

When, however, it has been removed by means of a spoon and the underlying bone exposed, it will be seen where the disease is making progress, and from where a scraping of bone should be taken. In my experiments I have inserted a fragment of tissue obtained as above described into a guinea-pig's hind leg just about the knee-joint, all hair having previously been removed by singeing with a platinum knife. A small pocket is now made with a sterilized needle, and the tissue carefully inserted. In a few weeks' time, should the tissue inoculated be tuberculous, the inguinal glands will be found enlarged, and as time goes on the tuberculous virus will be found to have spread over the animal's body, the glands and viscera being attacked in the following order, according to the results obtained by Professor Delépine:

During the second week after inoculation the lymphatic ganglia upon the same side of the body below the diaphragm and the spleen will be found enlarged.

During the third week, the liver, the mediastinal and the bronchial ganglia.

During the fourth week, the lungs, the cervical and the axillary ganglia.

After the fourth week some of the lymphatic ganglia of the opposite side of the body below the diaphragm become affected, but this takes place extremely slowly, and the sublumbar and popliteal glands escape for a considerable time.

Microscopic sections made from these glands, and stained for bacilli, will frequently be found to reveal their presence.

In this way a definite diagnosis of the actual character of the underlying lesion can be made, and the value of the knowledge thus obtained is naturally immense, both as regards prognosis and treatment.

The course of such tuberculous lesions is only too often a downward one, despite the most elaborate and painstaking treatment. The practical difficulties encountered in removing tuberculous deposits within bone are immense, and in no region of the body are these difficulties greater than when tubercle attacks the temporal bone, for reasons which must be obvious to all here.

The complications which have, to be feared are: (1)

meningitis, (2) tubercular enteritis, (3) general marasmus.

The treatment of such cases must be considered from two points of view, according as it is non-operative or operative. Cases will be met with, especially in infants, where any operative interference will from the first be seen to be hopeless.

Such are the cases where marked debility and emaciation are present, where advanced facial paralysis and masses of enlarged glands have been early symptoms, and where the discharge is abundant, fetid, and frequently blood-stained. In such cases palliative measures, antiseptic treatment, and, if possible, residence at the seaside, are indicated, but I am bound to say that in the majority of such patients whose cases I have followed an early death has been the usual history. The prognosis in such cases I believe to be essentially bad.

In other cases, however, where the present condition of the patient is good (and often enough it is so), and where the tuberculous lesion may be regarded as primary and local, much can be done by suitable operative interference. It is almost superfluous to say that the first and the main essential is to provide free drainage. This implies opening and cleansing the mastoid cells, and it is a remarkable fact how often in such cases, without any external and objective sign or indication, the mastoid cortex will be found extensively perforated, and a pultaceous mass immediately exposed to view. Under good illumination a very careful toilet of the part should be effected, and this can generally best be done by means of a sharp spoon. All softened and carious bone must be scraped away, and as smooth a cavity left as possible, even if this necessitates laying bare the dura and walls of the lateral sinus. The cavity thus obtained should be allowed to granulate from the bottom, and care must be taken to stimulate any sluggish area by means of applications of chlorid of zinc, nitrate of silver, etc. Frequently more than one scraping is necessary as fresh foci of disease appear. In one particular case which came under my treatment some years ago, and where the cause was proved to have been feeding with milk from a tubercular cow, five separate operations had to be undertaken before the morbid process was eradicated, which, however, it finally was, and the child has now grown up a healthy and sturdy boy. In very many of the cases the middle ear has been so extensively destroyed that its function as an organ of sense may be disregarded. Under such circumstances its contents should be freely curetted, and middle ear, antrum, and mastoid cells thrown into one cavity, and allowed to become obliterated by means of healthy granulation tissue. Where, however, a fair degree of hearing is present, efforts should be made to preserve the function of the organ as far as is possible.

An important point arises in connection with the treatment of the accompanying enlarged glands. Some of the glands may be enlarged purely as the result of septic absorption, and if the morbid cause be removed this enlargement will gradually subside, especially if aided by suitable treatment. But many of the glands are of a tuberculous nature, and are prone to undergo caseous degeneration, while at the same time they are a source of possible systemic infection. Hence I hold that after the mastoid area and the cavity of the middle ear have been attended to, and as soon as the condition of the patient admits of it, another operation should be undertaken with the object of removing these enlarged and tuberculous structures.

The facial paralysis which so often accompanies tuberculous disease of the middle ear is unfortunately usually permanent. Something may, however, be done by facial massage, and the internal administration of strychnia to assist in maintaining the tonus of the facial muscles.

General treatment, such as the exhibition of cod-liver-oil, iodide of iron, syrup of iodine, etc., is useful, as also is change of air and liberal diet. The general conclusions from a study of these cases may be summarized as follows:

1. That primary tuberculous disease in and around the middle ear is of fairly frequent occurrence, and that it most usually attacks the children of the poor, especially the poor of our larger cities.

2. That a generalized tuberculous infection may arise from a primary focus within or around the middle ear.

3. That the prognosis in such cases is not very favorable, at least 40 to 50 per cent. of the cases succumbing, even after operative treatment has been undertaken.

4. That in many of the cases operative interference is contra-indicated, owing to the extent of the existing disease and the asthemic condition of the patients.

5. That when operative interference is feasible, the main object should be to scrape away all available foci of disease and to provide efficient drainage.

6. That the best and the most reliable means of establishing the tuberculous nature of the disease is by means of properly-conducted inoculation experiments.

Dr. ARTHUR HARTMANN (Berlin) read a paper on *Congenital and Acquired Atresia of the Meatus Externus.*

Dr. Hartmann referred to previous reports on atresia auris congenita, which he considers should more correctly be regarded as absence of the external meatus.

He demonstrated two preparations with plaster casts of the rudimentary external ears of the same.

The first specimen was from a new-born infant, in which on both sides there was complete absence of the annulus tympanicus and membrana tympani, whilst the tympanic cavities and ossicles, though present, were not quite normally developed. In the second specimen, from an adult, the external meatus—*i. e.*, the pars tympanica and membrana tympani—was completely wanting. The articular surface for the jaw was on the anterior surface of that portion of the temporal bone which normally forms the posterior wall of the meatus. In this case also the tympanic cavity, the ossicles, and the antrum mastoideum were somewhat abnormally developed.

These specimens were important in their bearing on the question of the operative establishment of an external auditory meatus in cases of atresia congenita. They showed that this was not possible.

It is well known that even with both meatuses absent, hearing and understanding of speech can exist.

Reports of complete acquired closure of the meatus were rarer than those of congenital absence of the meatus. Dr. Hartman reported a case he had seen in which after diphtheritic-scarlatinal otitis the ossicles on both sides came away, and later complete bilateral bony occlusion of the meatus supervened.

Sufficient hearing-power remained to prevent the onset of deaf-mutism, loud speech being heard. On one side the meatus was restored by operation. After turning forward the auricle the new-formed bone was chiselled away, and the cavities of the middle ear laid bare, as in the radical operation. The meatus was covered by means of Körner's flaps. Healing was very slow. The hearing was considerably improved.

In the discussion which followed Dr. HOLINGER (Chicago) said that the paper was very interesting to him, because he was at present faced with the question whether to operate in such a case. In examining 510 children of the Institute for the Education of the Deaf and Dumb, in Jacksonville, he found a girl of fifteen with absence of both auditory canals. The girl was growing more and more deaf on account of constantly recurring attacks of otitis media. The first attack came on after scarlet fever, and the pus broke through the mastoid. The question of operation answered itself. He should operate in the following way. He should chisel behind the auricle down to the middle-ear, and remove the malleus and incus. He should allow the wound to granulate and then cover, according to Siebenmann, with Thiersch's grafts. Thus he should create a canal behind the ear. The operation would be to improve hearing and to stop the recurrence of the suppuration.

Dr. HARTMANN, closing the discussion on his paper, said: It is not advisable to operate on such cases as long as there is no inflammation. I do not believe that an operation according to Professor Siebenmann will improve the hearing power. If there is recurrent inflammation, as in the case of Dr. Holinger, we may proceed as he advised.

PAPERS.

Dr. T. BOBONE (San Remo) read a paper on *The Early Involution of Adenoid Growths on the Riviera.*

The paper was a contribution to the etiology of adenoid growths. Dr. Bobone said he had for some time been surprised at the fact that adenoids are excessively rare amongst the natives at San Remo. Moreover, he had observed that adenoids, in patients coming to the Riviera and for the removal of which the parents would not consent to an operation, began to slowly involute; so that some months afterwards nasal respiration was possible, speech was much improved, tendency to cold and cough with very slight provocation was lost, and normal development took place.

Dr. Bobone considers that pure and simple involution of the adenoid growths, although not generally admitted, is possible; and this involution he attributed to the same causes as the rarity of the vegetations amongst the natives. That cause must be looked for in the dryness of the climate and the clearness of the atmosphere on the Riviera.

The other etiologic factors in the causation of adenoid vegetations mentioned by authors, such as geographical latitude, diatheses, discharge from the nose, infectious fevers, etc., are also to be met with at San Remo, and notwithstanding, the adenoid vegetations, as already stated, are so rare.

Dr. Bobone believes the most important factor in the etiology of the vegetations to be the humidity of the climate of the country where they are observed, and that the greater the humidity, the larger the number of children with adenoids. He has been able to demonstrate this fact in observing the geographical distribution of the vegetations in Italy, where the frequency increases with the humidity of the climate of the different regions, as the following table shows:

Names of the Observers and Localities in which they are living.	Frequency with which they find Adenoid Growths.	Relative Humidity of these Localities.
Bobone (San Remo)	extremely rare	60°-65°
Cozzolino (Naples)	0.01 per cent.	65°
Massei (Naples)	0.3-0.5 per cent.	65°
De Rossi (Rome)	0.8 per cent.	65°-70°
Corradi (Verona)	5 per cent.	70°-75°
Poli (Genoa)	7 per cent.	{ moist climate
Kruch (Milan)	8 per cent.	{ (Weber),
Arslan (Padua)	very frequently	70°-73°
Ficano (Palermo)	very frequently	80°

Dr. Bobone is also of opinion that a factor which frequently complicates a simple case of vegetation is inflammation—*adenoiditis*. In the localities where the vegetations are most frequently met with, there also are attacks of adenoiditis more frequent, favored by the cold, the fog and the damp. Whereas on the Riviera the warm and dry climate is not favorable to the development of and frequently recurring attacks af adenoiditis; and when the vegetations are not irritated by inflammation, the involution can proceed.

Dr. Bobone added that in cases in which the parents would not consent to an operation, the good results he obtained he attributed more to the action of the climate than to other remedies.

Dr. ALLEN T. HAIGHT (Chicago) read a paper on *Naso-Pharyngeal Adenoids as a Causative Factor in Ear Diseases.*

Among the most interesting cases, he said, that came before the otologist were those relating to post-nasal vegetations affecting the hearing, and there were few patients to whom more satisfaction could be rendered than to those so affected. Adenoid vegetations seemed not to be restricted to countries, to climates, to sex, to color, or race of man. Naso-pharyngeal vegetations were a hypertrophy of the lymphoid tissue situated in the vault of the pharynx, bounded on either side by the orifice of the Eustachian tube, and presented on its surface several vertical furrows which partially subdivided it. It was his opinion, based on several years' experience in the Illinois Charitable Eye and Ear Infirmary and in private practice, that the main factor in producing both suppurative and non-suppurative inflammatory conditions of the tympanic and Eustachian mucous membranes was the presence of naso-pharyngeal adenoids, or the condition of the post-nares subsequent to their removal or absorption. Adenoid vegetation might produce inflammation of the middle ear (1) by constant irritation, on account of the obstruction to the circulation of the blood by pressure; (2) by blocking the orifice of the Eustachian tube partially or completely; (3) by their injurious effect upon the general economy of the child, and particularly upon the nerves of special sense; (4) by leaving as a sequela a post-nasal catarrh, which sooner or later establishes some form of middle-ear disease. In children who suffered from adenoid vegetations the hearing was generally very sensibly impaired, and it was the common thing for a child so affected to have questions repeated often and in a louder tone of voice. In many cases the Eustachian tube was completely blocked by dry secretions of the post-nares. He had observed diminution of power of hearing on the side where the adenoid existed. On the opposite side, where the post-nasal space was clear, the hearing was normal. He had seen cases where the hearing was seriously impaired, and the drum membranes normal in appearance, and yet with safety he assumed the faulty hearing to be dependent upon the growths in the naso-pharynx. Mouth-breathing, he believed, had an important otologic bearing on the subject. The mouth-breathing child was usually found shallow through the upper part of the chest, and with very small lung capacity. They frequently met with children affected with adenoids who were not mouth-breathers, and these children were plump, well developed, and of healthy appearance, although they usually had some ear complication. In his examination of twenty-six children for deaf-mutism, he found only four free from post-nasal adenoids; sixteen of those examined showed marked facial deformity from mouth breathing. He coincided with Harrison Allen and Lisson, who had expressed the opinion that there were many children in homes for feeble-minded and idiots all over the world who were affected with this disease, and who by a comparatively trifling operation could possibly be restored to usefulness and their families. It would be obvious to mention every analogous case reported of deaf-mutes who, after the removal of adenoid vegetations, gave evidence of hearing and began to speak some words. The general belief that adenoid vegetations were never present after the thirtieth year was contradicted by Couetoux, of Nantes, who operated upon a man of sixty-five to cure a marked unilateral deafness. Dr. Haight had found vegetations in ages above sixty, and frequently between thirty and forty. They did not differ histologically from adenoids in children. It was not uncommon to observe these formations in the aged who were hard of hearing. Notwithstanding all the writings of the past ten years, he did not think that the pathologic enlargement of the lymphoid tissue of the naso-pharynx had received sufficient attention in the world's text-books. If the symptoms of these growths were more generally recognized by the family physician, and their removal accomplished, they would not find so many chronic suppurative and non-suppurative inflammations of the middle ear, with the history dating back to an attack of diphtheria, scarlet fever, measles, or other fevers. As to treatment, he should say it was never too early nor was it ever too late. At the first recognition of existing growths the operation should be performed at once. He had found that curetting was the only true basis of treatment. He was not a believer in general anesthetics in children over the age of twelve, as local anesthesia after twelve made such an operation absolutely free from danger; but there were some cases where a general anesthetic must be administered, especially in refractory children and nervous adults. In children it was advisable to anesthetize in a sitting posture, and he preferred bromide of ethyl to any other of the numerous anesthetics.

Professor KNAPP also advocated the use of ethyl. There was absolutely no danger.

Dr. EEMAN (Ghent), Professor GRAZZI, and Dr. GRADENIGO also joined in the discussion.

DISCUSSION.

The Indications for Opening the Mastoid in Chronic Suppuration of the Middle Ear.

The discussion was opened by Professor POLITZER, (Vienna), Professor MACEWEN (Glasgow), Dr. LUC, (Paris), and Professor KNAPP (New York.)

OPENING ADDRESS BY PROFESSOR ADAM POLITZER (Vienna).

Professor POLITZER said it was a happy idea of the Organization Committee to have put on the programme a discussion on such an important question. There was no question of otology which had acquired more actual interest than the free opening of the middle-ear spaces for chronic suppuration of the middle ear. Experience had shown that the free opening of the middle-ear spaces was of the most vital importance, by which they were able to save the life of the patient and prevent other consequences to the middle ear hurtful to the organism. The indications were generally acknowledged, and in most cases with well-marked symptoms the surgeons were likely to be in perfect agreement; therefore there could be but little new to say in reference to the indications. The chief point in that discussion would be to decide whether it was justifiable without well-marked symptoms to operate as frequently as some operators maintained. Professor Politzer then enumerated all the indications for the so-called "radical operation," giving after his own experience a complete critical view on the subject.

In his enumeration, he classified the indications in two groups—objective and subjective.

The objective indications were briefly:

1. Caries of the walls of the tympanum.
2. Granulations and polypi in the neighborhood of the aditus, and recurring after removal.
3. Fistulæ opening into the mastoid cavities, and frequently leading to cholesteatoma.
4. Cholesteatoma.
5. Hyperostotic stricture of meatus.
6. Facial paralysis or paresis.
7. Painful swelling on the mastoid (indicating acute mastoiditis, fistula, cholesteatoma or sequestrum).
8. Obstinate long-continued fetid discharge, rebellious under all forms of treatment, especially when the perforation is in the postero-superior quadrant, and the remains of the membrane is adherent to the inner wall, and still more if pus, or especially crumbling masses of epithelium, can be sucked out by means of Siegle's speculum.
9. Symptoms of tuberculosis occurring in the course of chronic suppuration of the middle ear (the supervention of aural suppuration in the course of pulmonary tuberculosis is unfavorable for operation).

Further, high temperature preceded by rigor or oscillation of temperature, indicating sinus phlebitis or direct septic absorption. Also vomiting, with headache, and other brain symptoms, or changes in the fundus of the eye.

The subjective symptoms were:

1. Persistent or recurrent pain in the ear or mastoid process, especially with persistent and fixed pain in the parietal or occipital region, and increased by percussion, which frequently points to temporal or cerebellar abscess.

2. Vertigo, either permanent or intermittent attacks, which may be due to erosion of the external semi-circular canal, or extension of the disease to the interior of the labyrinth (as would be indicated by the usual tuning-fork tests for nerve deafness, and would call for a removal of the labyrinth, as advised by Jansen, over and above the original mastoid operation).

3. Well-marked brain symptoms, such as headache, heaviness, pressure, torpor, loss of consciousness, etc.

Operation was all the more called for if the objective signs were accompanied by any of the serious subjective symptoms, and the symptoms of serious brain complication, instead of being contra-indications, called for immediate operative interference. With regard to meningitis, it was now well recognized that the most pronounced symptoms of that disease might be due to a serous, as distinguished from a purulent form of meningitis, recovery from it being a frequent sequel to the thorough removal of the ear disease. Such symptoms, therefore, would not contra-indicate operation unless lumbar puncture showed the cerebro-spinal fluid to be infected.

Professor Politzer concluded that experience taught him that not rarely the clinical symptoms did not correspond to the pathologic changes found during the operation in the temporal bone. Sometimes only insignificant changes, such as a small quantity of granulation tissue in the attic or antrum, were found in cases where he had performed the operation on account of dangerous symptoms. On the other hand, he found grave changes where before the operation he would not have expected them.

These circumstances rendered it more difficult to draw strict lines in regard to the indications, and there would always be cases in which some surgeons, on account of the impossibility of predicting exactly the pathologic changes in the temporal bone, would hold that it was not advisable to wait for the appearance of well-marked symptoms, and decide to operate at once, while other surgeons would advocate more conservative methods. That many cases of the chronic suppuration of the middle ear could be healed by vigorous antiseptic treatment, by removing the granulations or cholesteatoma in the tympanic cavity and the attic, by partially removing the wall of the attic, had been shown by the daily experience of those surgeons who treated such cases by conservative methods. Although he was a strong advocate of the radical operation in suitable cases, he could not agree with those surgeons who performed it often for the mere purpose of the discharge—at least, until strenuous efforts had been made to stop it by other means. He thought that in these cases it was not justifiable to have recourse to an operation which, although not necessarily dangerous in the hands of a skilled operator, was still a serious one, especially when they considered (1) the many important structures in the vicinity which might be injured, (2) the possible permanent impairment of hearing in those who before the operation could hear fairly well, (3) the protracted healing process after the operation, which very often rendered the patient *hors de combat* for many months. It was his firm belief that these views would in course of time receive general assent, when further anatomic researches and more extended clinical observations had cleared up those points about which at present their judgment was still in doubt.

OPENING ADDRESS BY PROFESSOR WILLIAM MACEWEN
(Glasgow).

Professor MACEWEN said: Mr. President and Gentlemen,—I have to thank you for the honor you have conferred upon me by asking me to open a discussion on the indications for opening the mastoid in suppurative otitis media.

Instead of enumerating the individual indications for opening the mastoid, which may be found in more or less detail in most recent otologic works, and which may require to be supplemented or reduced as our experience ripens, it is thought desirable to regard the subject from a broader basis, and one which may be found more generally applicable. The following forms a useful practical rule:

When a pyogenic lesion exists in the middle ear, or in its adnexa, which is either not accessible or which cannot be effectually eradicated through the external ear, the mastoid antrum and cells ought to be opened.

As there are many ways of opening the mastoid, some

more, and many less complete, the observations made in this note cannot be equally applicable to all of them.

Some operators content themselves in "opening the mastoid" by sinking a narrow shaft into the antrum, through which they can inject fluid, and others perform a typical operation irrespective of the pathologic condition revealed.

The author does not follow the classic operations of Kuster, Stacke or Schwartze, but operates by first opening the mastoid at the base of the suprameatal triangle. From that point he follows the pathologic lesions anteriorly into the middle ear, especially exposing and carefully scrutinizing in all cases the attic of the antrum and tympanum, when, if found eroded, these plates are removed, along with the morbid contents of the middle ear. We then pass backwards and downwards, through the mastoid cells toward the sigmoid sinus, following the pyogenic erosions wherever they may lead in that direction, and when necessary exposing the knee of the sigmoid sinus. After opening the mastoid antrum and cells, the further procedure has a purely pathologic basis; if the disease revealed be extensive, so must be the operation. The greater part of this operative procedure is performed by means of the rotary burr, which is the safest instrument for such a purpose. One of the first objects of the operation is to secure the patient against subsequent pyogenic extension to the brain on the one hand, and the cerebellum and sinus on the other; and this may be done with a probable certainty, as far as the two most frequent localities for brain and sigmoid sinus invasion are concerned. It is to such an operation (with its pathologic basis) for "opening the mastoid" that the following remarks apply:

The ablation of the mastoid, while at once eradicating a suppurative process, chiefly located in the mastoid antrum and cells, affords at the same time ready access to the attic and inner wall of the tympanic cavity, and to the auricular extremity of the Eustachian tube. Immediately following the operation, one can initiate the formation of avascular tissue, and thus create an efficient barrier against pyogenic extension to the otherwise most accessible and most vulnerable parts of the brain, the cerebellum and the sigmoid sinus.

In persistent otitis media purulenta, the mastoid operation has at least three advantages over that of the treatment by way of the external auditory meatus; First, by exposing to ocular inspection all the affected area, and by thus enabling the operator to follow and eradicate all the recesses in the bone made by pyogenic invasion. In this way one does not act in the dark, as the whole pathologic field is open to inspection. Secondly, by being able to secure asepsis. Thirdly, by raising an efficient barrier against pyogenic extension, between the most vulnerable parts of the brain and the sinus.

Indications for Opening the Mastoid in Purulent Otitis Media.

1. There are many cases of purulent discharge of the middle ear, of such long standing, and so intractable to all remedies administrable through the external auditory meatus, that most surgeons would agree that in such the mastoid ought to be opened. When the symptoms are obtrusive, the pain severe, the discomfort great, the discharge profuse, and possibly foul-smelling, the patients themselves will probably demand relief, which the otologist will readily grant. It is not, however, to such pronounced cases that special attention is here directed. It is rather to those in which the decision is much more difficult, especially in the presence of very slight discharge, continuous, though apparently subdued by treatment. Many believe that very slight though persistent otorrhea can lead to no untoward result, the patient living a considerable number of years, possibly even a long life, with the discharge never properly away, and yet not sufficient to arrest attention. Its long duration causes the bearer of it to pay little attention to it, and by-and-by it may be disregarded, and even forgotten.

The pyogenic process may, however, proceed inward, giving rise to symptoms often misunderstood or attributed to other causes, and may eventually either prove fatal or, by undermining the constitution, thereby pave the way for the advent of other lesions. Many patients thus affected, though able to pursue their avocations, are yet subject to periods of malaise, with occasional recurrent slight febrile attacks, irritability, and nervous hypersensitiveness, exhibited in unevenness and irrascibility of temper, which attacks last from a few days to a week or more, leaving the patient slightly weaker, though relieved from the depression, and fit to enjoy life. These attacks are so frequent, and the patient becomes so used to them, that he comes to regard them as part of his ordinary habit, and often attributes them, with considerable plausibility, and sometimes with point, to colds, chills, biliousness, indigestion, etc.

When they occur, however, in the presence of pyogenic otorrhea of old standing, they may bear a different interpretation, and in the absence of other definitely assignable causes they may be considered as the result of slight absorptions. In some cases the cause and effect are a little more evident, as when patients have pyogenic pulmonary catarrh with organisms in the lung secretion similar to that found in the slight purulent otitis media, and when these pulmonary attacks are mainly coincident with the recrudescence of the otorrhea. In some such slight cases, after every other assignable cause was exhausted, and after treatment in other directions had failed, the mastoid was opened, when, in the midst of eburnation and sclerosis of the bone, marked osseous erosions, containing small quantities of secretion filled with pyogenic organisms, were found, and generally these led more or less directly to the sigmoid sinus, the coats of which bore evidence of long-standing irritation, and through which, no doubt, the pyogenic absorptions had taken place.

After the operation these patients became greatly improved in health, all their old general symptoms having disappeared along with the cessation of the otorrhea.

Cases with a history of an initial period somewhat similar to the above have been seen at a later stage by the author, coming under observation in a moribund condition from pneumonia, due to septic infections from thrombosis of the sigmoid sinus, originating in a purulent otitis media of old standing; the passage between the cells and the sigmoid sinus being in some instances very small and tortuous, and not unlike those apertures seen in the cases with slight symptoms just referred to.

When it is recollected that in many instances the otitis media purulenta is obscure and overlooked, and that the symptoms of the purulent absorption may be of a "typhoid" as well as of a "pulmonary" type, one can easily understand that death may be attributed to pneumonia or to enteric fever.

It is quite true that, with chronic otitis media purulenta, a fatal issue ensues only in a limited number of cases, a proportion, however, perhaps greater than is generally believed, but as one cannot, with any data obtainable at present, foretell which of these apparently slightly affected patients are to become the victims of a fatal issue, ordinary prudence dictates its removal even while it is slight.

It cannot be too often recalled that the virulence of the otorrhea cannot be measured by the quantity of the secretion, its odor, or the slightness of its initial symptoms, and that the pyogenic process may proceed insidiously until some slight exciting cause or accidental circumstance precipitates a dangerous or fatal crisis.

2. Another question arises, whether there be lesions in the middle ear, which, though it may be mechanically possible to remove them through the external auditory meatus, could yet be removed with greater safety through the mastoid. This must be answered affirmatively, while the middle ear and its adnexa are in a septic condition, and when by application through the external auditory meatus they cannot be made aseptic prior to the performance of an operation entailing the exposure of a fresh surface to the action of pyogenic organisms and their products. To operate through the external ear under such conditions is to court disaster. By opening the mastoid one can efficiently remove therefrom the suppuration, and can eradicate its cause, after which any operation involving exposure of a fresh surface can be preceded with in safety.

In numerous instances, cases of intracranial pyogenic extension have occurred in immediate sequence to the removal by way of the external auditory meatus of granulation tissue masses—so-called "aural polypi"—which were protruding into the middle ear. Some of these granulation masses protrude through the bone from the dura mater, which they serve to protect, *as long as they remain intact*, but when they are removed a fresh surface with open mouths of vessels is exposed, and absorption through the softened brain membranes is apt to occur.

Besides rendering the operation safe by asepsis, the

opening through the mastoid enables one to demonstrate the exact locality from which these granulation masses spring. This is difficult and sometimes impossible to do, by operating through the external auditory meatus. One must recollect that many of these granulation masses, presenting at the upper and back part of the middle ear, protrude through eroded bone, and that their presence is to be regarded as indicative of a diseased process which has attacked the osseous tissues as well as the soft parts; and therefore to an extent these granulation masses are symptomatic, and by removing them alone the disease is not removed, but only *one* of its indications.

As long as these masses are left *intact*, they may secrete, but they do not readi'y absorb, as they are destitute of lymphatics, and therefore in the midst of certain pyogenic organisms, not only may the granulation masses be left with safety, but they afford for the tissues from which they spring a definite protection from the invasion of certain pyogenic organisms. They are a provision thrown out by Nature in an attempt at repair.

In the presence of such granulation masses, one does not devise an operation merely for their removal, but for the eradication of the disease which has occasioned them. In removing them one has also to make provision that absorption will not take place through the wounded surface left thereby.

3. In many, if not all, of these persistent pyogenic otorrheas, the osseous tissue is involved, and it is very difficult, by means of treatment through the external auditory meatus, to eradicate the organisms that have housed themselves in the recesses of a minute particle of necrotic bone. In the interior of such harbors of refuge, situated in the mastoid, the pyogenic and other organisms are safe from any antiseptic wave or blast introduced through the external ear, and wait—and they have endless patience, even beyond that of the aurist—until the antiseptic has exhausted its energies, when they again sally forth, in the tide of a catarrhal effusion, disseminating themselves and affecting fresh areas. Erosion often steadily progresses within the mastoid cells, even when the middle ear has been rendered sweet. In such cases the surgeon would be deceived were he forming an opinion on the asepticity of

the mastoid cells from the condition of the discharge issuing through an external ear which he has rendered aseptic by chemicals, as a slight pyogenic discharge issuing through such chemicals would probably be rendered aseptic in transit.

In other parts of the body where a necrotic bone filled with pyogenic organisms is even exposed to view and of easy access, it is the greatest difficulty, and sometimes it is impossible, to entirely destroy these organisms by direct applications of antiseptics of such strengths as the neighboring tissues would withstand without themselves being destroyed. If this be so under such conditions, how much more difficult must it be by way of the external ear to eradicate pyogenic organisms through hidden, narrow, tortuous, and sometimes almost inaccessible passages which are often found in the mastoid process and cells.

4. In recurrent cases of purulent otitis media, one cannot pronounce the patient safe even when the otorrhea ceases—temporarily.

In one such instance, treated through the middle ear on the most approved principles, with great care, by an aurist of undoubted ability and experience, the patient, who had had a slight pyogenic otorrhea, was pronounced cured by the aurist, the discharge having disappeared, and the condition of the middle ear appearing to him in every way satisfactory. Within about three weeks of this time the patient came under my observation, suffering from pronounced symptoms of cerebellar abscess, and was plunged in profound coma, accompanied with great respiratory difficulty. He was operated on. two ounces of pus being removed from the cerebellum, after which he made a rapid recovery. The middle ear contained only a few drops of pus, the mastoid, antrum and cells contained more, and an erosion in the mastoid exposed the sigmoid sinus, which was thickened, the disease having spread to the cerebellum by continuity of tissue. With the data at the disposal of the aurist in this case it would have have been difficult for him to have acted otherwise than he did, and had he done so, it would have been at variance with the teaching of the day. This case, however, demonstrates that the information obtainable by inspection of the middle ear is not sufficient to reveal the pyogenic invasion of the recesses of the mastoid region.

Had the case been treated by opening the mastoid in the way described, the formation of the abscess in the cerebellum would have been prevented.

5. Cholesteatoma and tubercular processes with secondary pyogenic involvement are also conditions for which the mastoid requires to be opened, as it is only in this way that these diseases can be efficiently removed.

6. The problems connected with the question of operation upon recurrent attacks of purulent otorrhea are somewhat similar to those which arise in connection with appendicitis. Purulent otitis media and appendicitis have many analogies. They are both pyogenic, but while the latter is the result of the action of a well-known bacillus, whose course is definite, the former may be the result of one or other of a variety of organisms of greater or less virulency, and producing different pathologic effects. Both are apt to invade neighboring structures, the one the peritoneum, the other the intracranial tissues. Both are insidious in their action, and as long as they exist they are apt to undermine the health and reduce the vigor of the individual. Both tend to precipitate a sudden serious illness, and one which is often fatal. In both an early and complete operation not only at once relieves the patient from the depressing effects of the disease, but at once removes the possibility of a sudden and fatal termination. In both many, lulled into a sense of security by the apparent passivity of the disease and its long duration, and arguing from the fact, that as the patients have recovered from one attack they are equally likely to recover from another, postpone operation until the peritoneum in the one case and the brain in the other become involved, and a fatal termination is imminent, and then it may be too late to save the patient.

7. With regard to the fauna occurring in that perfect incubating chamber, the middle ear and its adnexa, and their relative pathologic significance, the time at our disposal prevents us dwelling at present further than to state that valuable indications may be derived from the identification of the particular form or forms of organism which may be present in such cases.

8. After what the author has elsewhere written, he presumes that it may be understood that the opening of the

mastoid must always be undertaken as a preliminary step to operating upon those intracranial lesions originating in purulent otitis media—abscess of the brain or cerebellum and sigmoid sinus thrombosis. To operate upon the several complications, and to leave uneradicated the paths by which pyogenic organisms enter, is to render the patient's recovery doubtful, and to expose him to fresh attacks.

9. Syme is credited with saying that diseases of the ear were of two kinds: the one which is curable, and is treated by the surgeon; the other which is incurable, and is treated by the aurist. Whatever be the special province of the present-day aurist or surgeon, let us hope that we relegate to neither many cases of incurable disease. The anatomy and pathology of the mastoid region were not understood in Syme's day, and the operation of opening the mastoid in its present conception was unknown. As the subject which you, Mr. President, have arranged for this discussion is the indications for opening the mastoid in purulent otitis media, we are precluded from entering into the consideration of the results attending that operation. The personal experience of the author leads him, however, to state that he regards the operation of opening the mastoid as the safest and most efficient way of eradicating otherwise persistent purulent otitis media. In conclusion, he adds that the more the pathology of purulent otitis media is studied, the more frequently the complete ablation of the mastoid recesses is undertaken, the fewer will become the so-called incurable cases of "ear disease." He regards the operation of opening the mastoid as substantially contributing to the well-being of human comfort and happiness, and materially lengthening life.

OPENING ADDRESS BY DR. LUC (Paris).

Dr. Luc said: The indications for opening the mastoid apophysis, which are simple enough in cases of acute suppuration of the ear, because they then consist of the combination of signs characteristic of purulent retention in the antrum and mastoid cells, are, on the other hand, numerous and varied in cases of chronic otorrhea.

Here the phenomena may suddenly appear at a given moment, after months and years of suppuration, nothing having occurred to interfere with the flow, which, causing the patient no pain, has been more or less neglected by him. But this is only a very limited aspect of the question, and in addition to this primary indication, which was the only one known at the commencement of aural surgery, others have been added. The progress of our methods of diagnosis has taught us that in the majority of cases the intractable nature of many otorrheas results from lesions seated in parts of the middle ear which are inaccessible to our means of treatment through the natural passages—the attic and antrum.

It is thus that we have learnt to open the mastoid, no longer simply to insure the flow of pus retained in the cavities, but to reach the extreme limits of the suppurative focus, and to dry up in a radical way a suppuration which is otherwise incurable. We must recall to mind that Schwartze of Halle, Zaufal of Prag, and Stacke of Erfurt, have been the principal promotors of this movement, which has since become so widely known. There is then a second cause of indication quite distinct from the first.

There is a third, since the numerous endeavors to open the cranium, which have been made within recent years, to circumvent the results of intracranial affection so frequent in the course of otorrhea, have led the majority of aural surgeons to the conviction that the antrum is the most certain track by which to reach the original focus, be it meningitis, encephalitis or thrombophlebitis arising from disease of the ear, and consequently this opening must be considered as the prelude or first step preceding the search for any of the foci in question. Our subject is thus divided into three chapters, which we will consider in succession.

1. *Indications for Opening the Mastoid in Chronic Otorrhea in Case of Purulent Retention.*

It is not our intention to dwell long on this subject. The indications for operation are almost identical with those present in the acute forms of otitic suppuration.

It is aroused by some intercurrent accident, occurring most unexpectedly, in the course of an old otorrhea, until then more or less neglected. Generally brought on by the development of exuberant granulations in the cavum, particularly in the neighborhood of the *aditus ad antrum*, the pus which had always easily flowed out of the attic into cavum and meatus finds an obstacle to its regular evacuation; the patient feels for the first time an ear-ache, localized generally at the base of the mastoid process. At the same time fever makes its appearance, while the general state is disturbed.

Palpation of the apophysis reveals a marked tenderness to pressure, predominant at or chiefly confined to its base, except where the cells extend almost to its point, this last region proving the chief seat of pain.

When the antrum is contracted, deep and separated from the surface by a broad band of eburnated tissue, retention causes the seat of disease to give rise to a sensation as of an otalgia, more or less intense, generally accompanied by sensibility to pressure at the base of the process, and a persistent febrile state, with the arrest or diminution of discharge. Consequently the persistence and accentuation of this state constitute an indication for intervention, provided it is necessary for a re-establishment of the flow of pus, a procedure such as removal of polypi that obstructed the meatus in such a case, growing, may be, from the cavum in the region of the aditus, not proving sufficient.

Pain, and that alone, gives, in our opinion, a reason to intervene when it has attained a certain degree of intensity, depriving the patient, for instance, of all sleep. Delay operation in such a case, other symptoms appear, such as swelling over the mastoid, edema, and the risk is run of finding signs that infection has entered the cranium.

Intra-mastoid retention of pus is happily not always so obscure in the course of chronic otorrhea. Quite often, in fact, pain and the sensitiveness of the region of the apophysis are not slow in exhibiting the classical local symptoms of mastoiditis, so-called mastoid retention; for it is well recognized to-day that all suppurations of the ear are accompanied by a purulent mastoiditis, and the local signs in question—swelling, edema, and redness of the skin—indicate not simple antral suppuration, but *imprisonment of pus in its interior*.

These signs always notify the natural advance of the pus to the soft external parts, which we consider as favorable and beneficial: favorable in the sense that the mastoid antrum is shown to be superficially placed, and removing at the same time the danger of eruption of pus towards the cranial contents; beneficial in the sense that it clears the mind of all doubt in those doctors most inclined to hesitate or temporize, and causes them to operate without delay.

What must happen in such cases? What must be the nature and extent of our intervention? We believe we can arrive at the principle. Whenever there is an indication to operate on the mastoid in the course of a chronic otorrhea, whether on account of the phenomena of retention, or that one proposes to discover and destroy the lesions maintaining the suppuration, the operation of opening the cavity and subsequent curetting ought to affect the whole of the cavities of the middle ear.

It is of slight importance whether we commence the operation through the antrum to reach the attic, along the aditus, after the method of Zaufal, or whether, following the teaching of Stacke, we enter the antrum *via* the attic. Yet, again, all the cavities, all the recesses of the middle ear, must be opened and cleaned out. Proceed otherwise, and the patient, notably in a case of retention, remains exposed to a persistence of this otorrhea, its reproduction, and finally to those accidents we seek to combat.

2. *Indications for opening the Mastoid with a view to the Radical Cure of Chronic Otorrhea.*

We believe we may express this principle, that any focus of suppuration, however inveterate it may be, is unable to resist the surgical treatment which realizes the triple result of exposing, then cleaning and draining, the whole of its suppurative surface.

What has for long justified the expression "intractable otorrhea" is that, up till recent times, apart from accidental retention of pus in the mastoid, indicating urgently the opening of the antrum, all treatment of aural suppuration was limited to the opening of the tympanum. This is only a very small part of the cavities of the middle ear. Above it is the lodgment of the ossicles, the attic, which is concealed from our inspection by the osseous ridge, which results from the difference of level between its roof and that of the auditory meatus. Behind it, and on a slightly higher level, is the petrous antrum, communicating with it by a narrow canal, the aditus, and extending more or less far in the direction of the base of the mastoid process.

The prolongations of the tympanic cavity upwards and backwards, the attic, and the antrum are the most often involved in the pathologic process, and especially in the suppurations. Further, certain peculiarities of the anatomic disposition render them peculiarly favorable for the retention of pus, and the perpetuation of suppuration at their level.

Indeed, both these peculiarities are explained by their situation, which renders them inaccessible through the natural passages both to our inspection and our means of treatment. Let us add, in regard to the petrous antrum, that as it dips more or less deeply into the mastoid process, below the level of the aditus, which is its natural outflow, and passes insensibly in many cases of old-standing osteitis, along with the mastoid cells, into a large suppurating cavity, which extends down to the tip of the process, it can only emit its pus into the tympanum when there is enough to overflow—a circumstance favorable for the development of fungating granulations, which in their turn can keep up the suppuration, and only disappear under the action of the curette. As regards the attic, it would seem at first sight that its situation immediately above the tympanum would favor in the best possible way the outflow of any pus secreted in its interior, but in its origin the mechanism of the retention of the pus is here of a special nature: namely, in the very middle of the cavity in question there is the bulk of the ossicles, which are held in their respective positions by complicated ligamentous apparatus, whose meshes allow the pus to circulate only with the greatest difficulty. We must add to this that the ossicles, which are generally affected by the spread of the fungating osteitis, contribute actively in keeping up the otorrhea.

The preceding considerations had not escaped the surgeons of the Halle school, where we must admit that modern aural surgery, which has become so fruitful in its results, had its birthplace. Schwartze was the first who endeavored to bring about the cure of several otorrheas by opening the antrum with a view to making a counter-opening for the outflow of lotions injected through the meatus.

On the other hand, one of his pupils, Ludewig, brought

about the cure of other chronic otorrheas by the extraction of the carious malleus and incus. The road was thus opened in the direction of genuinely rational treatment for the most intractable forms of otorrhea; but it remained for Stacke (of Erfurt) to give the complete solution of the problem, by proposing a new operative method of laying open and cleaning by one stroke the whole of the cavities of the middle ear.

Stacke, however, limited his first interventions to the attic, but soon experience taught him that this limited intervention ended most frequently in failure, as the lesions met with in this cavity existed almost always simultaneously in the antrum, and he arrived at this conclusion, to which we have ourselves been led by experience, and which we cannot express too strongly as a principle, that *the mastoid antrum, the actual prolongation of the attic, near the base of the petrous bone, participates in the great majority of cases in the suppurative lesions of the latter, and ought therefore to be opened and curetted at the same time, when sufficiently long attempts at local treatment of the otorrhea through the auditory meatus have failed.*

We are now naturally led to study the combination of signs, which in the course of a chronic otorrhea indicate that the attico-antral cavity participates in the suppuration, if, indeed, it is not the sole or principal source, and that consequently surgical interference is called for, answering the precise object of rendering accessible to the action of the curette and to drainage a region which is naturally beyond our means of treatment, owing to its anatomic situation.

Clinically, chronic otorrhea, originating in the antrum, presents itself under two distinct aspects, according as it is accompanied or not by a fistula. We should consider the symptoms of these two forms successively.

In most cases the mastoid fistula is on the external surface of the process, and as a rule at its base, but it may be further back or near the tip. Under these circumstances there is nothing easier or more instructive than to explore the opening by means of a probe. We are here, so to speak, brought directly in contact with the osseous lesion, and when the instrument has penetrated the antro-mastoid cavity, it gives us very valuable information as to its situation and dimension.

But the mastoid fistula does not always appear on the external surface. It may occupy other positions, and it is all the more important to recognize them, in proportion, as they are unusual.

Let us mention, in the first place, those not very rare cases in which it is found on the posterior wall of the auditory meatus. We have recently observed during the present year a remarkable example in a diabetic lady, about fifty years of age, who was sent on account of what appeared to be a simple otorrhea. The first peculiarity which arrested our attention was that insufflation of air into the tympanum was not accompanied by perforation sound, and further that the fundus of the meatus was obstructed by a large granulation-mass, growing from the posterior wall. After we had removed this granulation from the wall, we found a fistulous orifice, from which pus escaped in abundance. This led us to suspect an anterior mastoid fistula.

In this case we were unable, on account of the narrowness and obliquity of the osseous fistula, to confirm our diagnosis by means of the probe or exploratory irrigation by means of Hartmann's canula, but the lesions seemed sufficiently characteristic to justify the proposal to open the mastoid cavities, and we had an opportunity, a few days afterwards, of verifying the exactness of our diagnosis. In point of fact, the operation revealed that the process was converted into a vast suppurating cavity, from which the pus could only escape by an overflow through the narrow orifice on its anterior wall.

The diagnosis of chronic mastoiditis, which is difficult when a perforation occupies the position described, is much more so when, subsequent to chronic Bezold's mastoiditis, it is situated on the internal wall of the process, the pus escaping from there, along a deep track underlying the sterno-cleido-mastoid, to escape by an orifice often far away from the mastoid region, so that the first idea in one's mind is that one has to deal with a fistula arising from a cold, cervical, glandular abscess.

We published at the commencement of this year, in the *Archives Internationales d' Otologie*, a remarkable case, and we believe a unique one, of this clinical form, occurring in a young man of twenty years of age, who had come for the treatment of a cervical orifice in the right side of his neck, immediately behind the posterior margin of the sterno-cleido-mastoid muscle. This fistula had existed for four years, and was all that remained of a diffuse phlegmon of the neck, which had come on after an acute suppuration in the ear. On the first examination, we found that the tympanic membrane was destroyed, and that the ear still suppurated in a slight degree. The co-existence of a cervical fistula along with an old-standing otorrhea put us on the right line for the diagnosis, and our suspicions were confirmed by the exploration of the track by means of a probe pushed upwards, penetrating at first the mass of the sterno-cleido-mastoid, then the interior of the mastoid process. The young man recovered after a long and laborious treatment, which consisted, on the one hand, in the opening and curetting of the whole cavity of the middle ear, and, on the other hand, the opening in all its length of the fistulous track, in such a manner as to transform it into one deep groove; finally, in the resection of the bulk of the mastoid process, so as to reach the granulations which had developed in its interior around the perforations on its inner wall.

It remains now to consider the signs by which one may recognize the second category of cases of chronic mastoiditis, namely, those in which there is no fistulous track to offer more or less direct communication between the mastoid focus and the exterior.

It is in such cases, apart from accidental retention, that the diagnosis offers the maximum of difficulties. It seems to us that the term *latent mastoiditis* is peculiarly applicable to this clinical form.

Inspection and palpation of the mastoid region reveal absolutely no abnormal peculiarity; there is neither fistula nor redness of the skin, nor tenderness on pressure. The whole symptomatic expression of the affection is limited to an intractable otorrhea, in the customary sense of the word, that is to say, resisting the most varying therapeutic means, including the extraction of the ossicles.

What do we learn in such cases from the examination of the tympanic membrane?

We find it always perforated, that goes without saying, but its perforations may be referred to three distinct types, which are easy to classify:

1. There is first the type of *Shrapnell perforation*, situated above the short process of the malleus.

2. Then *circum-malleal perforation*, characterized by extensive destruction of the tympanum all around the handle of the malleus, which hangs in its middle. Often there are fungating granulations in the posterior region of the perforation, that is to say, in the neighborhood of the aditus, which indicate disease in this region, and irrigations in this direction made by means of Hartmann's canula confirm this presumption by bringing about the expulsion of cheezy and fetid pus.

3. Finally, the *postero-superior perforation*, characterized by a small loss of substance in the membrane, situated in the region of the aditus. This perforation, like, the preceding ones, frequently allows of the passage of small polypoid masses, which recur invariably every time they are removed, and irrigation directed towards them by means of Hartmann's canula produces the same result as in the preceding cases.

Independently of these polypoid growths, the different types of perforation which we have just passed in review permit us at times to discover whitish mother-of-pearl masses, which are nothing else but cholesteatomata occupying the attico-antral region. In such cases the Hartmann canula again plays its marvellous role, as a means of diagnois, by expelling these pathologic products, which are so characteristic, and placing them directly under the eye of the observer.

The different otoscopic confirmations which we have just enumerated would then offer strong presumtion in favor of the suppurative focus being in the attico-antral region, especially if in the case of large tympanic perforations the examination of the lower part of the drum revealed no lesion capable of keeping up the suppuration.

Does this diagnosis carry with it at once the indication for opening the mastoid and the attic? We do not think so. In a case which we have supposed of an intractable chronic otorrhea, but unaccompanied by indications of retention, or threatening complication, there is no urgency for interference. It is then our duty to avoid having recourse to the great surgical opening in question, until we have exhausted the rational local means applicable through the meatus. Among these means we would place in the first line irrigation practiced through the perforation in the direction of the attic and aditus by means of Hartmann's canula, this simple instrument of which we have now to speak strongly in praise as a curative means, just as we have already spoken of it as valuable for diagnostic purposes. In point of fact, these irrigations carried out regularly and followed by the insufflation of various antiseptic powders and by tampons plugged as deeply as possible, bring about a cure with considerable frequency in cases in which one would be tempted at first sight to consider the surgical opening the only possible means of treating them successfully.

If these means, employed methodically and regularly for several weeks, should fail, there is still one method of treatment to which we should have recourse, namely, the extraction of the ossicles through the natural passages, especially when we have to deal with a Shrapnell perforation, or when the ossicles appear manifestly affected with osteitis.

This operation ought to be followed by as complete curetting as possible of the attic by means of little curettes curved in different directions, after which the drum should be plugged with strips of gauze, which we should take care to introduce right up into the superior part.

After several consecutive weeks of this treatment we must be guided by our results. If the suppuration persists, with or without the regrowth of granulations in the region of the aditus, if the injections directed upwards and backwards by means of Hartmann's canula towards the antrum continue to bring about the expulsion of fresh quantities of pus, and particularly if this pus is fetid, laden with cheesy granulations and with necrotic scales, there should be no further hesitation; the extraction of the ossicles has only enabled us to touch a portion of the lesions, and there remains another focus to be opened and cleansed, which could only be done at the cost of an operative breach, necessitating an external wound; from this moment to open surgically the attic mastoid cavities becomes an absolute duty.

Indications for the Opening of the Mastoid in Chronic Otorrhea in Cases of Threatening of Intra-Cranial Complications.

We have just passed in review two classes of cases in which the operation of opening the mastoid was a necessity, but in these the urgency of their nature was quite different: in the first it was a matter of intervening without delay in order to insure the escape of pus supposed to be retained in the mastoid cavities, and at the same time to remove the focus of suppuration by opening it and cleaning it out in its whole extent; in the second class, on the other hand, this latter task was the only one which had to be fulfilled: we were in presence of an otorrhea which had proved its intractable nature under all methods of treatment directed against it through the natural openings, whence the conclusion that the focus keeping up the suppuration was inaccessible by these passages, and that it was necessary to attack it by an artificial route. But, however, there being no urgent call for the intervention, the operator can take his time and only decide to intervene after he is assured of the insufficiency of other methods of treatment.

We have now to consider a third class of cases, in which the urgent call for intervention is still more imperative than in the first; it is not merely a matter of allowing exit for purulent secretion which is prevented from escaping, and thus to put an end to the more or less severe pain, at the same time that one assures the patient against the possible danger of an extension of pus into the intracranial cavity, but it is necessary to combat without delay the real danger of a commencement of meningo-encephalic extension.

Before entering upon this subject there is one symptom, pointing to a well-known complication of chronic suppuration of the middle ear, with which we think we ought to occupy ourselves at this point, more especially in considering the indications for the opening of the mastoid. We refer to the occurrence of facial hemiplegia on the same side as the affected ear. It seems to us that we are all agreed in according to this eventuality, under the circumstances supposed, a very special degree of gravity. It marks, indeed, a further step in the progress of the destructive work of the osteitis, and it is not uncommon to see it followed more and more quickly by the explosion of an intracranical infection. In all cases, this occurrence being possibly the result of compression of the nerve, either by a sequestrum or by granulations blocked up in the region of the aditus, it is rationally indicated that we should at once go to the help of the nerve which is thus in danger, so as to be in time to prevent, if possible, a lasting facial paralysis. For all these reasons we ought to consider the occurrence of peripheral facial hemiplegia on the same side as the diseased ear, in the course of chronic otorrhea, if not as an indication sufficiently decisive to determine of itself the necessity for intervention, at all events as an argument of such a nature as to remove all hesitation regarding the necessity for operating without delay, in cases where the collection of signs presented by the patient would seem to render the surgical opening of the ear justifiable.

While there is only relative urgency in case of the occurrence of facial hemiplegia of otic origin in the course of an intractable otorrhea, this urgency to the opening of the mastoid becomes absolute in the presence of any symptomatic manifestation betraying the commencement of intracranial infection with or without concomitant retention of pus, the natural outflow of antral pus through the tympanum and the external meatus preventing in no way the fungating osteitis from carrying on this destructive work, and from exposing at any given moment the external surface of the dura mater to contact with the infectious germs of the aural focus.

We do not consider it our duty here to draw up a complete symptomatic table of intracranial infection arising from the ear, whether we are concerned with the commencement of meningitis, of encephalitis or of thrombophlebitis of the lateral sinus. It is quite evident that when the classical symptomatic combination of symptoms peculiar to one of these complications is produced, the opening of the skull and the search for the intracranial focus must be carried out without delay, as the only possible means of saving the patient.

Now, in such circumstances, even in the presence of certain symptoms called focal symptoms, pointing to the existence of a focus of encephalitis more or less distant from the petrous bone, our opinion (which we know besides to be that of most of our colleagues in aural surgery) is that instead of basing our choice of site for the cranial opening on considerations of cerebral localization, it is better to proceed straight away to carry out the antromastoid opening, pushing our resection of bone to the extent of laying bare the dura mater of the middle fossa of the skull, if we have reason for suspecting a lesion of the temporo-sphenoidal lobe, whereas we expose the dura mater of the posterior fossa and the lateral sinus if the symptoms observed suggest more probably a lesion of the cerebellum or an infection of the lateral sinus.

We have been supposing a case of confirmed intracranial infection; but before arriving at this, our patients pass often through a certain premonitory phase, the significance of which it is important to recognize, because in proceeding from this moment to the opening of the mastoid, we have a great opportunity for circumventing possibilities of accidents by means of disinfection limited to the osseous focus, or at least of preventing them from passing the barrier formed by the dura mater.

We cannot, therefore, too strongly insist upon the symptomatology of this period when the danger might yet be relieved by a simple operation, not risking the grave consequences attached to any interference beyond the limits of the dura mater.

In the first line of this symptomatic enumeration we would place pain, which is no longer limited to the deep part of the ear and the mastoid region, but diffused towards the forehead or the vertex, and taking on the character of severe headache. Often this is accompanied by a certain degree of photophobia, and the countenance acquires the contracted expression which is so peculiar to the initial stage of meningitis. Other symptoms may be added to these and accentuate their significance, even when there does not yet exist any established meningoencephalitis, a fact proved by the result of the opening of the mastoid at this period: it may be a vertiginous state, preventing the patient from standing erect, and even accompanied by nausea; there may be bilious vomitings, absolutely analagous to those occurring in confirmed meningitis; there is occasionally a shade of inequality between the two pupils. Lastly, the temperature does not always remain normal at this period, especially if the accidents in question are accompanied by purulent retention, or if there is the commencement of infection of the lateral sinus, in which case the fever may present the extensive oscillations which are so characteristic.

Once more we cannot insist too much on the urgency created in regard to the mastoid operation, not by the simultaneous appearance of all the preceding symptoms, but the occurrence in a decided form of even a single one of them.

Under such circumstances, we would formulate the principle *that the osseous opening ought not only, as in all cases of chronic otorrhea, to extend into all the cavities of the middle ear, but it ought to be carried to the denudation of the dura mater.* No doubt the operator is often saved this trouble, and in many cases in which a prompt intervention may have been determined upon, the occurrence of several *meningitiform* manifestations which we have just enumerated, but will find the explanation of the symtoms in question in the discovery of a perforation of one of the deep walls of the attico-antral space, leaving the dura mater bare, in direct contact with the pus of the focus.

The duty of the operator in cases of this kind is to carry out minute disinfection of all the walls of the focus, and especially of the denuded portion of the dura mater; also to leave the surgical wound sufficiently open in order to permit of a subsequent inspection of the osseous cavities which have been operated on; but, on the other hand, our confirmed opinion is that *we ought not at the time of this first intervention to open the dura mater*, because the simple extra-dural disinfection may suffice, and it suffices often to bring about the complete subsidence of even extremely anxious meningitiform disturbances. Now, those of us who have had any experience with cranial intervention know how different the prognosis after the operation must be according as the dura mater has been opened or not.

It ought only to be opened at a second operation, on which we must, however, decide without hestitation and without delay if after supervision for twenty-four hours subsequent to the first operation there is a persistence, or, still more, an accentuation of the symptoms of intracranial infection.

If the necessity for the deeper operation occurs, it is remarkably facilitated by the first one, which has had for its result to lay bare the region of the dura mater, behind which may be found the focus we are seeking for, either immediately on the surface of the pia mater, or, it may be, at a slight depth in the cerebral substance.

Under these circumstances, the mastoid opening will form the first rational stage of the intracranial intervention; it will have served to justify its further performance, and to simplify the process of carrying it out.

CONCLUSIONS.

A.

The opening of the mastoid is indicated in the course of chronic otorrhea under three distinct circumstances:

(1) When the object is to give vent to pus in cases of purulent retention.

(2) For the circumvention of conditions indicating the threatening or the commencing of intracranial infection of aural origin.

(3) For the cure of the otorrhea after it has been recognized that this has proved intractable to different methods of local treatment applied through the auditory meatus, including the extraction of the ossicles and the curetting of granulations accessible through this passage.

B.

The operation is only urgent in the two first cases.

C.

In all cases of chronic otorrhea the opening in the bone should extend from the antrum to the attic, or from the attic to the antrum, and be followed by curetting and complete disinfection of the whole of the cavities of the middle ear.

D.

In the case of threatening intracranial complications, the osseous breach ought to extend from the first to the suspected region of the dura mater; this membrane, however, not to be opened until a second operation, after a delay of armed expectation of as short duration as possible, if the threatening signs in question are seen to persist or still more to increase.

OPENING ADDRESS BY PROFESSOR KNAPP (NEW YORK).

Professor HERMANN KNAPP said we did not only want to be informed that under certain conditions, which his predecessors had so exhaustively and authoritatively dealt with, the mastoid should be opened, but also when, how and where, in particular how extensively, it should be opened, the description of the mere technique or the operation, however, lying outside the question. When acute purulent otitis media was on the border-line of becoming chronic, or had just become chronic, opening of the mastoid was indicated both as a curative and prophylactic measure. The indication for opening the mastoid was strengthened if tuberculosis, diabetes, syphilis, or some other constitutional disease, were present, particularly in the case of children. He thought the frequency of relapses in children was owing to the structural conditions of the infantile mastoid. He mentioned a case which had come under his own observation, to show that the suppuration may leave the tympanic cavity, attic, and antrum, but extend into and beyond the tip of the mastoid. The pus cells in this case travelled through the condensed bones in passages so small that they could not be followed with the naked eye. The indications for operation in advanced cases of destructive subacute chronic mastoiditis were absolute, and in the relapses of suppurative mastoiditis almost absolute. The prognosis in both cases was favorable. He had seen children recover who had a whole mastoid and a good deal of the adjacent temporal bone converted into gelatinous masses, and the dura extensively covered with soft discolored granulations. The best treatment of cases which from the beginning showed a disposition to long duration was to perform first the opening of the mastoid, and conduct the subsequent local and constitutional treatment with the utmost care and perseverence, so as to prevent the affection from becoming chronic. As particular requirements in such cases, he should lay stress on (1) a large, deep, and angular incision of the drumhead and the adjacent part of the posterior wall of the ear canal as soon as there was bulging, (2) opening the mastoid and thorough removal of all diseased tissue, (3) enlarging the antral canal by cautious scooping, (4) watching the course of recovery, using dry treatment rather than syringing. In chronic suppurative otitis media without symptoms of mastoid involvement that had resisted topical treatment and intratympanic operations, attico-antrectomy was indicated. In many cases it was difficult to determine when this should be done. During past years intratympanic operations had steadily lost ground. Many aural surgeons reported good results from the removal of the ossicles and cleansing the attic in cases of chronic otorrhea with or without cerebral symptoms. But, unfortunately, the good results in most of them had not proved permanent. He alluded to a patient who had long been treated by intratympanic procedures, but received only temporary relief. Such cases had determined him not to lose much time with intratympanic operations, although he would not go so far as an excellent otologist who told him that he had abandoned them altogether.

If the outer wall of the mastoid was perforated, and an abscess or a fistula present, it was indicated to evacuate the abscess and seek the perforation, and, guided by it and the fistula, open the mastoid freely and remove all morbid material. That was better than to let the patient take the uncertain chances of a spontaneous recovery, which was rarely complete and permanent.

If the disease extended beyond the mastoid process, the radical tympano-mastoid operation had to be followed by operations on the affected parts outside the ear.

If in chronic purulent otitis media the anterior wall of the mastoid bulges—which meant a suppuration in the cells adjacent to the posterior wall of the ear canal—a free incision down to the bone was indicated. We should then explore the wall with a probe, or, if the skin was swollen and painful, wait a few days to see whether the mastoid should be opened from the outer surface or from the anterior.

If the pus extended from the ear into the pharynx, forming a retropharyngeal abscess, he would open the mastoid and expose the tympanic cavity and attic clear to the tympanic orifice of the tube, and free it as far as possible from pus and disintegrated tissue.

The extension of the disease to the posterior cranial fossa was so important and so frequent that the removal of the posterior wall, in particular that part of it which formed the sulcus of the sigmoid sinus, had been recommended and practiced by some competent aurists in all cases. If the posterior wall showed no flaw on the closest search, and the suppuration was limited, he had left the wall alone; but when the contents of the mastoid had undergone extensive molecular disintegration, he considered the exploratory exposure of the sigmoid sinus and dura mater correct practice. Similar indications resulted from the extension of the suppuration into the middle cranial fossa, an occurrence less frequent than its extension into the posterior fossa.

Extension of the suppuration in the petrous bone might indicate opening of the mastoid as an initial step for removing carious and necrosed portions of the petrous or evacuating pus which had passed from the middle ear through petrous bone into the posterior cranial fossa, producing an epidural abscess on the posterior surface of the petrous bone.

Meningitis in the first stage might be recovered from by opening of the mastoid and posterior and middle cranial fossæ, exposing boldly the posterior surface of the petrous and liberating the pus.

Necrosis of the different portions of the temporal bone indicated the opening of the mastoid in most cases.

It was evident, Professor Kapp said in conclusion, that the opening of the mastoid in its recent development by the combined efforts of general and aural surgeons took rank among the most important operations.

Professor LUCAE (Berlin) then read a paper on *The Radical Operation in Chronic Suppurative Inflammation of the Middle Ear.*

At the outset I cannot sufficiently express my high estimation, he said, of the operation in question as an extraordinary means of cure in chronic suppuration of the middle ear. At times it has been only by means of this operation that I have seen healing brought about in a large number of cases. The following observations are intended to serve the purpose of diminishing the abuse of the operation as much as possible.

In the University Aural Clinic in Berlin under my direction there have been from April, 1881 (date of the foundation of the stationary clinic) up to August, 1899, 1,935 *operations for the opening of the mastoid process*, of which 852 were for acute and 1,083 for chronic forms of suppuration. At a superficial glance these numbers may appear large even in the chronic cases, but the experienced aurist will agree with me that the number of the chronic cases in which operation had been performed is by no means great in proportion to the acute.

It is obvious that in only a fraction of the chronic cases in which operation was performed was the operation such as is known as the "radical" one (opening of the whole of the cavities of the middle ear), this having only come into general use within recent years.

In order to form a more exact estimate of the frequency of performance of the "radical" operation in the chronic cases, I have calculated the percentage of operations to the total of cases of aural suppuration, and have selected for this purpose the last four years (the years are counted according to the customary prussian "State year" from April to April). This has been done particularly, because during this period the treatment by means of irrigation with formalin lotion has been adopted since 1895. These had a double advantage, because I was able to cure without operation the larger number of cases, or at least to improve them, and further that, if the remedy produced no good result, the indication for operation was all the more distinctly marked.

The following are the numbers arrived at during this period, namely, from April to April in each year:

1. 1895-96 in total, 2,061 middle-ear suppurations;
    648 acute, with 86 operations = 11·72 per cent.
    1,413 chronic, with 116 operations = 8·35 per cent.
2. 1896-97 in total 1,763 middle-ear suppurations:
    528 acute, with 66 operations = 12·5 per cent.
    1,208 chronic, with 85 operations = 7·03 per cent.
3. 1897-98 in total 1,700 middle-ear suppurations:
    581 acute, with 69 operations = 11·87 per cent.
    1,119 chronic, with 69 operations = 6·16 per cent.

4. 1898-99 in total 1,661 middle-ear suppurations:
   530 acute, with 61 operations = 11·51 per cent.
   1,131 chronic, with 90 operations = 7·95 per cent.

It must be mentioned that the number of new ear cases has by no means diminished, but, on the contrary, from 1895 to 1899 there has been an increase from 6,536 to 6,704. The percentage numbers speak for themselves sufficiently distinctly, and they show that the number of operations in chronic cases as compared with the acute ones is much smaller; in the year 1897 to 1898 they were only about half. Further, it is of interest to notice how small the absolute percentage of operations in chronic cases has been on the whole, and that the acute cases call for relatively few operations.

The statistical comparison of this period of four years of formalin treatment with the previous years would give no certain results, because most of the cases of chronic suppuration in the middle ear have been treated as outpatients, and only a few in the in-patient department, and, as happens unfortunately in every polyclinic, many of them fail to return with any regularity. We have, however, the general impression that the results of the formalin treatment have been better than those of any other. It is particularly in cases running a "cold" course without any threatening symptoms, and where it was only on account of the fetor of the discharge that there was any suspicion of deeper-seated affection that the formalin treatment was most remarkable. The general rule was that when the treatment was carried out carefully several times a day for four or six weeks, and no improvement in the fetor was brought about, the subsequent operation always revealed severe affection of the temporal bone (empyema, caries, cholesteatoma, etc.). Formalin has the double advantage that it is not only a powerful disinfectant, but it is very cheap. The strength of the solution used by me for irrigation is fifteen or twenty drops to one litre of boiled water. I have never observed severe or lasting irritation produced thereby. The only unpleasant effect, especially in frightened children, is that the remedy runs through the Eustachian tube, and produces occasionally transitory pain in the pharynx. This, however, is soon overcome by means of gargling with cold water.

In such cases a weaker solution may be employed, as the effect of the formalin is very powerful.

Gentlemen, as I think in German, I have spoken in my native tongue. But I now only wish to say some words for the British ears not understanding German. I am of the opinion that the opening of the mastoid, or at least of the cavum tympani, is a very important help in the treatment of chronic otorrhea. But also one may get on in plenty of cases by non-operating. I beg to add that instead of being proud of saying, "I have operated on so many patients," one should be prouder of saying, "I have cured so many patients also without operating."

Professor GUYE (Amsterdam) said the mastoid operation was a very great boon to the patient and to humanity in general, as Professor Macewen had so well said, but, nevertheless, as to finding the indication for mastoid operations only in discharge which did not give rise to dangerous symptoms he could not agree. He was with Professor Lucæ when he said that one should be prouder of having cured cases of chronic suppuration without an operation. He considered that the important thing in a case of chronic otorrhea was to keep the meatus as clean as possible, the using of carb. glyc., and, thirdly, to have great care for the keeping open of the Eustachian tube. His practice was, to patients who could bear the expense of a menthol insufflator, to get them to blow methol into the nose and through the tympanum, after Politzer's method. The operations ought to be reserved for really dangerous cases.

Dr. MOURE (Bordeaux): I am quite of the same opinion as the openers of the discussion, who do not hesitate to open the mastoid whenever a discharge from the ear resists medical treatment, followed or not by the extraction of the ossicles, when this treatment has been properly carried out. It is certain, however, that surgical treatment ought to be limited to some otorrheas, and not practiced in all, as seem to think the partisans of surgical treatment à outrance. When a purulent otorrhea is complicated by local pain; when irrigation directed towards the attic washes out cheesy matter, or, still more, mother-of-pearl pellicles; when the otorrhea continues to be fetid, in spite of regular irrigation; when, finally, we see the granulations recurring, in spite of ablation or cauterization, still more if there are spots of caries towards the superior or posterior parts of the meatus—in all these cases we must not hesitate to interfere surgically. Moreover, it may be said that all those who have had occasion to perform a certain number of operations of this kind have a tendency the more they operate to be more and more ready to operate. They recognize the necessity for operating, as also the efficacy of surgical treatment, which alone affords the means of curing certain otorrheas which are intractable under ordinary treatment.

Dr. McBride (Edinburgh) joined views with Professor Politzer, Professor Lucæ, and Professor Guye in their conservative methods with regard to mastoid operations. Professor Macewen had laid down that a persistent discharge alone from the ear without other symptoms was an indication for mastoid operation. Under certain circumstances it might be so, but by no means generally. The question came to be, What could they promise to their patients from a mastoid operation? In chronic cases they could promise the patient nothing. A certain proportion did not do well after the operation, the discharge remained, and the patient was exactly where he was before. But he agreed with Dr. Knapp that they did not do quite enough operations in acute cases just beginning to become chronic. Here the discharge usually ceased, the membrane healed, and hearing was restored after draining through the mastoid.

Dr. Jansen (Berlin) was prepared to accept, as his own, the statement of Professor Macewen, that frequently disease of the mastoid process did not show itself by outward signs. The question with regard to operative treatment was easier when, instead of making the diagnosis simply of suppuration of the middle ear, they designated beforehand the region of the middle ear which was affected. Then cases with suppuration in the lower section of the tympanum did not come into the question, but, nevertheless, it was only with great difficulty that they could effect a cure of disease in the large sinus between the fenestra rotunda and the facial. Further, the rare form limited to the attic was also to be excluded from consideration, as it did not require to be exposed through the mastoid process. On the other hand, the complication of abscess in the tube, which was very rare, called for an opening through the mastoid process. There only remained the conditions localized in the antrum and mastoid process.

It was desirable to differentiate between antrum and mastoid suppuration, because suppuration limited to the antrum was often cured without operation. When the discharge was slight, and always about the same in quantity, there was a great probability of there being uncomplicated antrum suppuration. A more exact description of the symptoms which indicated retention and increased pressure in the antrum and mastoid cells, as Dr. Luc had described, was possible, and it enlarged the circle of cases in which the indication for operation was urgent.

Professor GRADENIGO (Turin) said that, having performed a great number of middle-ear operations by the retro-auricular method in cases of chronic suppurative otitis media, he had come to the conclusion that the indications for the operation, such as had been generally stated in the discussion, were exaggerated. For the purpose of healing simple chronic pathologic conditions of the tympanic cavity, the extraction of the ossicles, or even of the hammer only, and removal through the external auditory canal of the posterior bony wall, were for the most part sufficient. In such cases the retro-auricular method did not give better results, and even exposed the patient to risks of various kinds. It required a long aftertreatment, difficult to be carried out, especially in children, and the final result often compromised the success of the best performed operation. Amongst the decided indications for the retro-auricular operation, with the opening of the mastoid, must be considered the cases of cholesteatoma antri, and all cases where symptoms existed pointing to mastoideal pathologic conditions or to intracranial complications. Regarding the technique, he preferred the Zaufal-Stacke method.

Dr. NOYES (New York) said while he fully agreed with the advisability of operative treatment for cases where there was any bone disease, he recommended the dry treatment. There was a class of chronic cases in which the acute process might have already considerably subsided, for which the treatment by dry powdered boracic acid was most effective.

Professor KÜMMEL (Breslau) said: One class of cases has not been mentioned—hysterical girls; they are able to imitate any kind of symptoms. I want to illustrate this by reporting the case of a girl who has been operated upon for the fifth time, and never anything has been found. The skull has been trephined over and over, until there is a defect of the size of the palm of the hand. Her brain has been punctured in at least twenty places. Still, about every six months she becomes ill with the same symptoms; she reproduces all the appearances of dizziness; she shows facial paralysis by contracting the one side of the face or the other; she has temperature up to 40.2° C., or 104½° F. This girl is quite well now, with her over twenty punctures of the brain and seven or ten narcoses.

Professor EEMAN (Ghent), on the subject of opening the mastoid, said that, speaking generally, he was a very warm advocate for the radical operation, but he thought it was a duty in many cases to try *at first*, and *before* performing a radical operation, *all* the other means which science possesses against chronic otitis purulenta.

He particularly wished to direct attention to the cases in which the extraction of the malleus can be sufficient to effect a complete and lasting recovery. He said that in his clinic the extraction of the malleus had been performed very often, and with splendid results, about 15 per cent. of the cases being entirely cured. Some of these cases came under treatment with conditions which would certainly have led other surgeons to an immediate and radical mastoid operation, such as fever, intracranial symptoms, inflammation and narrowing of the external auditory canal, etc. In these cases, under appropriate treatment, inflammation subsided in a few days, and then it was possible to ascertain that there was perforation of the membrane of Shrapnell, and caries of the head of the malleus. Extraction of this ossicle gave a perfect cure; some of the cases had continued under his observation for years after the operation, and he was able to state that the results had been lasting. Professor Eeman desired to warmly advocate the extraction of the malleus in cases of chronic purulent otitis presenting perforation of Shrapnell's membrane and caries of part of the malleus, and the postponement of the radical operation until it has been practically demonstrated that the removal of the malleus was insufficient to cure the patient.

Moreover, he said that he could not agree with the assertion of Schwartze, that isolated caries of the malleus was rare, and that as a rule both incus and malleus were affected at the same time; in his clinic isolated caries of the malleus had been found to be frequent.

Dr. OSCAR BRIEGER (Breslau) expressed himself as follows: Among the indications for radical operation we have included the failure of local medical treatment to produce a cure. According to the present standard of our knowledge this indication will have to be admitted to some degree. But it would be erroneous to deduce that the operation would render subsequent treatment superfluous. On the contrary, after the operative opening of the cavities of the middle ear, the alterations of the mucous membrane, which may become manifest, besides the morbid foci in the bone, may require further local treatment. It is occasionally possible to shorten the after-treatment by combining it with local treatment of the mucous membrane. It is, for instance, advisable in processes which reveal lasting maceration of new-formed or implanted epidermis to plug with gauze soaked in alcohol. Formalin also answers those purposes, as well as combinations with other drugs—for instance, weak solution of nitrate of silver in alcohol—according to the intensity of the process in question.

Luc recommends especially the evacuation of the cavities of the middle ear. If it is o be understood by this that after each radical operation the ossicles ought to be removed, it must be objected that in the interest of the function the preservation of these has been advised. In general this advice is superflous, because the connection of the columella is in those cases interrupted by destruction of the long process of the incus. It is quite true that the function is sometimes remarkably good after this, in general, complicated method. But it happens that even after complete skinning over of the cavities of the middle ar fetid secretion continuous from carious points of the remaining malleus. And this is less accessible for treatment, and more dangerous, because the local conditions are altered to a variable extent by adhesions, etc. It is necessary, at least, to make a careful selection of those cases where the ossicles are to be preserved. With regard to the contra-indications against the operation, Dr. Brieger is inclined to exclude meningitis. There are cases where marked symptoms of meningitis are present, and nevertheless there is only circumscribed suppuration, which may heal if new infection from the cavities of the middle ear is excluded by means of an operation; but recovery may take place in spite of diffuse meningitis, as ascertained by lumbar puncture, if the primary centre of infection is destroyed by radical operation, and if by this puncture more favorable conditions are created, recovery may in those cases be effected by removal of the infected material, by the lumbar tapping, or perhaps at the same time by the production of the new transudating lymph, which may have some bactericidal propensity. Of course successes of this kind are rare in extensive meningitis, but are sufficient to justify us in rejecting extensive meningitis as an absolute contra-indication against radical operation, the more so as the operation itself is harmless in those desperate cases.

Dr. BARR (Glasgow) regretted that the subject of their discussion excluded the methods of operation, and the results of operations, especially the latter, because he thought that one of the most important considerations with regard to the subject was the results of operative measures in chronic suppuration of the middle ear. Probably the most interesting class of cases was that for which there was no immediate demand for operation—cases where there were no objective or subjective symptoms demanding speedy operation. They were indebted to Professor Macewen for uttering a warning about continuing the treatment by the external meatus too long before adopting operation. They must not, however, be too much discouraged by certain dangers of ordinary treatment referred to by Professor Macewen, such as the removal of granulation tissue or polypi, as the experience of otologists showed that these were not great. Still, it was well that a surgeon of Professor Macewen's vast experience should utter those words of warning. Although the question of attic treatment had been rather disparagingly referred to by Dr. Knapp, Dr. Barr believed that the attic syringe was of great value, although many of the attic syringes in use were too narrow in the bore. He had found that in many cases after the attic treatment, including the removal of the malleus and incus, and the efficient use of a proper attic syringe, no radical mastoid operation was required.

Professor FARACI (Palermo) thanked Professor Gradenigo for approving of his osteotomy forceps. In the majority of cases he had found the removal of the larger ossicles and the resection of the outer wall of the attic and antrum sufficient. He thought it non-justifiable to open the mastoid as a whole till the ossicles had been removed through the meatus. As regards endocranial threatenings, there were two categories: (1) If the complication had occurred, the mastoid was a small part of the whole operation. (2) If the complication was only threatening, the operation through the meatus sufficed, as in a case quoted with meningitis symptoms.

His conclusions were that the mastoid should be opened:

1. When it was invaded by the morbid process in whole or in part.

2. When all the other methods of treatment, including the ablation of the larger ossicles and the resection of the outer wall of the tympanic attic and mastoid antrum, had been found fruitless.

3. In cases of manifest intracranial complications, the mastoid operation being followed by the further interference the complications demanded.

Dr. SUAREZ DE MENDOZA (Paris) thought that pain alone was not necessarily an indication for opening up the cavities of the middle ear in their totality. Sometimes in such cases the mastoid was found almost or quite healthy, and the pain was due to eburnation of the mastoid cells. A simple gouging of the mastoid or its erasion by means of an electric burr might be sufficient in such cases. With pain as the sole indication, they might cease operating deeper if the condensation of the bone and the absence of pus or granulations allowed them to attribute the pain to the condensation of the osseous tissue.

Dr. MILLIGAN (Manchester) said that in cases where local treatment had been faithfully tried for a period of twelve months, and where suppuration persisted, he was in favor of performing a mastoid operation. By local treatment he included the ordinary methods of antiseptically cleansing the parts, the removal of granulation tissue, the removal of diseased ossicles, etc.

Where such methods failed he thought recourse to an exploratory operation justifiable. By its means the paths of infection could be followed up, concealed foci of sepsis could be attended to, and extension to more deeply seated parts frequently arrested.

He desired to associate himself very largely with the opinions expressed by Professor Macewen.

Mr. T. MARK HOVELL (London) said that the mere fact that a discharge had existed for a long time was not a sufficient reason for the mastoid process being immediately opened up. He considered that the operation should not be undertaken in chronic suppurative inflammation of the middle ear until the ordinary methods of treatment had been fairly tried. About ten years ago he saw a lady who had a discharge from one ear which had lasted for forty-three years. It ceased entirely after about six weeks' treatment by the usual method with an antiseptic lotion and dry boracic powder. The discharge had not returned.

Mr. Hovell was of opinion that when the attic was cleared out the mastoid antrum should be opened up at the same time, otherwise a second operation might become necessary.

Dr. C. R. HOLMES (Cincinnati) said he had practiced, and was likely to continue to practice, the lines laid down by Professor Macewen. Dr. McBride had said that they could not promise results to mastoid cases. He certainly wished to put himself against that statement. He believed that in almost every case they could promise the patient a cure. They should save the patient the possibility of two operations when they knew one thoroughly performed would cure.

Dr. DENCH (New York) said each case must be treated according to the local conditions present. When the mastoid operation was involved a complete mastoid operation was imperative. If during the operation the surgeon found that infection of the lateral sinus had taken place, he must not hesitate to remove every source of infection. In one of the speaker's cases a second operation was necessary, owing to jugular involvement.

Mr. CRESSWELL BABER (Brighton) thought that most were agreed that in chronic suppuration of the middle ear, accompanied by mastoid symptoms, the bone should be opened. The interesting point to consider was whether the mastoid should be opened in cases of chronic suppurative otitis media without any symptoms except the discharge. In those cases he considered that, as a general rule, first of all, every means of arresting the discharge through the meatus (such as careful cleansing, curetting, removal of ossicles, etc.) should be tried, and if the purulent discharge from the tympanum still continued, the risks of pyogenic infection from this focus should be put before the patient or his friends, and the possibilities of an operation on the mastoid placing him in a safer position explained, although, of course, no certainty of a cure could be promised until the parts had been exposed by operation, and the full extent of the disease ascertained.

Dr. RUDLOFF (Wiesbaden) read a paper *On the Operation for the Removal of Adenoid Growths with the Head hanging over the Table, while the Patient is under the Influence of Chloroform.*

In his opening remarks, Dr. RUDLOFF drew attention to the method of performing operations on the hanging head, in cases in which there is danger of blood-suction. He then described his method, of which he had made use during the last eleven years. His experience included over 700 cases. He advocated the free administration of chloroform, and employed Boecker's and Hartmann's curette in performing the operation. In describing the method of operating he drew attention to the following points:

1. Adenoid growths occasionally have their origin in Rosenmüller's fossa. In removing them it is important (*a*) to avoid injury to the pharyngeal orifice of the Eustachian tube; (*b*) to bear in mind that the tissue surrounding the carotid artery extends into the lateral wall of the fossa, and that danger of injury to this artery is to be guarded

against. How necessary this warning is is proved by the case recorded by Schmiegelow.

2. Adenoid growths must be thoroughly removed (*a*) in order to avert as far as possible the danger of recurrence; (*b*) because a certain percentage of the cases which occur are tuberculous.

3. If the tonsils are enlarged it is advisable to remove them some time previously.

Dr. Rudloff illustrated his method by means of a specimen (sagittal section through the head), and exhibited the instruments he employed. He further showed casts illustrating the varying dimensions of Rosenmüller's fossa, and the relation existing between this fossa and the orifice of the Eustachian tube, and referred to a specimen showing the relation between the carotid artery and the lateral wall of Rosenmüller's fossa, which was exhibited in the Congress Museum. His statistics recorded a recurrence of 3½ per cent. In concluding, he remarked that he did not necessarily confine himself to the method he had described, but adapted it to the individual requirements of the cases which came under his care.

Professor V. Uchermann (Christiania) read a paper on *Rheumatic Diseases of the Ear*.

Mr. President and Gentlemen—Rheumatic diseases of the ear are but little known and seem to be rare. The symptoms are apparently not sufficiently distinct, nor the etiology so clear as to establish a safe conclusion with regard to cause and effect. Still, I am of opinion that a closer investigation of the matter will enable us to recognize certain common features, symptomatic and pathologic, by which a clinical diagnosis of the special cases can be made or rectified. To attempt this, and at the same time to draw the attention of my colleagues to an interesting group of ear diseases as yet little known, is the aim of this paper. At the outset we are met with the old difficulty, What is rheumatism? The answer from an etiologic point of view appears to be more unsatisfactory than ever. Infection admitted, is it a specific infectious disease, or is it only a kind of pyemia dependent upon one or more pyogenic bacteria? Whatever the case may be, we have the clinical picture, which cannot be dispensed with. As we are well aware, the characteristics of the disease are—its tendency to attack the connective tissue (fibrous or muscular) and the endothelial-lined cavities, and to form fibrinous exudates and infiltrates. In this way it appears in the joints, muscles, heart, skin, etc. In addition to this there is its painfulness in certain localities, also its being acted upon by salicylic acid in acute forms, by atmospheric changes in chronic forms. It is necessary to set aside all cases whose only claim to being rheumatic is that they appear to have arisen after a rheuma—that is, a cold or catarrh. To this class belong, for instance, many of the so-called rheumatic cases mentioned by Gradenigo in his labyrinth diseases (Schwartze's Handbook). It is also necessary to differentiate between acute and chronic forms. .Among the former the best known are the polyarthritis acuta (rheumatic fever), acute muscular rheumatism and erythema nodosum; among the latter, the chronic rheumatic muscular and joint diseases. All the rheumatic ear affections that have up to the present been described belong to the acute forms of rheumatism appearing as complications of rheumatic fever. Ménière (*Revue Mens. d' Otologie et Laryngologie*, November, 1883) mentions a case where otalgia, in the form of severe intermittent pain, preceded by four days the attack of ordinary acute polyarthritis. A similar case is given by Wolff (*Verhandl. der Otiatrischen Section der Wiesbadener Naturforscher Versammelung*, 1887), who also adds that the joints of the ossicles can be affected. The clinical or pathologic proof, however, is not given. In both cases the appearance of the drum does not seem to have been altered. Moos has observed a case of apoplectiform (Ménière) deafness during the period of convalescence after acute rheumatic fever, complicated with endocarditis (perhaps embolic). In a second case various cerebral hyperesthetic symptoms appeared with attacks of pain and hyperacusis in the eighth and ninth weeks, hardness of hearing ending in total deafness (Schwartze's Handbook, tome i., p. 544). Among the deaf-mutes in Norway is a case where an examination of the ear points to the existence of a combined middle-ear labyrinth affection caused by this disease (Uchermann, "The Deaf-Mutes in Norway," vol. i., p. 446).

I have seen two cases where ear affection preceded ordinary rheumatic fever. Both cases were of adults; one a lady twenty-five years of age, who had had rheumatic fever several times before, the other a gentleman of thirty-five, of very rheumatic disposition. In both cases there was an acute inflammation of the middle ear, with marked injection of the drum, abundant secretion of serous or sero-fibrinous fluid, together with quite an unusual amount of pain, both spontaneous and when touched, which continued even after the opening of the drum. In the case of the lady, during the fourteen days before the beginning of the fever, an infiltrate formed on the posterior wall of the bony meatus, involving the adjacent part of the drum, of the size of half a pea, red and very sensitive. In the man's case there was a swelling of the posterior part of the drum, also a more diffuse swelling and sensitiveness of the septum cartilagineum nasi on the same side, with superficial (catarrhal) erosions. In both cases the ear affections healed after eight days with the beginning of rheumatic fever, possibly the result of paracentesis and salicylic acid, though the swelling of the septum did not disappear for several months, and caused considerable impediment to the nasal respiration.

But there are also other cases where the rheumatism from which the ear affection arises is of a chronic character, and where the ear disease itself runs a course less acute and violent, but sometimes for the organ itself more fatal. In the case of a young man about thirty, with a marked rheumatic history, I have seen without any apparent cause, and alternating with rheumatic affections of the throat, a bilateral, so called, otitis media serosa, that is, a collection of serous or sero-fibrinous yellowish fluid in the tympanic cavity, with the slightest inflammatory signs. The case ran a slow course, but finally yielded to repeated incisions of the drum. I venture the hypothesis that many of the cases of serous middle ear affections, especially those marked by yellowish or amber-colored exudate, are rheumatic in origin or foundation, and that treatment with salicylic acid should be tried before any surgical intervention is resorted to. In another case, that of a young, plethoric man, about thirty-four, the symptoms when I first saw him (February, 1895) were the following: he complained that for a year he had suffered from tinnitus aurium and deafness of progressive character, which latterly had greatly increased. He experienced no dizziness, and hitherto he had enjoyed good hearing and freedom from ear-troubles. Occasionally he had felt rheumatic pains, but otherwise had never had a disease of any consequence. On examination the right drum revealed a small round cicatrix (as big as a shot); in the upper and hindermost quadrant there was a little dullness, but no retraction, the left drum being also dull and not retracted. Both the drums were movable by Delstanche. By auscultation the left ostium tubæ Eustachii was found narrower than the right, otherwise nothing was abnormal. From the left ear the hearing of speech was gone. He could neither hear No. 64 of Appun's set of tuning-forks (64 double vibrations in a second) nor Galton's whistle. Rinne was $-5'''$, Schwabach much shortened ($-$). On the right ear Rinne was $5'''$, Schwabach was $-$. The deeper tuning-forks were heard more distinctly than the higher, the Galton not at all. On this side he heard words spoken in a loud voice at a distance of from 3 to 4 inches. In spite of internal treatment with salicylic acid and iodid of potassium, together with local treatment (leeches, injections of iodid of potassium and pilocarpin, massage (Lucae, Delstanche)—after a couple of months he was completely deaf. At his repeated request at last I tried a stapedectomia on the left ear. On probing, the stapes at first gave the impression of immobility, but by traction became loosened, and then was immediately replaced. The only result was considerable giddiness for a month, during which time he had to lie quite still on his back. At the same time he had rheumatic pains in the right shoulder. About a year later there appeared a reddish, fluctuating swelling of the left eyebrow and upper eyelid, with its seat in the periosteal tissue. By incision I removed about a teaspoonful of serofibrinous fluid, upon which the swelling disappeared. A year after, however, it reappeared in nearly the same place, and yielded to the same treatment. On this occasion there was also a swelling over the left tuberositas frontalis. Last year he called on me for a nose affection. There was a dry catarrh of the anterior part, with a formation of crusts and a dry perforation of the cartilaginous septum of considerable size. It had developed since the last time I had seen him, and proved very stubborn under the ordinary treatment. In connection with this case I might mention two similar affections of the nose that have come under my notice; one the case of an elderly man, very rheumatic, who eventually died of rheumatism (articular, etc.), owing to general exhaustion. The other a case now under my treatment, where there is no perforation, but the pale, swollen mucous membrane is specked with white fibrous (sclerotic) spots.

It is then a case of what is commonly called secondary sclerosis, with involvment of both the labyrinthean bony capsule and the nervous elements. The history of the case and its accompanying symptoms make it fairly certain that it is of rheumatic nature, and, like the affections elsewhere, bound to the connective tissue. For instance, a swelling of the lining of the canaliculi for the N. cochlearis and the lining of the vestibulum, with the result of more or less fixation of the stapes, will easily account for the acoustic phenomena. While with regard to the bone (labyrinth capsule) the result may be an eburnation (though with the preservation of the greater cavities vestibulum, scalæ, etc.), or may be, in some cases, the apparent reverse, a rarefication ("spongiosirung," Siebenmann). To sum up:

1. Rheumatic fever is sometimes preceded, sometimes accompanied, by otalgia, alone or together with an acute swelling and injection of the drum and the adjacent bony meatus, followed by a serous or sero-fibrinous secretion of the middle ear (otalgia, myringitis, otitis externa, otitis media *rheumatica*), or it may be complicated during its progress with affections of the middle ear and the internal ear (labyrinth, perhaps the auditory nerve).

2. There are other more independent rheumatic ear diseases with persons of a rheumatic constitution or tendency (previous rheumatic fever, etc.). The ear affection appears as an otitis media serosa with yellowish, half-fibrinous exudate, or as a (secondary) sclerosis with progressive character.

3. The characteristics of the different forms are: In the *acute* forms—painfulness, excessive injection, and the

tendency to the formation of fibrinous exudates. In the *chronic* forms—the tendency to the formation of fibrinous exudates, and the tendency to affect the bony capsule, with severe tinnitus and slow but steady progression. Salicylic acid seems to influence the acute forms but not the chronic. These latter, judging from the experience of a case at present under my treatment, are perhaps more influenced by a general rheumatic treatment.

In the discussion which followed Dr. HARTMANN said: The paper of Dr. Uchermann reminds me of one patient who probably comes in this line. A man slept one very cold and wet night in the woods; when he awoke he found he had completely lost his hearing.

Dr. UCHERMANN, closing the discussion, said: It is possible that Dr. Hartmann's case comes in this line, but we will have to differentiate between acute catarrhal inflammation of the ear and rheumatic inflammation of the middle ear. One is easily accessible to treatment with salicylic acid, the other is not. Furthermore, in rheumatic cases we always find other manifestations of rheumatism; exceptionally, rheumatic otitis shows infiltration and exudation in the ear alone.

www.ingramcontent.com/pod-product-compliance
Lightning Source LLC
Chambersburg PA
CBHW022148300426
44115CB00006B/398